India's
Manufacturing Sector
Policy Framework

Comments...

"The Competition was marked by excellent participation, indicating painstaking research by the groups of people behind each individual effort. Although they generally followed a middle of the road approach adopting well established trends of competitiveness and recognizing the inevitability of globalization, there were occasional flashes of original thinking, which distinguished some of them from the rest.

"I am deeply impressed by the social awareness exhibited by WTI and EMPI Business School in organizing the Competition. It is my earnest desire that the good work may be continued in future as well."

— Shri Prabhat Kumar
Former Cabinet Secretary, Govt. of India
Chairman of the Jury, National Manufacturing Sector Policy Competition

"These prize winning reports are marked by sheer volume of pertinent data, incisive analysis and a high degree of creative thinking. The jury's task was therefore pleasurable but by no means easy. We Think for India has come up aces with the idea of this competition."

— Smt. K.J. Udeshi
Executive Director,
Reserve Bank of India

"It is high time that Industry and Government got together to shape a manufacturing policy for India; otherwise, in the new post-WTO era, fifty years of effort at building an industrial base will be undermined. The initiative taken by WTI and EMPI has to be commended for bringing some of India's best minds together to define various policy alternatives."

— Shri A.P. Gandhi
Chairman, FAG Bearing (I) Ltd.
and Former President, Hyundai India

"The reports are remarkable in that for the first time an attempt has been made to elicit 'out of the box' thinking on a major policy issue by accomplished professionals. The ideas presented ... ...will be carefully scrutinized by policymakers throughout the country with a view to their actual implementation."

— Dr. Prodipto Ghosh
Addl. Secretary, Prime Minister's Office,
Government of India, New Delhi

"All the reports reflect exceptionally high standards. This initiative will greatly strengthen India's thrust towards becoming a world class manufacturing source. I congratulate the winners and also the WTI for having organised the event."

— Shri R.K. Krishna Kumar
Vice-Chairman, Tata Tea Ltd.;
Managing Director,
The Indian Hotels Company Ltd.,
(The Taj Group of Hotels)

"A good effort towards public-private academic interaction on policy matters. More than the content of "policy drafts" what is to be appreciated is the involvement by teams in putting together diverse ideas, models, facts and projections for a very crucial engine of development - Manufacturing. A useful example for researchers and students of Development Studies, Industrial Economics, Management and Business."

— Dr. Arvind Virmani
Director and Chief Executive,
Indian Council for Research in International Economic Relations (ICRIER), New Delhi

"Here we have an unusual exercise in sensitizing academia to the problems of industrial development. Far too long has the intellectual community in India kept aloof from issues of the real world, and the needs of the country. This exercise by WTI and EMPI has elicited substantial interest from many top institutions, and hopefully that will continue for mutual benefit of academia and policy makers.

"This exercise should not be a one-off affair, but an annual exercise."

— *Padma Bhushan* **Prof. P.V. Indiresan**,
Former Director, IIT-Madras

"WTI (We Think for India) has made a signal contribution by inviting Academics, Industry Professionals, Consultants and Government Policy-makers to think of the Manufacturing Sector in a strategic context. The submitted entries display a considerable enthusiasm in projecting this sector as crucial for long-term industrial and economic development."

— Dr. R. Natarajan
Chairman, All India Council for Technical Education (AICTE), Government of India

we think for india

India's Manufacturing Sector Policy Framework

eMpi B-School

...Presenting 3 top ranking independent reports
outlining a comprehensive manufacturing policy framework for India...

- Andersen Business Consulting
- IIM Lucknow
- IIT Madras

Academic Foundation
NEW DELHI

Editors :

Pankaj Saran
Head of Research & Vice President,
EMPI B-School, New Delhi.

Prof. S.P. Verma
Sr. Director & Dean (Emeritus),
EMPI B-School, New Delhi.

Published in 2003 by
ACADEMIC FOUNDATION
4772-73 / 23 Bharat Ram Road,
(23 Ansari Road), Darya Ganj,
New Delhi - 110 002. INDIA.

Phones : 23245001, 02, 03, 04.
Fax : 011-23245005
e-mail : academic@vsnl.com
www : academicfoundation.com

India's Manufacturing Sector :
Policy Framework

ISBN 81-7188-311-7

Printed at Print Perfect, Mayapuri Industrial Area, New Delhi.

This publication, released by
Hon'ble Shri A.B. Vajpayee
The Prime Minister of India
on 24th May, 2003
is dedicated
to
the youth of India
who represent our collective hope
of making India a great power

Contents...

Foreword...

SOMEONE told me about a remark that the legendary chairman of General Electric, Jack Welch made in a board room discussion. He said Mexico with the "M" in it stands for Manufacturing while India with its "I" stands for Intellectual Capital, as far as GE is concerned. How prophetic this statement has been. Subsequent experience has borne out so clearly that India is fast becoming a Software-Services provider in the world as compared to the rest of East Asia which is becoming more a Product-Hardware supplier to the world.

The above development has to be looked at within the wider context of :

- The role of "Manufacturing" in the developmental process of a developing nation.
- The importance of manufacturing to India in the light of its size, population, skill level and employment.
- The role of Indian manufacturing in a QR-free, post-WTO Trade Regime.

Specifically, national consensus is required for creating a dynamic Public/Private partnership to make manufacturing retain its importance for national development without sacrificing the natural competitive advantage India has generated in services. Thus in short, we need a National Manufacturing Sector Policy.

We think for India (WTI) was set up with a unique mission to make an attempt, perhaps for the first time in India, to create a structured dialogue platform, wherein Corporates, Academia and Government can come together to formulate public policy. WTI, as one of its first endeavors, took up manufacturing as a theme issue, because of the crisis Indian manufacturing is facing, by announcing a national level competition. This competition, an EMPI initiative was supported by FICCI and Businessworld. Response to the competition was overwhelming with as much as 167 Teams from leading corporates and academic institutions registering. Thereafter, a two staged selection process was followed, which finally resulted in three top teams emerging from the nine short-listed teams. The final winning entries were Andersen Business Consulting, IIT, Madras (Chennai) and IIM, Lucknow.

These teams demonstrated the wealth of knowledge in this country and brought together a number of perspectives to promote a genuine 'democracy of ideas'. Not only have these teams worked painstakingly for almost six months to put together these "Policy Drafts", but also demonstrated sensitive understanding about the uniqueness of India's development context wherein the highest technology is to work with the lowest, the large corporates have to work with the small- and the cottage-scale, the knowledge worker has to work with the unskilled amongst so many other contradictions that abound in India. Another interesting characteristic of all the winning teams has been a sharp focus on devising workable models for implementing the suggested manufacturing policy. This dimension assumes greater importance in the context of India, where a number of well-meaning schemes have got grounded because of poor implementation.

It is hoped that these "Policy Drafts" will influence the policy landscape in India apart from strengthening Indian Democracy to become a true 'democracy of ideas'. This exercise not only holds out directions for manufacturing in India but a direction for promoting the freedom of expression constructively from the youth of a developing country, who stand out uniquely as the builders of new India in the twenty-first century, which will be the century of hope for India.

— **R.A. MASHELKAR**, FRS
Patron, WTI;
Director General, Council for Scientific and Industrial Research (CSIR), Government of India

About WTI...

IT was on 21st March 2001 when I was at the EMPI Business School for addressing the students at their invitation, I entered into a conversation with Shri Pankaj Saran, the young and dynamic Vice-President of the Institution regarding the hopes and aspirations of the youth of India and the way in which they would like the 21st Century India to be transformed. Naturally during this conversation, the question arose ' **Who thinks for India.'** Since we did not have an answer for this question either in the metaphorical or contemporary context, I suggested to Pankaj that we should form a Forum called "Who thinks for India" and invite eminent personalities to guide us in its activities. To this, there was instantaneous response from all the members of the Academic Council of EMPI and it was decided then and there to take this idea forward.

The next day, when I approached and put forward this idea to Dr. Abid Hussain, my benefactor, friend, philosopher and guide, he reacted in his characteristic style, "Babu, why do you leave the question open-ended and doubtful? We should positively affirm our faith in India and work for its progress and I would therefore suggest that the name of the Forum should be **WE THINK FOR INDIA**." That was how WTI was born.

Judged from various angles, WTI is a thought process, an idea, a concept, a movement, an institution and framework for a plan of action. A good number of distinguished personalities who came forward to bless this forum and associate themselves in its establishment have decided not to stratify its structure and get bogged down in rigid procedures, but allow it to blossom gradually as a concept about which every Indian will be proud. Taken forward as a movement, WTI should address issues relating to nation building in all its dimensions in order that there is scope relating to nation building in all its activities and deliberation on thought provoking and action oriented lines. As far as the institution and framework are concerned, it was felt that sincerity, commitment and genuineness of the persons involved and the way they would deliberate upon issues will enable WTI to emerge as a movement for the benefit of people of different cross sections of the society.

The WTI journey has started with the first important initiative relating to an all India competition on National Manufacturing Sector Policy, details of which are contained in this document. WTI believes that more than **Vision** one should proceed in the direction of **Implementation**. WTI believes that in an atmosphere in which very often articulation is construed for action we should build up leaders, entrepreneurs and executives who would ensure delivery of results within stipulated periods. In this context WTI is looking for —

- Who these personalities would be,
- When they would arrive, and
- How they would deliver.

WTI would endeavour to motivate the younger generation of India to achieve results in all fields and demonstrate that they should not only dream and visualise but also cross milestones and set benchmarks for excellence.

WTI earnestly appeal to every member of the society to appreciate the larger canvas on which pioneering efforts are to be made and ensure that success should be the goal.

— L.V. SAPTHARISHI

Member and National Co-Convenor, WTI;
Additional Secretary, Ministry of Commerce,
Government of India, New Delhi

WE THINK FOR INDIA
e-mail : wti@indiatimes.com

WTI & EMPI

a special relationship

EMPI (Entrepreneurship Management Processes International) B-School Group, New Delhi-a globally recognised management education centre, conducts AICTE approved Post Graduate MBA level (PGDBM) program and an unique Under Graduate package- Graduate Studies in Entrepreneurial Management (GSEM). In addition, EMPI has an Institute of Advertising, Communication and Management, a Centre for Global Business Management Studies which provide parallel Post Graduate and Under Graduate studies. EMPI has an association with Academic Associates and Researchers, the Global Leadership Program, Wharton School, the Tokyo based Japanese Multinational Enterprise Study Group, MOSSAICC Academy - Mombusho SAI (Indo-Japan) and the European Union under whose auspices EMPI has setup the EU-India International Management Centre (EUIMC) in partnership with the University of Turin, Italy and Helsinki School of Economics and Administration, Finland.

A major concern for EMPI in its educational process of training managers with a social awareness, has been the importance of the context and environment of business, which assumes far greater importance for a developing country like India. Not only is business just a profit maximization activity, but it has a much larger responsibility to become an engine of progress for the country's development which has to be just, equitable, ethical, widespread involving the weaker and marginalized sections of society. This challenge is all the more daunting in that, this responsibility has to be accomplished without compromising Business and Economic efficiency for Global competitiveness. A crucial ingredient in the above is the importance of Public Policy and its interface with Business, which again in a developmental state has a much larger role to play than unbridled free enterprise.

It was within this back-drop that EMPI was privileged to have one of India's most capable bureaucrats, Mr. L. V. Saptharishi as a Chief Guest for its Annual Day. Expectantly the discussion veered around to WTO and India's response to it ? The answer was characteristic of Mr. Saptharishi and which also matched with what EMPI wanted to do, was : why don't we set up a forum ? Thus was born WTI or "We Think for India." WTI was very fortunate to have within its fold some of India's most distinguished personalities under the Chairmanship of Hon. Justice Ranganath Misra, Former Chief Justice of India. From the beginning, a key idea of both Mr. Saptharishi and myself, echoed by all the members - particularly Dr. R.A Mashelkar, DG, CSIR was that WTI must not become just another NGO, but should add to making India into a more vibrant democracy. It was then a new concept, emerged, perhaps for the first time in India, for creating a participative public policy platform which will bring together Academia, Industry and Government for structured dialogue and deliberation on various policy themes. With consensus on what WTI should do, the suggestion came that WTI should as a first activity look at the current crisis facing Indian manufacturing. Consequently with support from FICCI and Businessworld, a National Manufacturing Policy Competition - 2002 was announced. The success of the Competition more than anything else has also seeded a new working model for the involvement constructively of Indian Intelligentsia for nation building.

EMPI looks forward to its special relationship with WTI in all its future endeavors with the hope that what started out as a small idea will snowball into a national movement in the years to come.

— PANKAJ SARAN

Member and National Co-convenor, WTI;
Vice-President, EMPI B-School, New Delhi

e-mail: empi@del2.vsnl.net.in

Acknowledgements...

THE present publication is the culmination of a whole sequence of events, activities and thought processes involving a great deal of human application and teamwork at various levels. It all started with the National Manufacturing Policy Competition. The event which turned out extremely successful would not have been possible without the active involvement and support of so many 'pillars of support' — Distinguished members of the jury, who spent their valuable time in going diligently through the short-listed reports, Board of Patrons and Fellow Members of the WTI, who have most enthusiastically supported at all times and in all ways by creating both the intellectual fervent for new ideas and pragmatism for action; supporting organisations FICCI and Businessworld for coming along with their unstinted support when the idea was a 'fledgling'; Management, Faculty and Students of EMPI who provided the entire organisational back-bone to the initiative; Academic Foundation (our publishers) who put together the entire publication in a record time; our sponsors for providing us with the finances without which no initiative can succeed; but above all the participants who brought life to the whole concept by creating wonderful "policy drafts" distinguished by their perspective, creativity, diligence, analysis and projection for what a policy framework for Indian manufacturing should be.

— L.V. Saptharishi, Pankaj Saran
National Co-convenors, WTI

ANDERSEN

Indian Manufacturing – Towards Global Competitiveness

Policy Imperatives

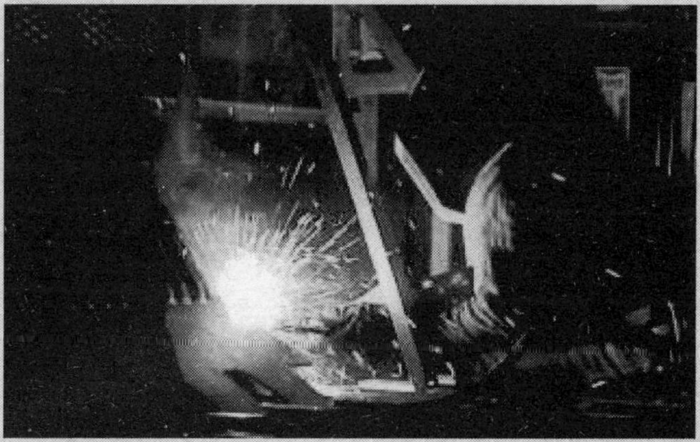

Andersen Business Consulting

Authors : Sanjay Upendram (Team Leader), Vikram Prabhu, Arun Krishnan, Kaushika Madhavan, Eric Pagdiwalla.

Contents

Indian Manufacturing – Towards Global Competitiveness
Policy Imperatives
(Andersen Business Consulting)

Executive Summary

"There is a tide in the affairs of men,

Which, taken at the flood, leads on to fortune;

Omitted, all the voyage of their life

Is bound in shallows and in miseries"

William Shakespeare - Julius Caesar

Shakespeare's observation is equally pertinent to nations. Indian manufacturing is at crossroads today. It faces great challenges, as well as great opportunities. India today has the potential to emerge as a global manufacturing hub, but is not in a position to turn this potential into reality. Key and timely decisions need to be taken to set it going on the path to growth and global competitiveness. The need of the hour is to enable radical and discontinuous growth, to bring India on par with other key developing and even developed economies. Otherwise, there is a grave danger of India getting marginalized in the world market, despite its huge potential.

Discontinuous Growth In Manufacturing

Growth in value added for different economic activities normally follows a continuum - starting from basic, low-value added activities like agriculture, going on to low-end, labour intensive manufacturing, to high-end manufacturing, and services. (Figure 1).

FIGURE 1

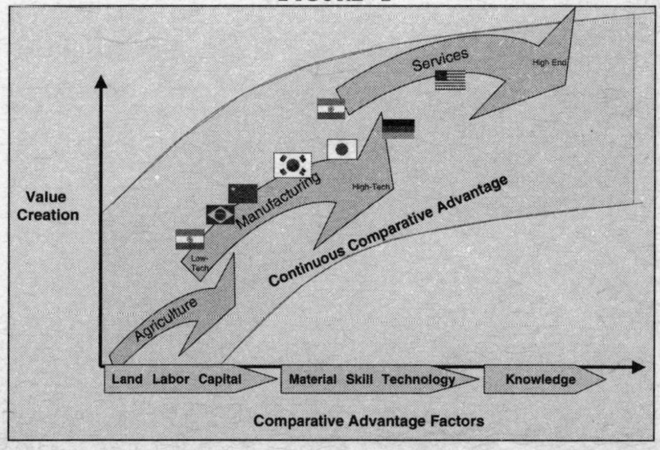

India is currently at the labour intensive, low-end manufacturing position catering primarily to repetitive manufacturing, with marginal participation in the high value added activities such as concept development, R&D and design. Its comparative advantage also lies primarily in availability of raw material and low cost of labour. While it is not unrealistic to aspire for India to move up the value chain and the continuum, given the large pool of engineers, for achieving radical or discontinuous growth there is a need to leapfrog over the normal evolution cycles, to get directly to high-end manufacturing and related services.

Given the relatively low level of development, and the fact that the bulk of India's workforce is engaged in low-end manufacturing, it will be a challenge for the entire manufacturing industry to make this jump.

However, specific sectors and firms can and have achieved such discontinuous growth. Examples include:

a) Moser Baer India - World's third biggest supplier of CD/DVDs

b) Sundaram Fasteners - Global supplier to GM for radiator caps

c) Taiwan emerging as a global producer for high tech electronic products

Such growth has been achieved either by catering to global volumes and maintaining world class performance standards (Moser Baer, Sundaram Fasteners) or by exclusive focus on specific industries (Taiwan's focus on electronics). This growth has been achieved in spite of constraints that are existing. However, we need to ensure that such successes are not an exception but the rule, through pro-active and focused policies that get implemented effectively.

Therefore, propelling India along the path to becoming a global manufacturing hub through radical growth, would require:

1. Identification of specific industries and sectors, where India can provide a comparative advantage, and putting them on the 'fast track'. By achieving radical growth in these industries, the economy as a whole can grow faster than the normal growth path.
2. Putting in place basic requirements like infrastructure, financing and a rational tax structure through policy initiatives, so that :
   a) The capabilities of the manufacturing industry as a whole can be brought on par with world standards.
   b) India's large labour force is leveraged effectively by enabling small and labour intensive industries to become more competitive.

With this background, we believe that any policy aimed at achieving higher manufacturing growth in India should focus on enabling three key goals:

1. Achieve global levels of excellence in Indian manufacturing.
2. Improve standard of living of the people through manufacturing growth.
3. Develop India into a strong player in the global market.

Achieving the above goals would require policies addressed at not only the manufacturing sector, but also the macro-economy as well as individual firms. To this effect, our policy recommendations aim to:

1. Identify specific industries for the 'fast track' and provide them all support for radical growth
2. Remove reservations in the SSI sector.
3. Promote collective industry efforts to transition the SSI sectors and leverage India's large labour force.
4. Promote industry - academia - government collaboration.
5. Rationalize tax structure.
6. Aggressively promote FDI inflow.

7. Develop bilateral trade relations with ASEAN countries and China.
8. Reform bankruptcy and labour laws.
9. Reform and privatize the power sector.
10. Privatize ports.
11. Improve road infrastructure.
12. Restructure government departments.

While the above policies are to be addressed by the Central government, they need to percolate down to the state level as well. It is to be noted that the government has already initiated policies in many of the above areas. However, the key to success is to follow through and ensure that the implementation is speedy and effective.

Good policies and even effective implementation alone cannot make Indian manufacturing globally competitive. Indian manufacturing firms need to look at their individual operations, benchmark themselves against global standards and improve their performance to match them. Policies can only provide support - it is up to individual firms to capitalize on the support and become world class.

1

CHAPTER ONE

Indian Manufacturing at Crossroads

INDIA'S manufacturing sector is undergoing a transformation, from a protected environment, to one of open trade and global competition. Despite progressive liberalization of policies over the last decade, the sector's performance has not been in proportion to the potential. Slow pace of reforms, mismatch between policy intent and implementation and inadequate development of key enablers like infrastructure, utilities, R&D and labour have combined to keep India lagging behind most other developing economies in industrial and manufacturing growth. As the world is moving towards more open and free trade, India needs strong policy decisions and effective implementation to emerge as a strong global economy.

The major contributor to India's economy today is the Service sector, followed by Industry. Nearly 75% of India's GDP comes from these two sectors. Of this, services constitute 45% and industry, 30%, of which manufacturing accounts for 19%. Manufacturing is a key building block of the economy, since growth in this sector will have a complementary effect on the services sector as well. For example, an increase in the demand for cars will also lead to demand for related services like software and design services (for manufacturers) and financing, repair and maintenance services (for consumers). Growth in production of consumer goods, textiles, etc., (manufacturing) will also aid growth in retailing (services). Therefore, any future growth in the Indian economy will greatly depend upon growth in manufacturing.

Government policy in India, from Independence till 1991, was focussed primarily on protecting the domestic industry and limiting private sector monopoly. During this period, Indian manufacturers were shielded from global competition and as a result, performance levels in India considerably lagged global benchmarks. While the economy has been opened up to global competition over the last decade, manufacturing in India is still a long way behind global standards in key areas like quality, productivity and process efficiencies. Further, in terms of infrastructure development, literacy rates and standard of living, India still lags behind not only the developed countries, but also developing countries like Thailand and Malaysia.

Global competition is intensifying, and no country or market is insulated from this phenomenon. Reduction in trade barriers and improvement in information technology and communication now allow companies to source products and services from the cheapest and most efficient sources across the globe. The Indian software industry has seized the opportunity and grown tremendously over the last 5 years; however, Indian manufacturing is yet to realize its potential in the global market.

India is today a strong regional power, which has the potential to become a significant global power in the future. On the one hand, there is an opportunity to leverage India's strategic location, engineering and design skills and workforce to emerge as a global hub for manufacturing and services. On the other hand, India faces stiff competition from China and other SE Asian countries, which enjoy most of the advantages of India, and also have healthier economies. These developments have placed Indian manufacturing at crossroads today. Several key questions need to be answered:

- What are the key opportunities for India in the global market?
- What are the essential key features of Indian manufacturing?
- Does Indian manufacturing have a comparative advantage in the global manufacturing arena? If so, how can India pro-actively leverage this advantage?
- What are the policies required and how quickly and effectively can we implement them?

We begin our analysis by understanding the global manufacturing landscape and trends.

2

CHAPTER TWO

Global Manufacturing Landscape

2.1 Trends in Maufacturing

Britain led the world in manufacturing output for most part of the nineteenth century. However with increased industrialization, other western nations like USA, Germany and France surpassed UK in terms of industrial output. In 1970, USA, Japan, Germany, France, UK, Italy and Canada accounted for around 75% of the world's manufacturing output.

Exhibit 2.1

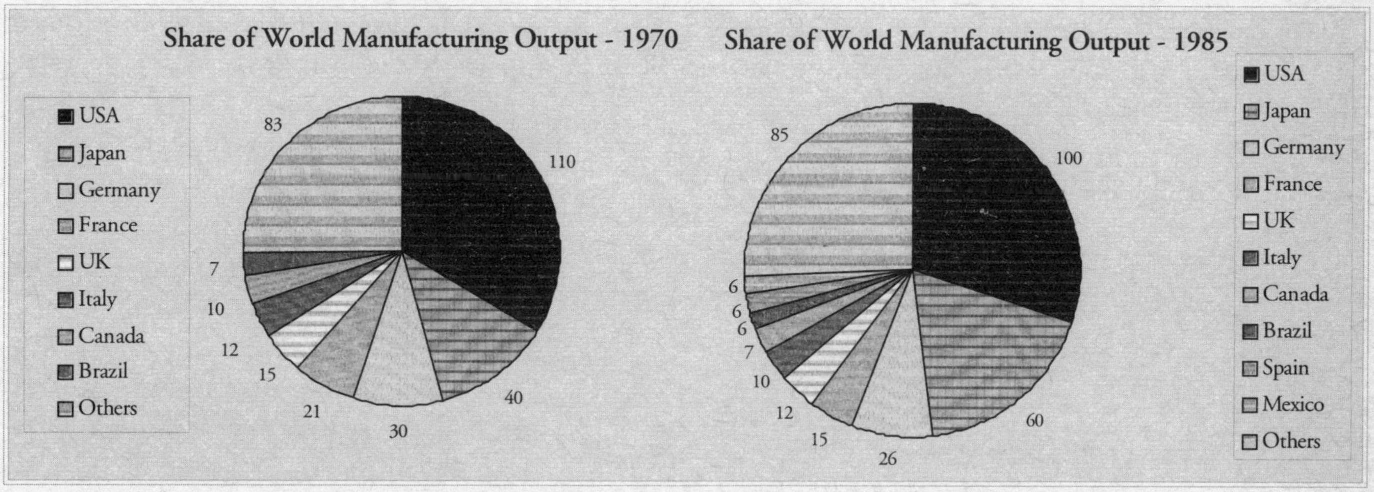

From 1970 to 1985, the share of these countries dropped to around 67%. During this period, with the exception of Japan, whose share grew by 50%, the shares of the other countries contracted. During the same period Brazil, Spain and Mexico increased their industrial output to be amongst the top ten industrial nations in the world.

In the developing world, the major catalysts for growth have been the economies of South East Asia. In 1963 Hong Kong, Singapore, Taiwan and South Korea accounted for only 0.35% of the world's manufacturing output while their share had grown to 1.55% by 1980.

EXHIBIT 2.2

Manufacturing as percentage of GDP 1973-2000

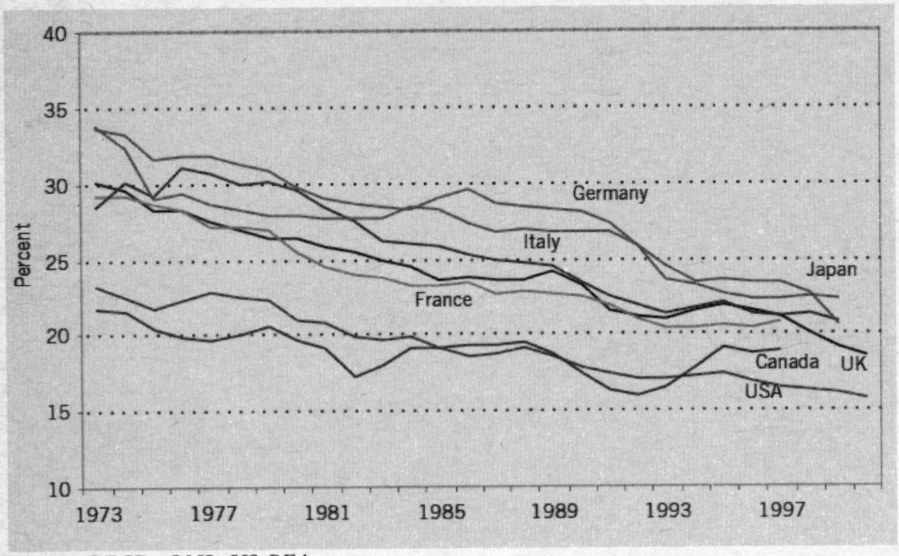

Source: OECD, ONS, US BEA

Source: The Government's manufacturing strategy for UK

From exhibit 2.2, we see that in the developed countries the share of manufacturing in total output has tended to decline, and with it the proportion of workers employed in the manufacturing sector. UK's manufacturing output as a proportion of GDP fell from 32% in 1970 to 19% in 1999. In USA, the share of manufacturing fell from 25% to 16% in the same period and in Germany from 36% to 22%.

Industrial production in developed countries is decreasing with a higher percentage of low value added manufacturing activities being outsourced to countries that possess favourable conditions for factors such as labour costs. Thus we see industries like textiles moving from USA (where USA had a share more than 15% in 1975) to countries like China, India and Pakistan.

This change has been brought about as the balance of economic activity has shifted from manufacturing to services. In the USA and UK, the proportion of workers in manufacturing has shrunk since 1900 from around 40% to barely half of that while in Germany and Japan manufacturing's share of jobs is now below 30%.

For developing countries the industrial value added as a percentage of GDP has either increased or remained stagnant as seen in Exhibit 2.3. The growth for countries like China is due to the increased share in low value added manufacturing activity.

Exhibit 2.3

Industry value added as % of GDP

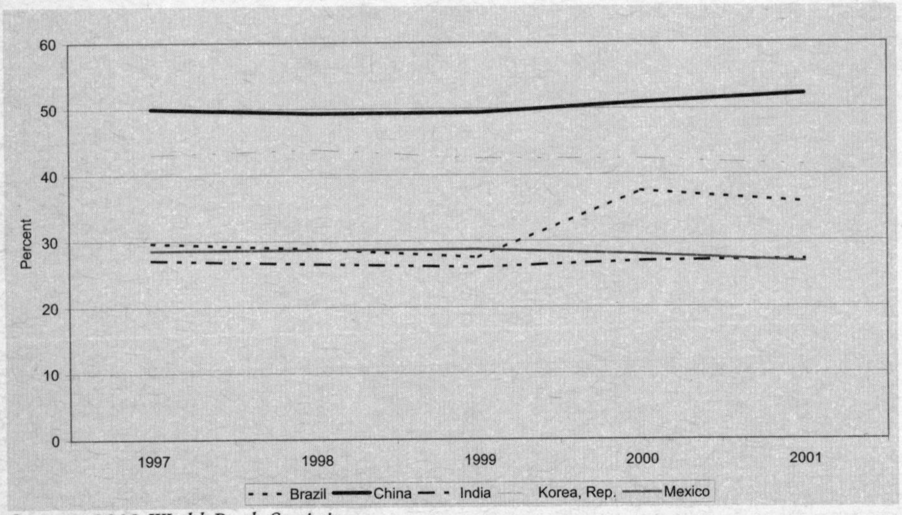

Source : 2002 World Bank Statistics

Supporting this trend is the growth rate of manufacturing for the periods 1980-1990 and 1990-2000 shown in Exhibit 2.4. The figure shows that developing nations have shown robust growth in manufacturing activity over these two decades though on a smaller base.

From the above graph we see that the growth of manufacturing is concentrated in developing countries with China exhibiting one of the highest growth rates of 13.4% between 1990-2000. The other countries that exhibited healthy growth rates during the same period are Malaysia (9.8%), Singapore (7.1%), India (7%), and South Korea (7.5%).

2.2 Trends in Exports and Trade

Trends in trade over the last three decades have been characterized by a growing participation and share of developing countries. Between 1970 and 1999 merchandise exports of these countries grew at an average annual rate of 12%, compared to 10% for the world as a whole, resulting in their share in world trade increasing from less than one fourth to almost one third. During this period, developing countries also became important markets for each other's products, the share of trade among them reached 40% of their total exports at the end of the last decade.

Manufactured items accounted for 70% of developing country exports at the end of the 1990s. During the 1970s and early 1980s the share of manufactured exports was around 20%, while the share of agricultural commodities fell from about 20 per cent to 10 per cent during the same period. Earnings from mineral and oil exports fluctuated considerably due to sharp changes in prices, but their overall trend was in a downward direction.

Although the share of developed countries in world manufactured exports fell from more than 80% in 1980 to about 70% in 1999, their share in world manufacturing income (value added) rose. Among the developing countries, it was mainly the East Asian economies that improved their share in world manufacturing income. The increase in exports of these developing countries can be attributed to greater share of labour-intensive, low value-added manufacturing. Thus an increase in export from these developing countries may have taken place without commensurate increases in income and value addition. Exhibit 2.5 compares the evolution of manufacturing trade and value addition in the G-7 countries with a group of seven of the more advanced developing countries (D-7). The developing countries included in the figure are Hong Kong, Malaysia, Mexico, South Korea, Singapore, Taiwan and Turkey. This clearly shows the trend currently followed in manufacturing. The high-value-added activities like manufacturing design as well as manufacture of high value added products still takes place in the developed countries where repetitive labour intensive, low value added work is contracted to the developing countries.

Exhibit 2.4

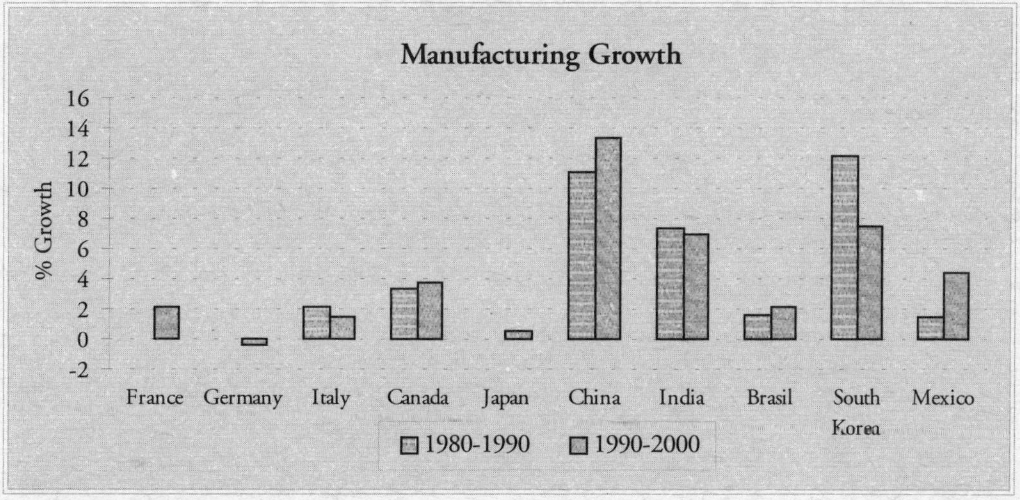

Source : 2002 World Development Indicators

Given these trends and increasing opportunities for developing countries, how is India placed to capture its share of the manufacturing pie? We seek to answer this by studying the manufacturing landscape in India.

EXHIBIT 2.5

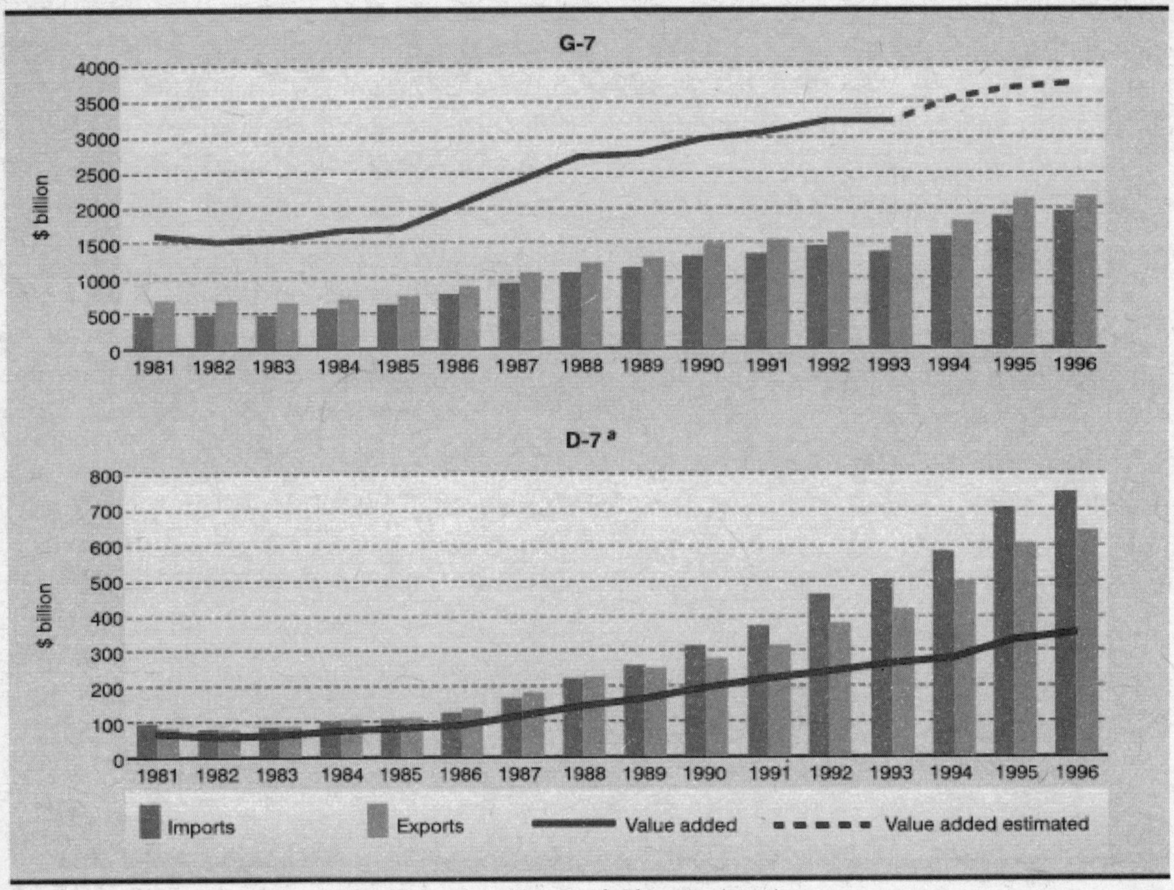

Source : UNCTAD secretariat calculations, based on Nicita and Olarreaga (2001)

KEY INFERENCES:

- Developed economies are increasingly outsourcing manufacturing activities to the developing countries.
- Most of the outsourced manufacturing is labour intensive, relatively lesser value added work
- This provides an opportunity for developing economies to increase manufacturing exports
- Countries like China and India are the key focus areas for such manufacturing growth in the future.

3

CHAPTER THREE

Manufacturing Landscape in India

3.1 Indian Manufacturing Lags Behind Other Key Developing Economies in Asia

Manufacturing plays a significant role in countries' economies. In less developed countries, the focus is more on agriculture and fulfilling of basic necessities. As countries become more prosperous, consumers spend more on value-added goods and services, and the role of manufacturing and services in the country's economy increases. In highly developed countries, services will be the dominant sector, since much of manufacturing in these countries is outsourced.

India being a developing economy, the importance of manufacturing cannot be under-emphasized. Indian manufacturing constitutes nearly one fifth of the economy. It employs nearly 1.8 million people in the organized sector alone, and contributes more than 75% of India's exports (Refer Exhibit 3.1)

Exhibit 3.1

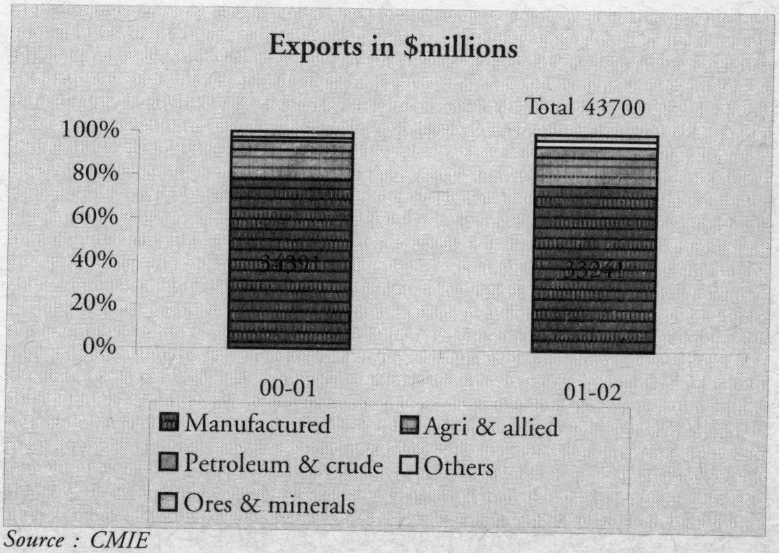

Source : CMIE

Thus the manufacturing sector has a significant role to play in India's prosperity and growth. India's manufacturing sector has been growing steadily over the years. Manufacturing value added grew at a rate of 7.4% during the period 1980-90. From $ 29.8 billion in 1980, the Gross Value Added (GVA) had increased to $ 72.8 billion in 1995. In the process, the share of manufacturing in India's GDP had gone up from 16% to 19%. (Refer Exhibits 3.2 and 3.3).

However, when compared to other key developing countries, India's performance is quite low. For example, against growth rate of 7.4% for India, the GVA growth rate of China

EXHIBIT 3.2

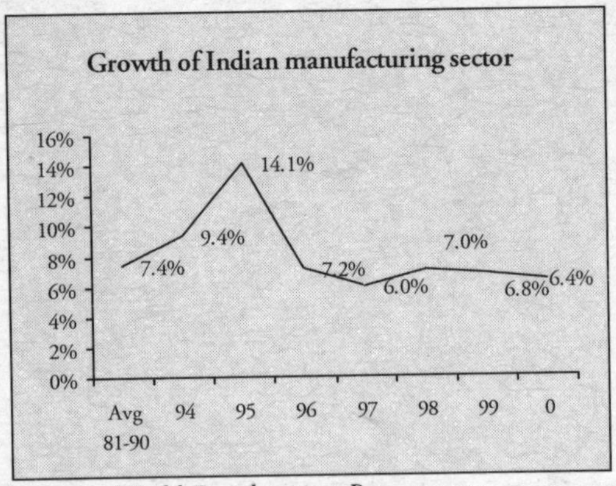

EXHIBIT 3.3

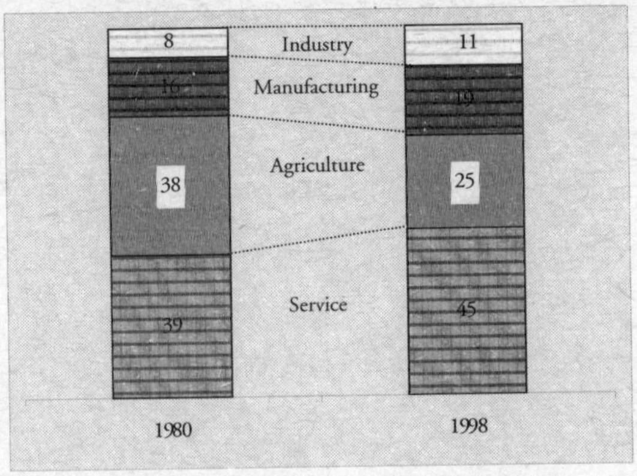

Source : World Development Report

was 10.7%, that of Korea 12.1%, and other ASEAN countries, 7.5%. The comparative value of GVA for China was $ 369 billion (nearly 5 times that of India) and Brazil was $ 179 billion. Even as a percentage of GDP, Indian manufacturing, at 19%, falls far below China (37%), Brazil (23%) and other SE Asian countries (ranging from 22% to 34%) (Refer Exhibits 3.4 & 3.5).

EXHIBIT 3.4

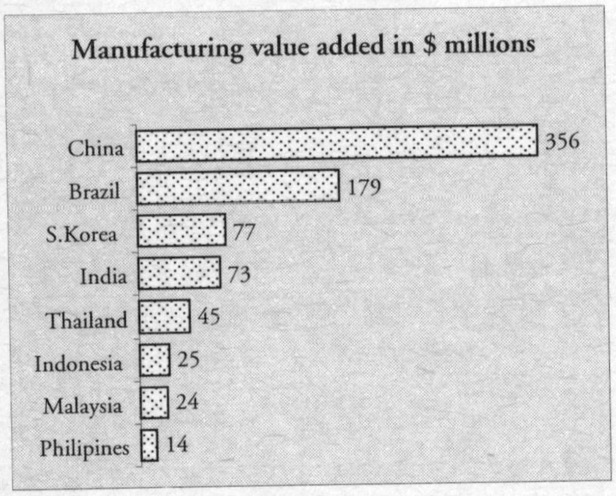

EXHIBIT 3.5

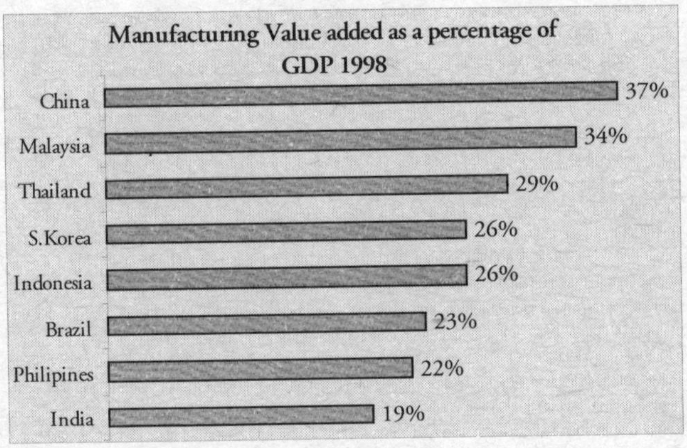

Source : World Development Report, 2000

Manufactured goods constitute 75% of India's exports (excluding services) - for 2001-02, out of total exports of $ 43.7 billion, manufactured goods constituted more than $ 33 billion. Most of the goods exported are low value, low tech items. High tech products constitute only 11% of India's exports, as compared to 67% for Malaysia, 21% for China and 18% for Brazil (Refer Exhibit 3.6).

EXHIBIT 3.6

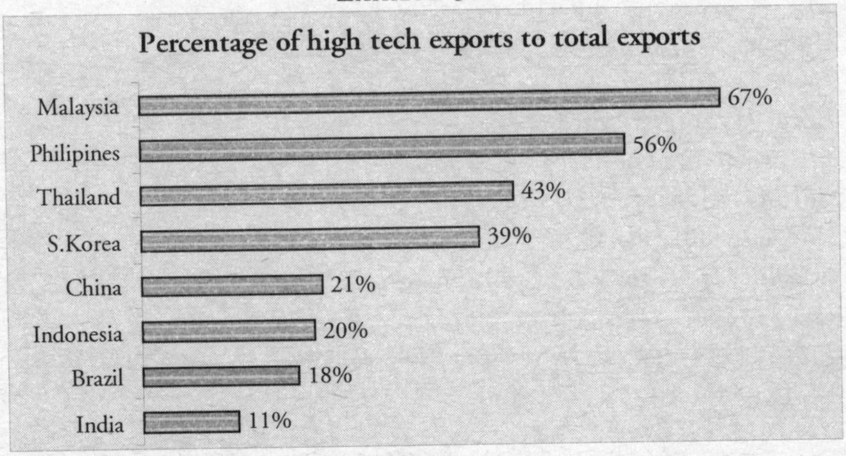

Source : World Development Report, 2000

It is evident that while Indian manufacturing has been growing in absolute terms, it is in grave danger of getting marginalized in the world market by the SE Asian Tigers and China. Thus, efforts at promoting growth in Indian manufacturing must also be aimed at making it more competitive in world markets.

3.2 Development of Indian Manufacturing Over the Years[1]

3.2.1 *Growth*

India's manufacturing sector has been registering healthy growth right from the 1950s to the present. In fact, growth in manufacturing has consistently outstripped the overall growth in GDP, indicating the increasing significance of manufacturing in the overall economy.

Table 3.1

Rates of Growth of India's GDP and Manufacturing - 1950 to 1998 (at constant 1993-94 prices)

Time period	Manufacturing	Registered manufacturing	Unregistered manufacturing	GDP at factor cost
1950s	6.4%	7.8%	5.1%	3.7%
1960s	4.2%	4.9%	3.4%	3.1%
1970s	5.0%	5.0%	5.0%	3.9%
1980s	7.3%	8.0%	6.2%	5.5%
1990s	9.3%	10.2%	7.5%	6.7%

Source: National Accounts statistics from EPW Research Foundation

Manufacturing growth started accelerating in the 1980s and got a further fillip in the 1990s. Several reasons are cited for this - investment in public infrastructure, gradual reduction in government controls and higher inflow of private investment in the 1990s being the key ones.

Manufacturing growth has not always led to a corresponding growth in employment generation. The 1980s can be called the period of 'jobless growth', since the high manufacturing growth rate of 7.4% during the decade translated into just 4,84,000 additional jobs in the registered manufacturing sector (in a total labour force of nearly 8 million factory workers). The key reason for low employment increase was the adoption of capital intensive production techniques by manufacturers. In the 1990s, the increase in investments and the setting up of more small and medium sized factories led to a resurgence in employment generation–nearly 1.8 million new jobs were created in the registered manufacturing sector during this decade. However, some studies have found that in the unregistered manufacturing sector, growth was actually negative during the mid 1980s to 1990s. Given that 80% of manufacturing employment is in this sector, this trend is significant.

In the later part of the 1990s, manufacturing growth has declined marginally. From a growth of 7% in 1998, the growth rate dropped to 6.4% in 2000. This may be a reflection of the overall slowdown in the economy–in fact, despite the slowdown, manufacturing growth continues to be higher than the growth in GDP.

In terms of employment, the manufacturing sector employed nearly 30.5 million people in 1999, out of which the majority (nearly 78%) were in small scale industries, khadi and village industries (Refer exhibit 3.7).

Manufacturing in India can be broadly classified into two categories–registered manufacturing and unregistered manufacturing. Registered manufacturing refers to the factory, or organized sector, and includes all factories using power and employing ten or more workers, or factories not using power and employing twenty or more workers.

[1] *Data presented in this section has been based on the Annual Survey of Industries data for 1979-80 to 1997-98, presented in 'A review of Indian Manufacturing' by JJ Thomas*

EXHIBIT 3.7 (A)

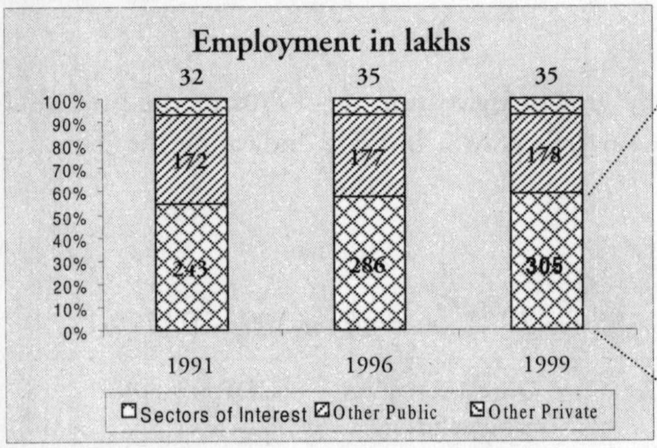

EXHIBIT 3.7 (B)

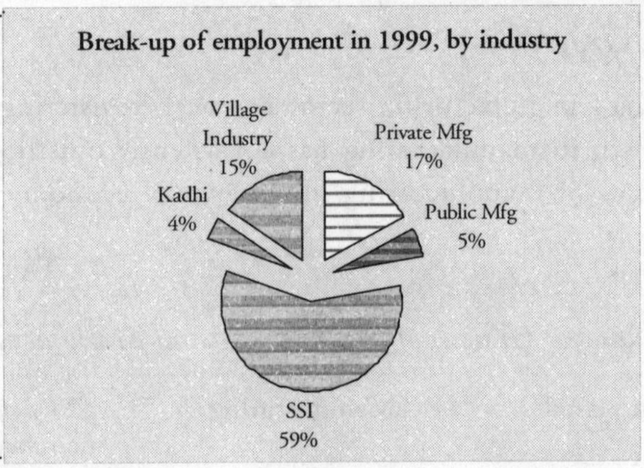

Source: *KVIC Annual Report, Ministry of Labour (DGE&T)*

EXHIBIT 3.8

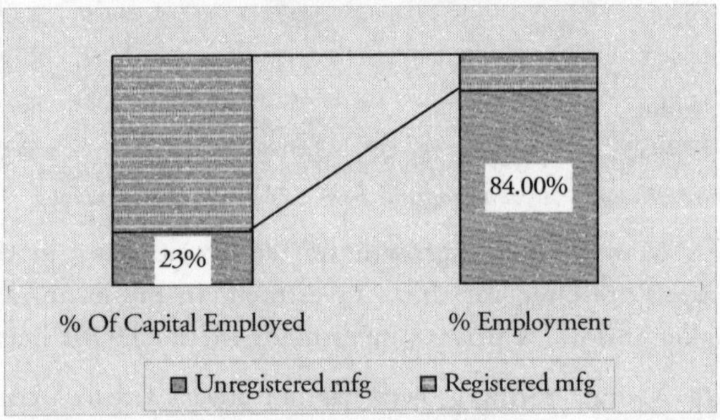

All other types of manufacturing activities fall under the unregistered category. Registered manufacturing has been growing at a consistently faster rate than unregistered manufacturing, and today accounts for nearly 2/3rds of India's manufacturing output. However, unregistered manufacturing accounts for nearly 80% of the manufacturing labour force (Refer Exhibit 3.8)

3.2.2 Composition and Growth Trends

The Annual Survey of Industries (ASI) identifies 24 manufacturing industries, which are classified into seven categories, as given in Table 3.2.

TABLE 3.2

Sl.#	Category	Industries in this category
1	Food and related	Food, Beverages, Tobacco
2	Textiles and related	Cotton textiles, Wool, Silk, Jute, Textile products
3	Wood, Leather and Paper	Wood, Leather, Paper and paper products
4	Chemicals and related	Chemicals, Rubber, Petroleum
5	Minerals, Metals and Metal products	Non metallic mineral products, Basic metals and alloys, Metal products and parts
6	Machinery and Transport	Machinery and equipment, Transport equipment
7	Others	Other manufacturing, Repair of capital goods, Electricity

Of these categories, Textiles and Food are highly labour intensive–each of these two categories accounted for nearly 18% of manufacturing employment in 1995-98. Chemicals, Metals and Machinery use relatively lesser. In terms of Gross Value Added (GVA) however, Chemicals, Minerals and Machinery are much higher than Textiles and Food (Refer Exhibit 3.9).

3.2.3 Trends

During the period 1979-80 to 1995-98, Chemicals and related industries have shown the maximum growth–their GVA contribution has grown from 18% to 26% over this period. In terms of employment, the

EXHIBIT 3.9

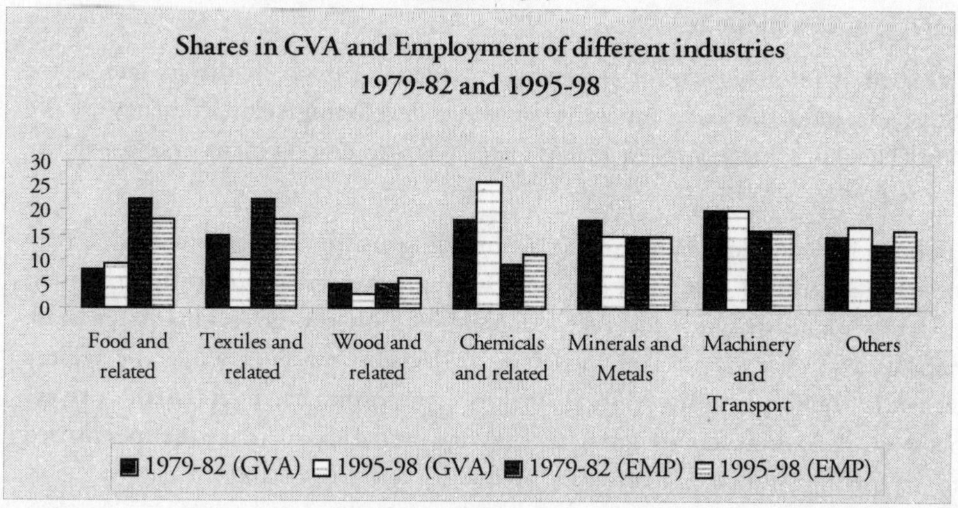

growth has been marginal, from 9% to 11%. Therefore, Chemicals and related industries have improved their productivity levels over this period. Food related industries have also improved in productivity, having grown marginally, from 8% to 9%, while their share of employment has dropped from 22% to 18%.

In contrast, GVA share of Textiles have dropped from 15% to 10%, while the drop in employment has not been proportionate (22% to 18%). Minerals and metals have dropped in GVA from 18% to 15%, while in terms of employment, their share has remained at 15%. Machinery and Transport have maintained their share of 20% in GVA and 16% in employment.

In terms of specific industries, Rubber and Petroleum were the high growth industries during the 1980s, having growth rates as high as 15%, while in the 1990s, Transport equipment had the highest growth rate of 21%, driven by the growing demand for passenger cars.

A key reason for the skew in growth rates across categories could be the level of investment in different industries. During the 1980s and 1990s, Chemicals, Metals and Transport equipment received the maximum investments, while labour intensive industries like Jute, Wood, Leather and Textiles received very low investments - just 0.5% to 1% of total investment went into these industries.

The above categories are all in the registered or organized sector of manufacturing. However, as already stated, 80% of manufacturing employment comes from the unregistered sector and SSIs, which continue to suffer from lack of scale economies and low productivity levels.

3.2.4 Conclusion

The trends indicate a gradual shift in growth, within the registered sector, from low value added, labour intensive industries to high value added, capital intensive ones. However, the bulk of employment continues to be in the former. There is also a clear dichotomy between registered and unregistered manufacturing in terms of growth rate and labour force. While the registered sector has cornered most of the investment and value add, the unregistered sector contains most of the labour pool, which is a key competitive advantage for India.

In order to leverage labour resources effectively, any policy aimed at future growth in Indian manufacturing has to focus on improving competitiveness of labour intensive industries in the registered sector, as well as the unregistered sector.

3.3 SLOW PACE OF REFORMS AND INEFFICIENT IMPLEMENTATION HAVE MADE INDIAN MANUFACTURING UNCOMPETITIVE

The Indian government's policies with regard to industry, foreign investment and trade have been evolving over the years. In the early years of independence, the focus of the government was on attaining self-sufficiency in all sectors of the economy, protecting and nurturing indigenous firms, generating employment, promoting small industries and preventing private sector monopolies.

To this end, the government increased state ownership of industries, restricted growth in the private sector and controlled foreign investment. Such measures did have inherent sense in developing self-sufficiency in the case of essential goods like food grains. They also helped to an extent in generating employment and providing basic social security.

Over the years, however, these policies had the detrimental effect of breeding inefficiencies, low productivity and lack of professional management in most Indian companies. Wide ranging government controls prevented companies from developing independent capabilities and increased dependence on the government. In state owned enterprises, government ownership and bureaucracy led to delays in decision making, while the regime of reservations, price controls and subsidies removed motivation to improve performance. Even in the private sector, protection from global competition, lack of access to latest technology, and lack of investments ensured that Indian manufacturing remained far behind global benchmarks in quality, efficiency and costs.

In terms of infrastructure and utilities too, government monopoly and lack of competition led to inefficiencies, high costs and poor service levels. In turn, these factors have increased the operating costs of Indian manufacturers, making them still less competitive in the world markets. In a recent survey conducted by KPMG, some industry heads ruefully referred to these as 'India costs' - the cost of doing business in India!

The absence of a free market and competition gave the Indian consumer a raw deal. Customers had limited choice and often had to be content with sub-standard products and services.

In 1991-92, faced with twin crisis of industrial slowdown and mounting imbalance in balance of payments, the government initiated a process of structural reforms. These were aimed to 'evolve an industrial and trade policy framework, which would promote efficiency, reduce the bias in favour of excessive capital intensity and encourage employment-oriented pattern of industrialization'.

As a result of the reforms, Indian industry has been opened out to competition from global players. Indian manufacturers are also freer to grow, invest and compete in global markets. However, the pace of reforms has been slow, especially with regard to freeing government controls. Also, procedural and bureaucratic delays in approvals and decision-making remain to a large extent. As a result, Indian manufacturing has not developed rapidly enough to keep pace with competition.

Tables 3.3 to 3.5 summarize the evolution of India's industrial, trade and FDI policy from 1947, into the 1990s.

TABLE 3.3
Industrial Policy

1947 - 1970	**Increase Government Ownership** • Reserve specific industries for state ownership • Identify specific industries where state will take dominant position • Licensing for specific industries
1970s	**Limit Free Market Development** • Compulsory licensing for all industries with investment above a certain level • Specify industries with compulsory licensing • SSI reservations • Specify industries to which private business houses and foreign companies will be confined

1980s	**Limited de-control** ♦ Raise minimum asset limit for defining business houses from Rs.200 million to Rs.1 billion ♦ Remove restrictions on business houses on certain industries if they enter industrially backward areas or make specific export commitments
1991	**De-licensing and de-control** ♦ Abolish licensing in all except 18 industries ♦ Abolish MRTP ♦ Open oil exploration, refining to private sector ♦ Open power to private investments ♦ Overhaul excise structure, revise excise tax rates
1995	**Further relaxation in licensing, controls and reservations** ♦ Attract private investment in infrastructure ♦ Liberalize policy for drugs and reduce span of price control ♦ Remove licensing on consumer electronics ♦ Remove price controls on coal ♦ Further reduce industries requiring licensing to 9 ♦ Remove reservations selectively in SSI sector

TABLE 3.4

Trade Policy

1947 - 1970	**Restrict imports** ♦ Designate government agencies as sponsoring authorities for imports ♦ Allow imports only if certified under 'essential' or 'indigenous clearance' (domestic procurement not possible) ♦ Government monopoly in import of certain goods ♦ Set up DGTD to monitor indigenous capacities and production, assist in 'indigenous clearance' certification

1970s	**Import segregation**
	♦ Do away with indigenous clearance
	♦ Categorize imports into four groups:
	  ♦ Banned goods
	  ♦ Restricted goods - imports allowed if not available in the domestic market
	  ♦ Open General License - import permitted without license by actual users
	  ♦ Canalized goods - imports allowed only to government monopolies

1980s	**Limited de-control** ♦ Encourage import substitution through phased manufacturing programs ♦ Increase list of goods under OGL
1991 onwards	**Opening up of free trade** ♦ Free trade of exim scrips ♦ Reduce peak tariffs ♦ Remove negative list of imports ♦ Remove user requirements for imports of capital goods, raw materials and components under OGL ♦ Reduce basic import duties

Contd...

TABLE 3.5

FDI Policy

1947 - 1970	**Promote Indian ownership** • Restrict operations of managing agencies • Set high corporate tax on foreign companies • MRTP to handle all applications for licensing from big business houses or subsidiaries of foreign companies
1970s	**Discourage foreign investment** • Specify industries where foreign firms would be allowed to operate • Do not allow any foreign investment in industries reserved for the government • Reduce foreign ownership to less than 40% in foreign companies, unless engaged in specific core activities
1980s	**Limited de-control** • Set lower limit for referring a company to MRTP, to Rs.1 billion
1991 onwards	**Promote FDI** • Raise limit on foreign equity holdings from 40% to 51% • Set up Foreign Investment Promotion Board to promote FDI on a case by case basis • Allow automatic approval for FDI in specific industries • Reduce taxes on royalties

Based on the analysis, key questions that arise are:

- What are the comparative advantages that Indian manufacturing enjoys?
- What are the key issues that need to be addressed for leveraging the advantages?
- How can Indian manufacturing become globally competitive?

We address these by analyzing the key enablers and barriers to manufacturing growth in India:

KEY INFERENCES:

- ☞ Indian manufacturing faces stiff competition from other developing economies in Asia and elsewhere, in the domestic and global markets.
- ☞ There is a need for balanced growth between capital intensive and labour intensive manufacturing, and to make the unregistered sector more competitive.
- ☞ The pace of policy reforms needs to be stepped up, and government functioning needs to be more efficient, to make Indian manufacturing globally competitive.

4

CHAPTER FOUR

Making Indian Manufacturing Globally Competitive - Key Enablers And Barriers

India has several inherent strengths, which have contributed to its steady growth over the years, and which need to be effectively leveraged for the future. At the same time, there are significant barriers, which stand in the way of the manufacturing sector growing to its full potential. In this chapter, we discuss the key enablers and barriers, and also look at examples of China and Taiwan, to see how India can utilize its strengths more effectively.

Figure 4.1 shows the framework we used to understand the enablers and barriers

FIGURE 4.1

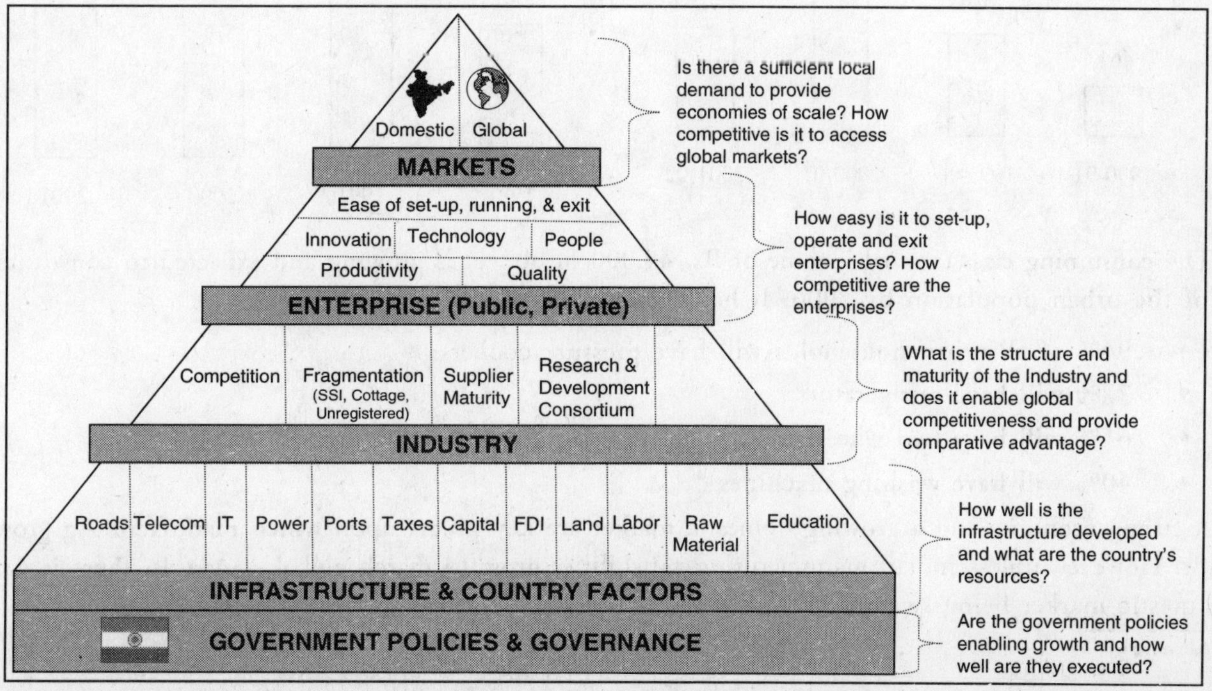

4.1 India Enjoys Several Key Advantages that can Contribute to Manufacturing Growth.

4.1.1 Availability of Raw Materials:

India has large domestic resources of metals, minerals, rubber, wood, oil and gas, which power domestic manufacturing. It is the third largest producer of cotton in the world. It also has about 120 billion tons of coal reserves, making it fourth largest in the world. There is abundant availability of iron ore, manganese, mica, bauxite, titanium, limestone and natural gas. India is one of the world's largest producers of milk, tea and leather. With its long coastline, fisheries and marine products are another area where it has resources in abundance.

Availability of a wide range of raw materials is a competitive advantage for India, as it reduces cost of inputs for manufacturing. Also, production of cash crops like cotton and rubber directly assist manufacturing growth as they act as immediate inputs for industry. India's coal reserves can go a long way in providing for ever-increasing power needs. However, the key to leveraging these resources lies in efficient utilization and avoidance of waste.

4.1.2 Increasing Local Demand, Powered by a Growing Economy

India's economy has been growing steadily over the last two decades. This has led to an increase in disposable income levels, as well as private consumption spending (Refer exhibits 4.2 and 4.3)

Exhibit 4.1

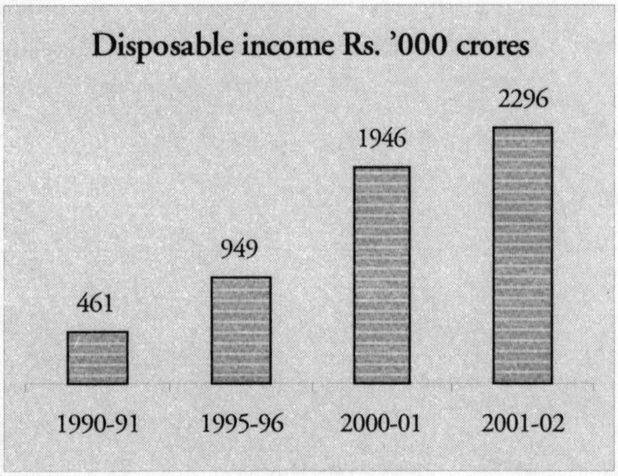

Exbibit 4.2

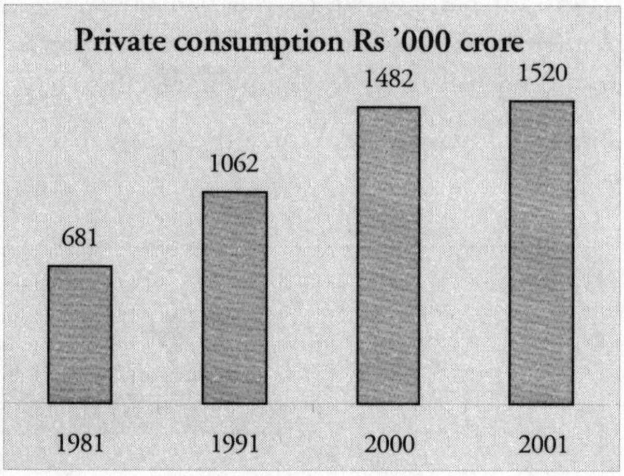

The consuming class (annual income of Rs. 45,000 or above) is growing and expected to constitute nearly 50% of the urban population by 2006. It has been estimated that by 2007[2],

- 94% of all urban households will have pressure cookers
- 77% will have refrigerators
- 61% will have two wheelers
- 40% will have washing machines

A strong economy and a growing domestic market are key pillars upon which manufacturing growth can be built. However, unless Indian manufacturing capabilities grow to match global standards, there is danger of the domestic market being swamped by imports.

[2] *NCAER data*

4.1.3 Growing Service Sector

India's service sector, as already mentioned, has experienced explosive growth and constitutes nearly half of India's GDP. Growth in services will have a complementary effect on manufacturing as well. For example, growth in demand for software services will also lead to demand for computer hardware, peripherals and electricity. Growth in financial services and easy availability of finance will aid the purchase of consumer durables.

Continuing growth of India's service sector will, in the long term, stimulate demand for manufactured goods.

4.1.4 Low Labour Costs

India has one of the lowest costs of labour among developing countries. Labour cost as a percentage of value added is only 21% in India, as compared to 23% for China and 30% for Taiwan (Refer exhibit 4.3). Lower cost of labour will make Indian products cheaper in world markets and is a competitive advantage that can be leveraged for manufacturing growth.

4.1.5 Growing Population of Indian Diaspora

Indian Diaspora, composed of non-resident Indians (NRIs), people of Indian origin (PIOs) and their descendents, living in other countries around the world, are currently estimated to number around 20 million[3]. They are over a million strong each in 11 countries. Many of them have strong ties with India and can be a huge potential source for FDI for Indian industry. In the case of China, for example, more than 50% of FDI into the country flows from non-resident Chinese.

EXHIBIT 4.3

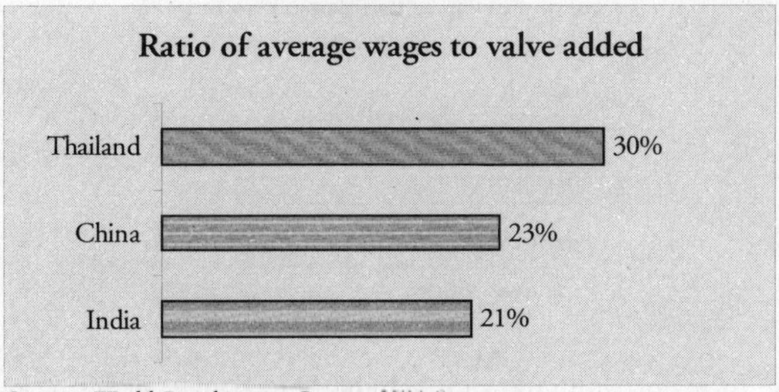

Source : World Development Report, ICU firm surveys

However, despite the above favourable factors, Indian manufacturing is yet to develop its potential, as several barriers hamper it. These barriers exist at the macroeconomic, sectoral as well as individual firm levels.

4.2 SEVERAL MACROECONOMIC BARRIERS HAMPER THE GROWTH OF INDIAN MANUFACTURING

4.2.1 Poor Infrastructure

- **Roads:** Despite having one of the most extensive road networks, totaling about 3 million kilometers, India's roads are mostly in poor condition. Most of the roads are still to be paved (Refer exhibit 4.4). Poor road conditions hamper transport and distribution of materials and add significantly to freight costs, besides disallowing easy access to different areas. As such, lack of good road infrastructure is a significant deterrent to manufacturing growth. The government's initiatives in Highways development, such as the Golden Quadrilateral project, are steps in the right direction, and need to be completed quickly and efficiently.

[3] *Data from Ministry of External Affairs*

EXHIBIT 4.4

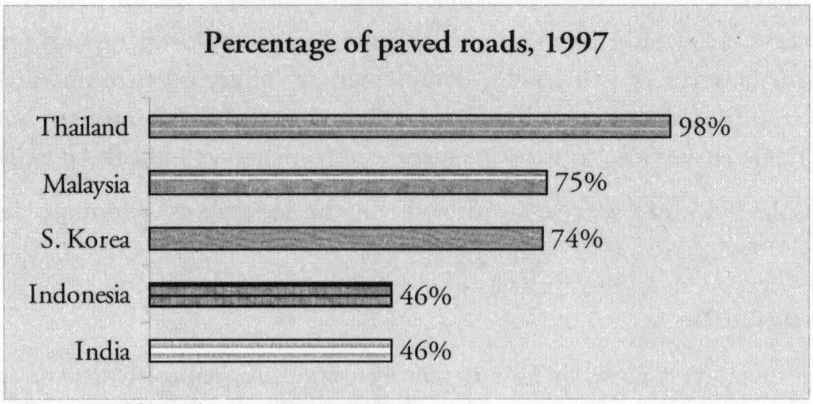

Source : World Development Report 2000

- **Ports:** India has 12 major ports and a number of minor ports along its long coastline. However, performance of most ports is highly inefficient, leading to long lead times and high costs of operations. Average time taken to clear goods at Indian ports is about 10.6 days, as compared to 7 days for Taiwan and less than 24 hours for Singapore. Shipping and transport costs form 8.5% of exports in India, as compared to 5.4% for China (Refer Exhibit 4.5).

EXHIBIT 4.5 (A)

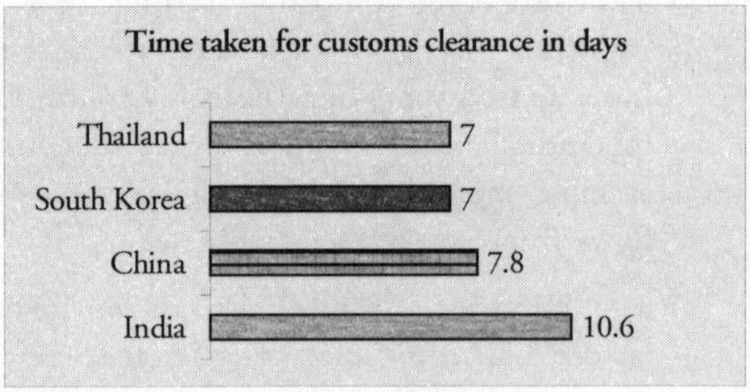

EXHIBIT 4.5 (B)

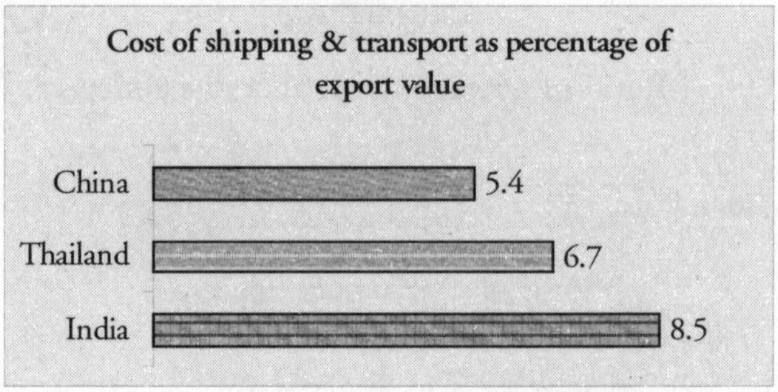

Source : World Bank CII firm level analysis 2000

Despite being strategically located close to major shipping routes, India is in danger of losing out in terms of trade, due to its inefficient port infrastructure.

- **Power:** Electricity generation and transmission in India is state owned, and most electricity boards function inefficiently, with unreliable power supply, inefficient metering and large transmission losses. India also has one of the highest costs of electricity to industry. As a result, most Indian manufacturers are forced to depend upon their own generators to fulfil their power requirements (Refer exhibit 4.6)
- **Communication:** India lags behind its Asian counterparts in terms of penetration levels of basic and mobile telephones. Number of telephone mainlines per 1000 people is just 32 in India, vis-à-vis 92 in Thailand, 112 in China and 199 in Malaysia. Number of mobile phones per 1000 is just 4 in India, as compared to 50 in Thailand, 66 in China and 213 in Malaysia. India has just 45 computers per 1000 people, as against 59 for China, 243 for Thailand and 1031 for Malaysia. In terms of Internet penetration, India, with 5 million Internet users, is still far behind China 22.5 million)-refer Exhibit 4.7.

Absence of a world class communication infrastructure leads to slow and in-efficient communication, which affects information flow and decision making and hampers business development.

EXHIBIT 4.6(A)

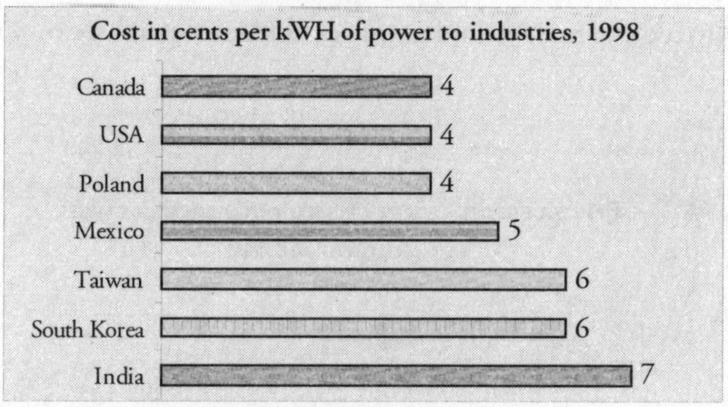

Source : World Energy Outlook 2000

EXHIBIT 4.6 (B)

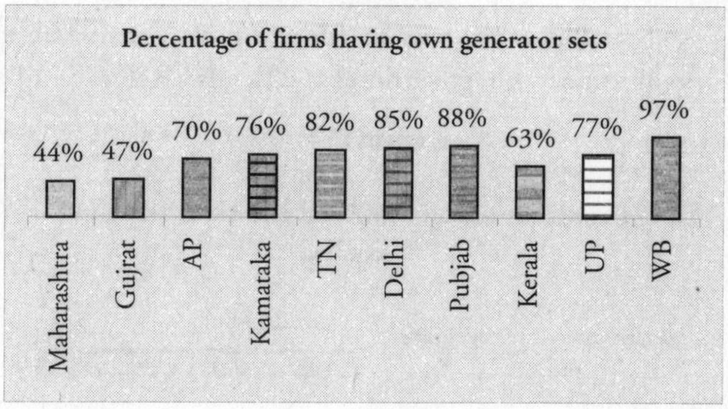

EXHIBIT 4.7

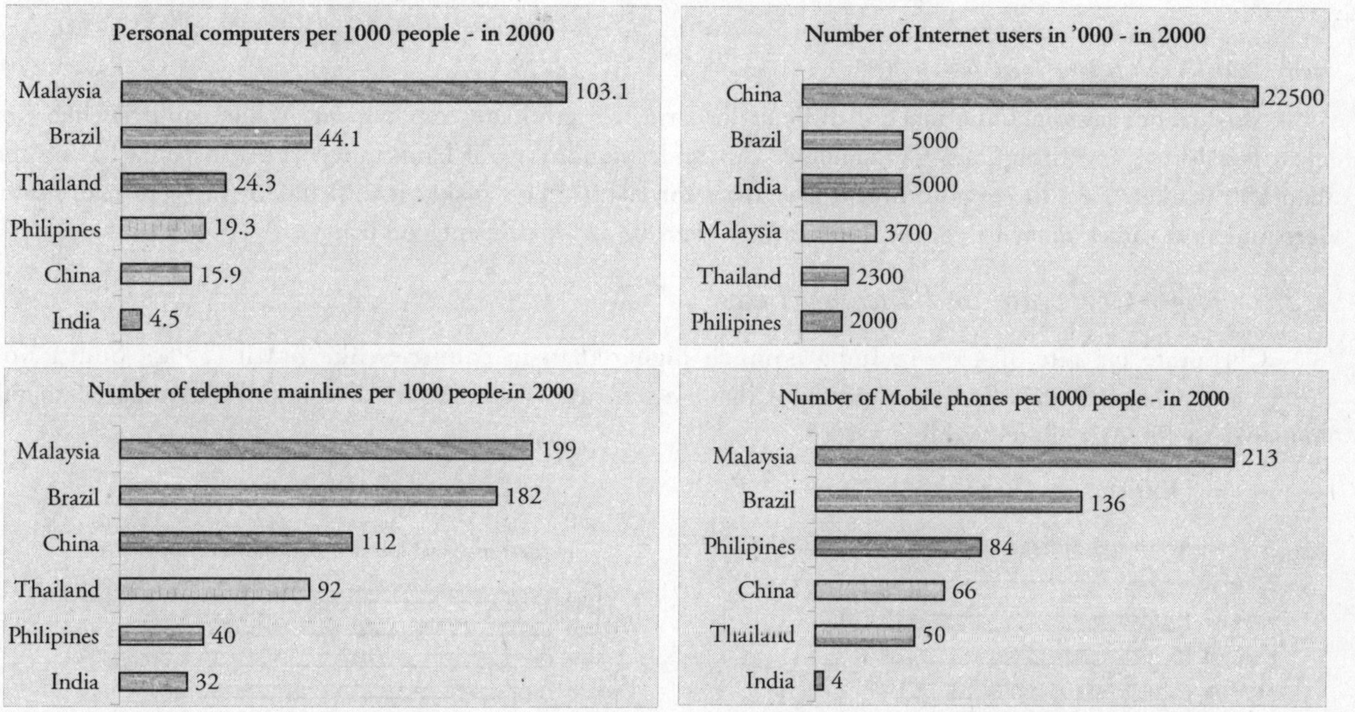

Source : World Bank

4.2.2 *High Level of Inefficiencies in Government Services*

Apart from poor infrastructure, poor service levels, delays and excessive bureaucracy in government departments are other problems faced by companies seeking to do business in India. Most firms rate the quality of service rendered by public agencies as poor. (Refer Exhibit 4.8).

EXHIBIT 4.8

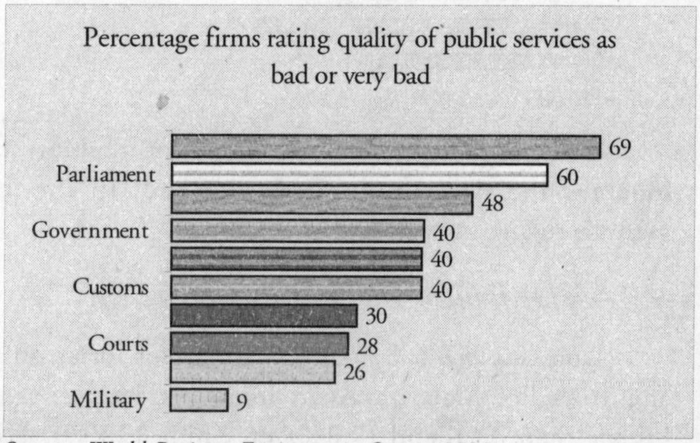

Source : World Business Environment Survey 2000

4.2.3 Difficulties in Starting and Running Businesses, Lack of Flexibility in Labour

Starting a business in India entails considerable procedural delays and requires the management to spend a lot of time with government officials (Refer Exhibit 4.9).

EXHIBIT 4.9(A)

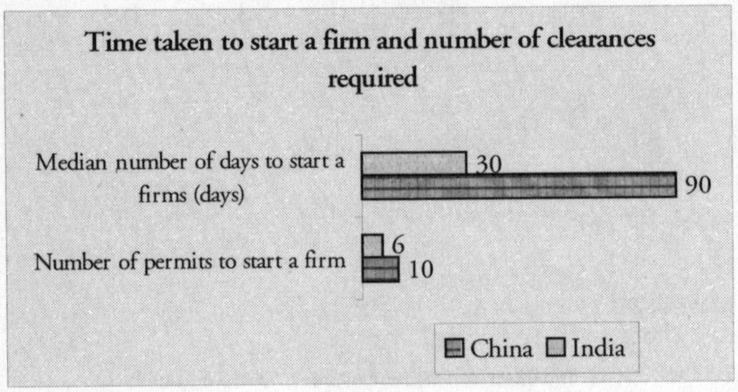

EXHIBIT 4.9(B)

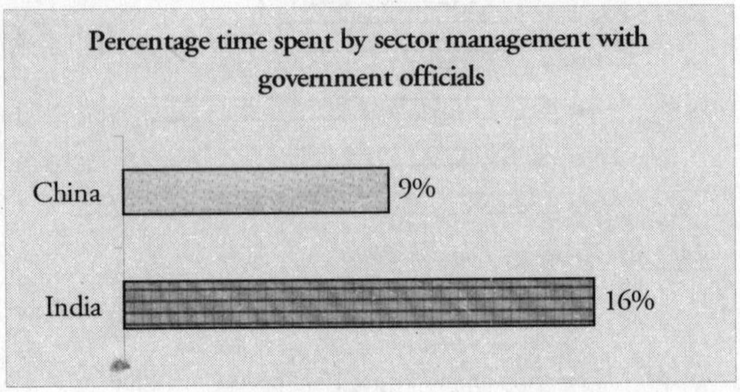

Source: World Bank CII Firm Level Analysis 2000

Also, labour laws in India make it difficult for firms to terminate employment. While countries like Korea allow free hiring and firing, and China allows this under conditions of bankruptcy, firms in India having more than 100 workers have to get government clearances for layoffs. This makes it difficult for firms to take business decisions and forces them to continue operating, even under inefficient conditions.

4.2.4 High Corporate and Income Taxes

Corporate tax rate of 40% in India is much higher than in countries like Brazil (15%), South Korea (28%) and China (30%). Similarly, personal income tax rates in India are also high, as compared to most countries in SE Asia. Refer Exhibit 4.10.

EXHIBIT 4.10(A)

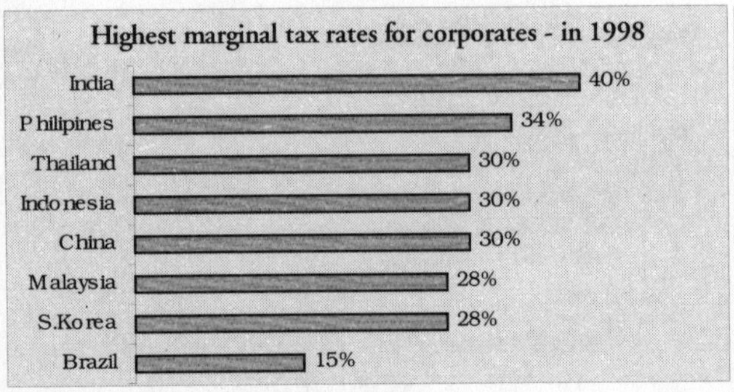

EXHIBIT 4.10(B)

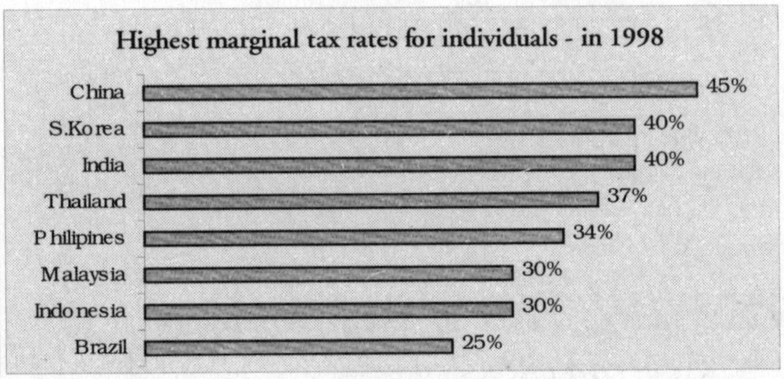

Source: World Development Report 2000

While high corporate taxes affect profitability of businesses, high individual tax rates reduce disposable incomes and dampen domestic demand. In the long run, both these factors can thwart industrial and manufacturing growth and profitability.

4.2.5 Difficulty In Obtaining Credit.

Domestic credit by the banking sector forms only 48% of GDP in India, as compared to 120% for China and 162% for Malaysia. Apart from this, interest rates in India, though reduced recently by the RBI, are still higher than Thailand (7.8%) or China (5.8%). High interest rates increase the cost of funds and affect investment. In fact, high interest rates, collateral requirement and paperwork in getting credit are cited as amongst the biggest obstacles in doing business in India. (Refer exhibit 4.11).

EXHIBIT 4.11 (A)

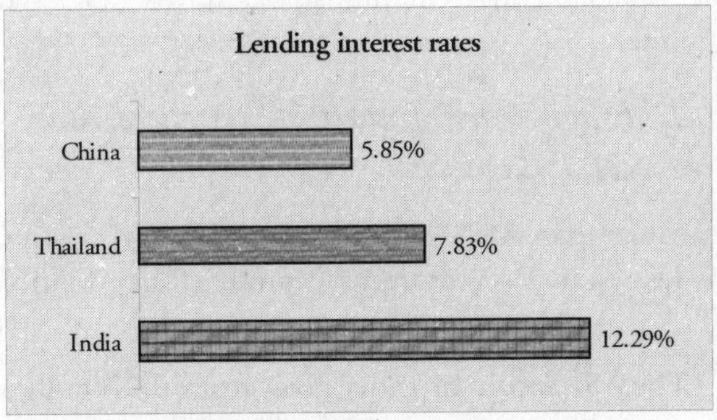

EXHIBIT 4.11 (B)

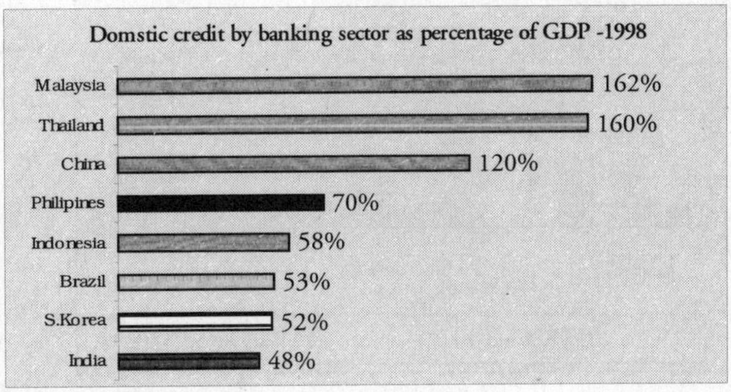

EXHIBIT 4.11 (C)

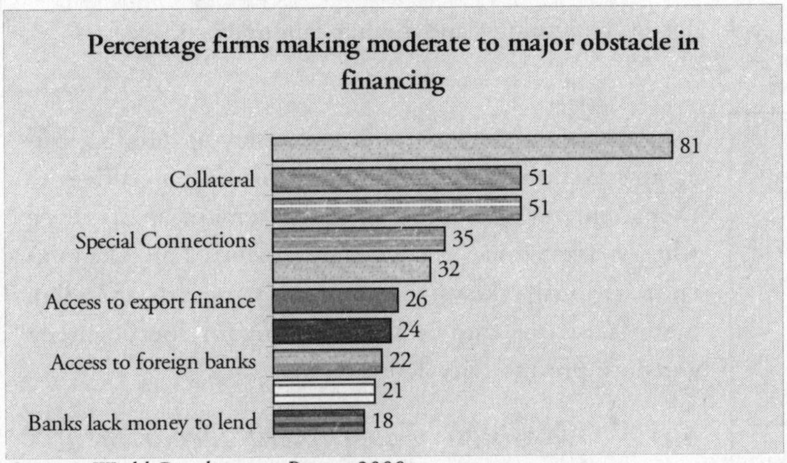

Source : World Development Report 2000

4.2.6 Minimal Foreign Direct Investment

Apart from difficulties in obtaining credit from domestic banking sector, investment in India is further affected by relatively low inflow of foreign direct investment (FDI). FDI inflow into India is less than a tenth of that into China. As a percentage of GDP, FDI accounts for only 0.5% in India, as compared to 4.1% for China and 4.7% for Malaysia.

Unless India is able to attract a greater share of the FDI flowing into Asia, it stands to lose out to China and other Asian economies.

EXHIBIT 4.12(A)

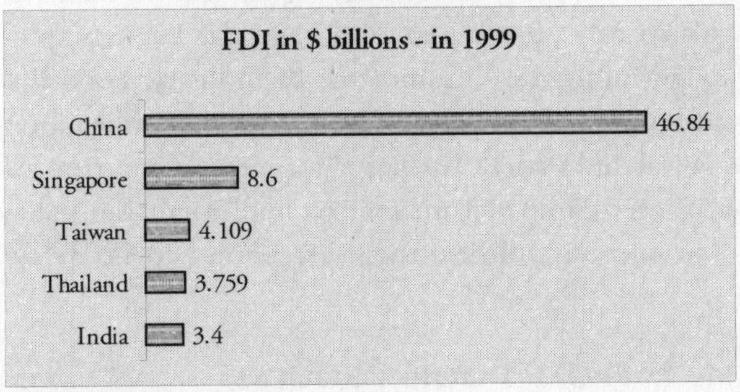

EXHIBIT 4.12 (B)

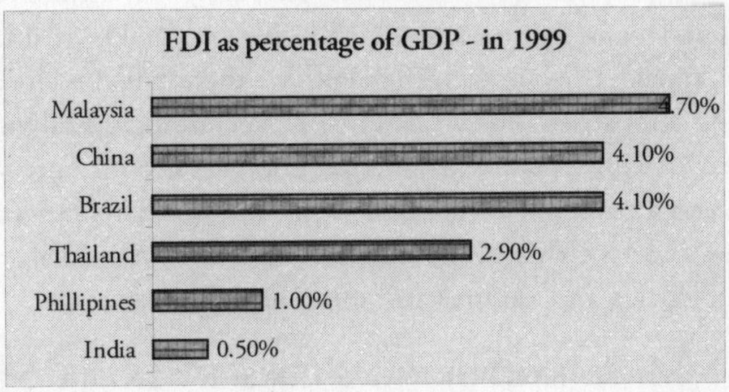

Source: World Development Report 2001

4.3 APART FROM THE MACROECONOMIC BARRIERS, THERE ARE ALSO A FEW KEY SECTOR LEVEL BARRIERS AFFECTING INDIAN MANUFACTURING

4.3.1 Low Labour Productivity

India's advantage of low labour costs is almost negated by low productivity levels. This is mainly due to

EXHIBIT 4.13

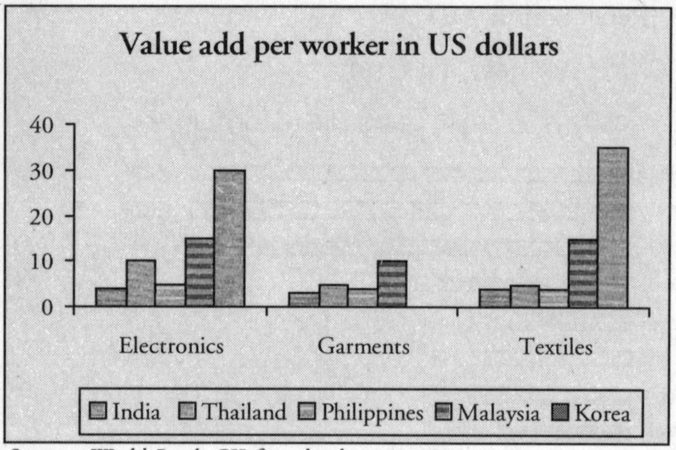

Source : World Bank CII firm level survey 2002

the fact that majority of the labour force in India is employed in the unorganized sector, where the value added is low. Exhibit 4.13 shows the comparison of value added in different industries, across key Asian economies.

4.3.2 Fragmented Industry Structure, due to the SSI Policy

More than 85% of Indian manufacturing units have less than Rs.1 crore as capital. (Refer Exhibit 4.14).

The SSI sector in India grew from 0.87 million units in 1980-81 to over 3.2 million units in 2000. Most of these units are sick and unable to compete with competition from international players. They suffer from lack of scale economies, poor quality levels, lack of new technology and higher labour costs.

EXHIBIT 4.14

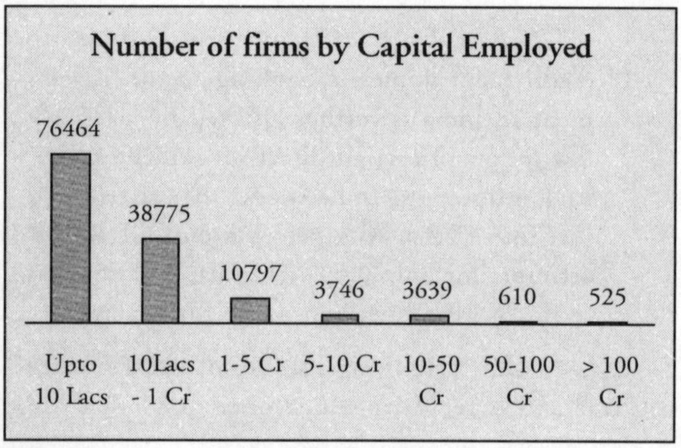

The government's policy of reserving product categories exclusively for SSIs, and providing incentives to them, encourages these firms to remain small, even when market forces dictate they expand. This phenomenon seriously skews the industry structure of Indian manufacturing, and is a key reason for inefficiencies and low productivity levels.

4.3.3 No Single Government Department in Charge of Manufacturing

The central government's structure impedes effective policy implementation. There are 40 different ministries reporting directly to the Prime Minister. Such a structure dilutes responsibilities and makes it difficult to drive forth a common agenda. For example - if we consider 'Manufacturing Policy', there will be close to 14 ministries - Commerce & Industry, Food Processing Industries, Power, Science & Technology, Railways, Shipping, Steel, Heavy Industries & Public Enterprises, Labour, Road Transport & Highways, SSI, Agro & Rural Industries, Textiles, Finance & Company Affairs and Petroleum & Natural Gas - that need to function in a co-ordinated manner to implement the policy. Given the fact that each ministry has its own priorities and accountabilities, there are bound to be delays, inefficiencies and failures in taking key decisions.

4.4 MANUFACTURING FIRMS IN INDIA SUFFER FROM POOR PROCESS EFFICIENCIES, LACK OF INNOVATION AND LOW QUALITY LEVELS

4.4.1 Lack of Innovation

India does not aggressively promote innovation, which is the key to manufacturing growth in the world markets. There is very low emphasis on promoting innovation, either in education or in business. School curricula focus heavily on classroom teaching and do not promote innovative teaching methods to kindle creative thinking and working. Even so, the percentage of scientists and engineers in the population is also lower in India compared to countries like China (Refer exhibit 4.15 (a)).

Indian businesses give relatively low importance to developing capabilities in indigenous innovation and development. Indian automotive companies, for example, spend just about 0.8% of their turnover on R&D, as compared to 3% to 6% for countries like Japan and Germany respectively (Refer exhibit 4.15 (b)).

Exhibit 4.15 (a)

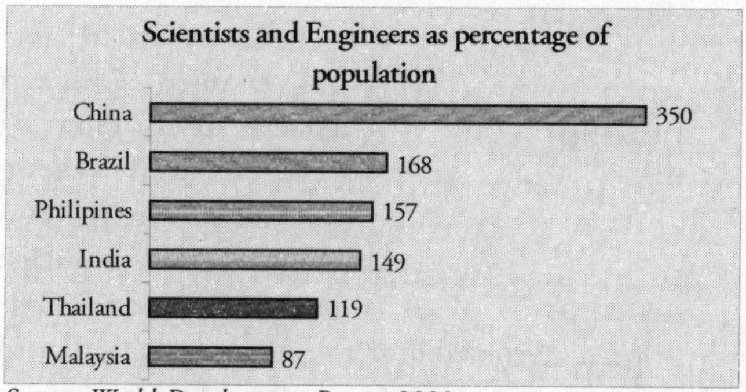

Source: World Development Report 2000

Exhibit 4.15 (b)

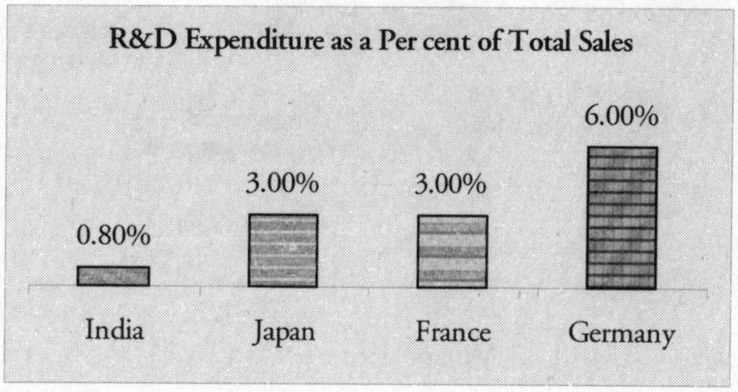

4.4.2 *Low Quality Levels*

Average quality levels achieved by Indian manufacturing firms would rule them out of competition in most world markets. In the automotive ancillary industry, for example, quality levels achieved by Indian firms are still at 3000-5000 PPM levels, whereas most global players are aiming for sub-100 PPM levels. In a recent survey conducted by KPMG, poor product development capabilities and inadequate quality levels were cited among the key factors that adversely affect Indian exports. This also leads to the 'Made in India' brand having a poor image abroad.

There are many more barriers that affect manufacturing in India, and we shall not delve into all of them in detail. However, the following factors also merit mention:

1. Lack of focus on process efficiencies and IT.
2. Lack of co-ordination among government, research institutes and academia.

In summary, there are several factors that need to be addressed for Indian manufacturing to grow and realize its potential. These factors encompass initiatives to be taken at the government, industry and individual firm levels. A co-ordinated effort across all concerned agencies would be required to achieve this.

4.5 Comparison with Other SE Asian Economies

4.5.1 *China*

China's economic liberalization started in the late 1970s. Right from the beginning, China followed a 'dual track' approach, which was aimed at softening the shocks of de-regulation.

Fig. 4.2 gives the schematic of China's liberalization policies.

The dual track policy has enabled China to transition smoothly from a planned economy to a market driven one. China's GDP growth has consistently been higher than India's - in fact, the growth has been over 8% for the last 25 years. Compared to the Indian economy, manufacturing plays a much stronger role in China's economy. While the contribution of industry in India's GDP has been gradually growing and is currently at

FIGURE 4.2

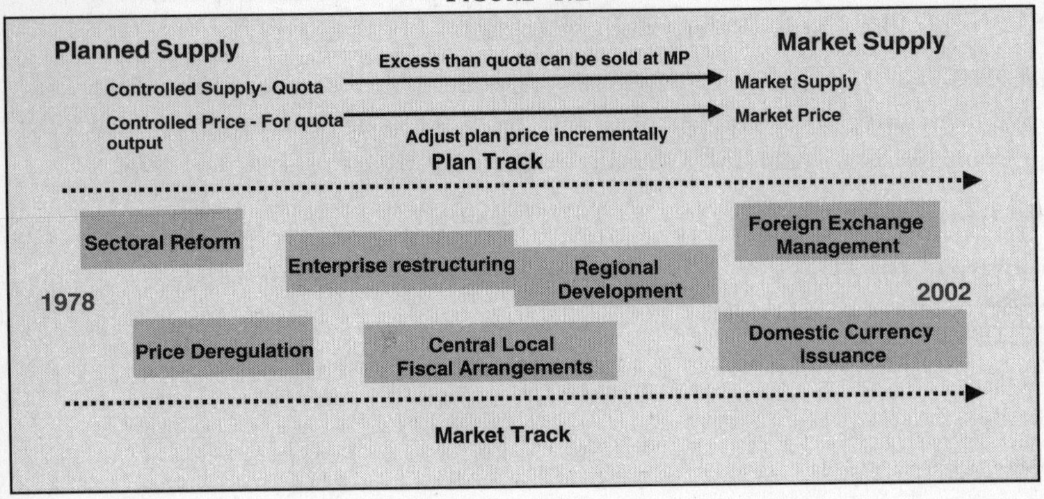

about 30%, China has constantly had a strong industry sector, contributing to nearly half of the country's GDP, over the last two decades (Refer Exhibit 4.16).

An export led focus has enabled China to achieve steady growth in exports, and exports constituted nearly 30% of GDP in 2000. Similar to India's SSI sector, China has a large Town and Village Enterprise (TVE) sector, which is a major contributor to industry. TVEs contributed largely to the expansion in China's domestic market and also formed about 20% to 40% of China's exports in the 1990s (Refer Exhibit 4.17).

EXHIBIT 4.16 (A)

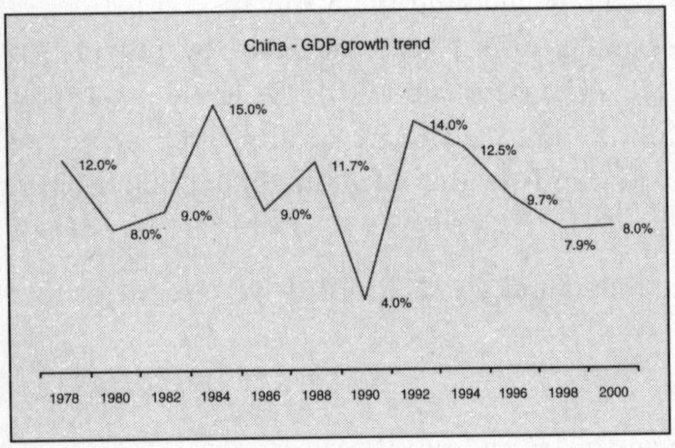

EXHIBIT 4.16 (B)

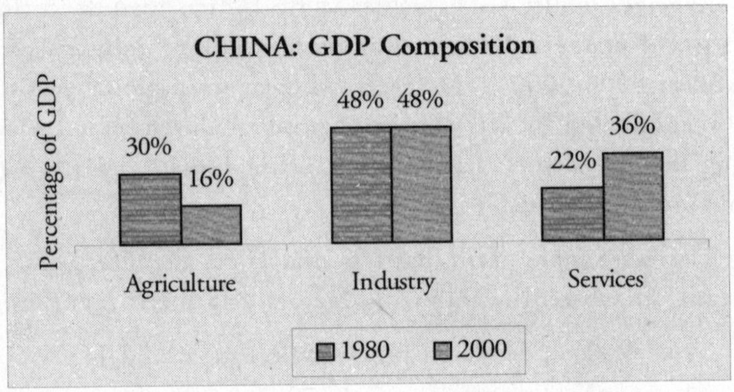

Source : EIU country reports

EXHIBIT 4.17

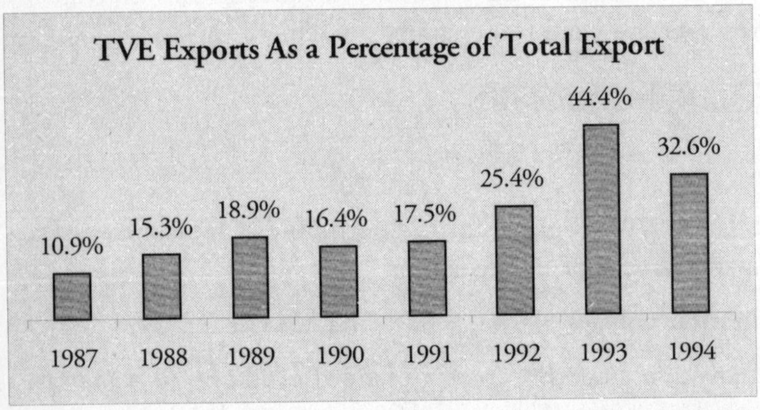

However, unlike Indian SSIs, the TVEs in China are more liberalized and open to respond to market forces. Their share of employment is also lower than in Indian SSIs.

China has been able to attract foreign direct investment (FDI) inflows much more than India. The bulk of the inflow (more than 50%) is from Chinese residing overseas. In contrast to India's policy of curbing foreign investment, China historically encouraged the formation of joint ventures (JVs) between foreign investors and Chinese firms, as a means of growing the domestic industry while encouraging FDI. Figure 4.3 indicates how China's laws concerning FDI have evolved over the years.

FIGURE 4.3

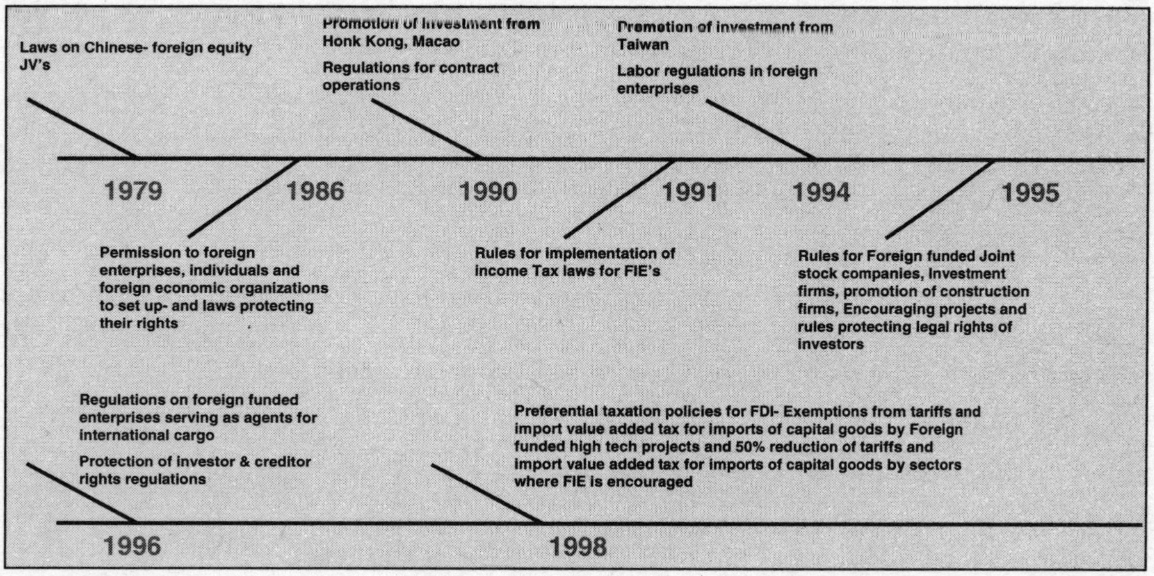

India can definitely benefit by adapting and applying key features of the Chinese model, in areas like:

- De-reserving the SSI sector
- Improving FDI inflow

We shall return to these themes in more detail in later sections.

4.5.2 Case Study: Taiwan

Taiwan's GDP has grown rapidly during the 1990s, driven mainly by exports. Exports have consistently been at over 40% of GDP during this period. Almost the entire export value (98.5%) comes from manufacturing, out of which the bulk (41%) is accounted for by Electronics. (Refer Exhibit 4.18).

EXHIBIT 4.18(A)

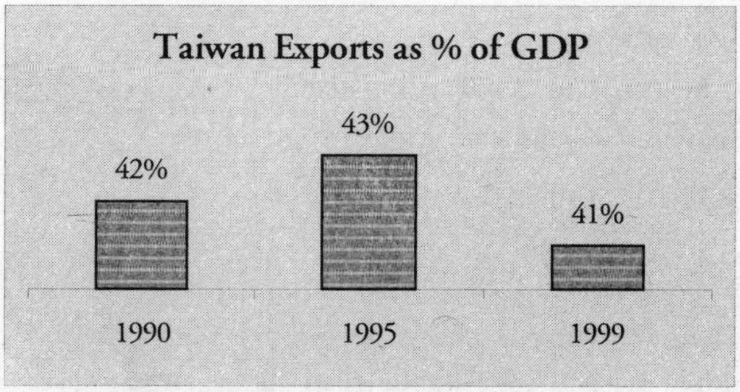

EXHIBIT 4.18 (B)

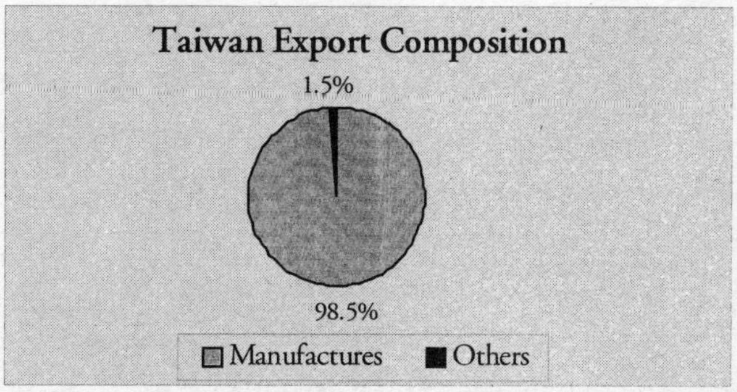

Source : World Bank, unido.org

Taiwan's development as a base for high-tech electronics manufacturing is a result of the co-ordinated efforts of the government (National Science Council - NSC and the Ministry of Economic Affairs - MOEA), industry, research institutes and academia.

The government has a seven point agenda to promote priority industries:

- Provide R&D and plant investment assistance for selected products.
- Encourage technology transfer to ensure self-reliance in key technologies.

- Create a favourable environment for high-value-added and low cost R&D, production and marketing.
- Strengthen upstream industries to support middle and downstream development systems and vertical industry integration.
- Utilize government purchasing projects to obtain foreign expertise and technology.
- Encourage fair competition in the domestic market with updated laws and regulations.
- Co-ordinate agencies to establish development plans for key industries.

Examples: ITRI

ITRI is a government sponsored non-profit organization dedicated to developing the high-tech industry in Taiwan. It acts as a driver for government industry co-ordination. Many small firms in Taiwan depend entirely on ITRI for their R&D. It employs 5,700 people, and has a 5 point agenda:

1. Promote development of high-tech industry.
2. Improve domestic product design.
3. Improve processing technologies.
4. Provide technical assistance to industry.
5. Assist in development of defence technology.

Taiwan's case study also provides valuable lessons that can be used to promote manufacturing growth in India. Key action areas that India can focus on, based on Taiwan's experience, include:

- Focussed thrust on identifying priority industries.
- Strong collaboration among government, industry, research institutions and academia.
- Favourable, market friendly government policies.
- Focus on developing human capital.
- Macro-economic stability.

Having analysed the global trends, India's manufacturing capabilities and the key enablers and barriers for India's manufacturing growth, we need to address the following questions:

- What are the essentials of a Vision for Indian manufacturing?
- What are the key policy decisions that need to be taken to achieve this vision?
- How should these policies be implemented and what will be the expected outcome?

We address these questions in the remaining parts of this document.

KEY INFERENCES:

- Efforts to improve and grow Indian manufacturing should aim to leverage India's inherent strengths -natural resources, growth in domestic market, low labour costs, large labour force and growing population of NRIs.
- However, key barriers to manufacturing growth need to be overcome. For this, the following areas need attention:
- Government needs to focus on:
  - Improving infrastructure
  - Removing reservations and allowing SSIs to compete in the open market
  - Reducing procedural delays in government

- Improving efficiency of public services
- Reducing tax rates and interest rates
- Attracting foreign direct investment
- Improving focus on manufacturing, by reducing the number of ministries concerned with it
- Greater interaction with industry, institutions and academia
- Promoting a culture of innovation, manufacturing and craftsmanship

☛ Firms need to focus on:
- Improving process efficiencies and benchmarking against global players for quality and productivity levels
- Promoting innovation and R&D

5

CHAPTER FIVE

Vision For Indian Manufacturing and Guidelines For Policy Recommendation

IN this chapter, we outline our vision for manufacturing in India, and explain the key considerations we took into account in developing the policy decisions. Based on the vision, we discuss the guidelines we evolved for arriving at the various policy recommendations.

A Vision for Indian Manufacturing

Based on our analysis of the global developments in manufacturing, the evolution and current status of manufacturing in India, as well as the state of key enablers and barriers, we feel that the manufacturing vision for India should be anchored upon three key foci:

1. Achieve global levels of excellence in Indian manufacturing.
2. Improve standard of living of the people through manufacturing growth
3. Develop India into a strong player in the global market

While each of the above can be an achievement in itself, we feel that they are deeply interconnected. To achieve any one of these, we need to address the other two as well. Our manufacturing policy recommendations are based on this premise, and address all three.

These foci form the basis for our recommendations. The key guiding principles we used are elaborated below:

1. Achieve global levels of excellence in Indian manufacturing

Indian manufacturing capabilities should be developed to a level where Indian products are competitive across global markets in terms of price, quality, technology, delivery and service. For this,

- Indian firms should be enabled to access the latest technology from across the globe.
- Indigenous R&D, innovation need to be encouraged and a passion for manufacturing needs to be created.
- Infrastructure, public services and utilities should be improved and made more efficient to assist manufacturing growth. From 'India costs', firms need to experience 'India benefits'.

- Government, industry, research institutions and academia should be facilitated and encouraged to work in collaboration to improve industry capabilities.

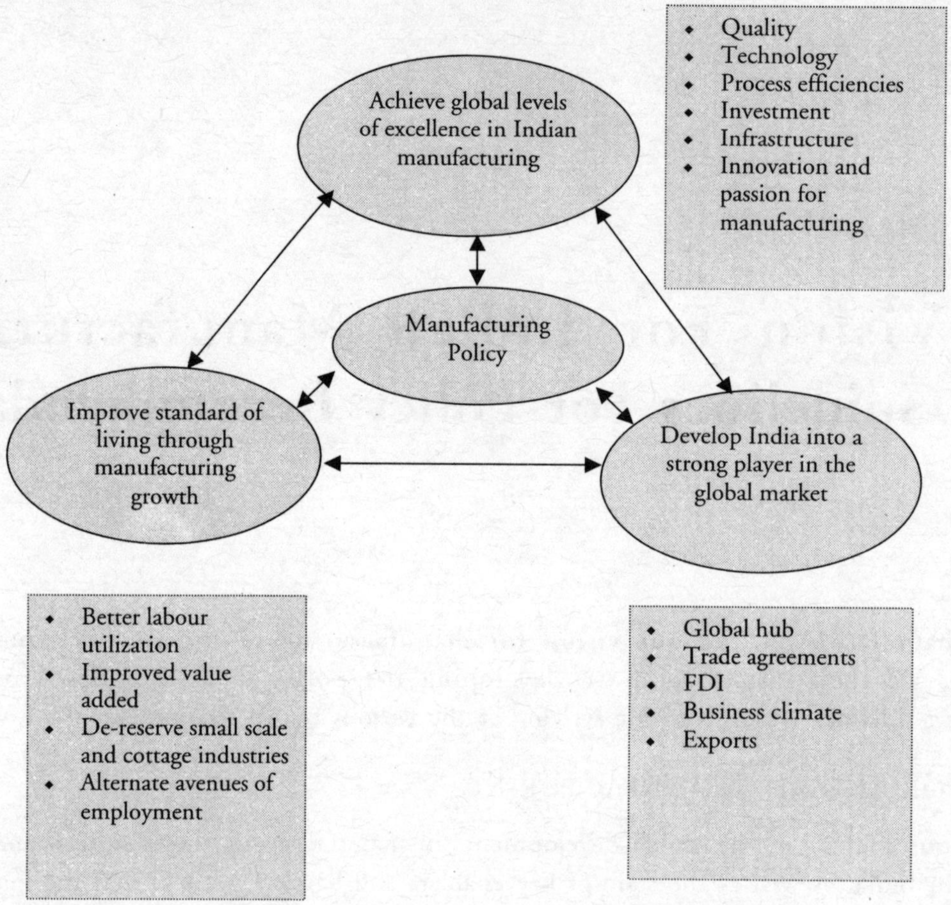

- Firms should be able to obtain funds easily and cheaply, and be encouraged to invest in developing technology.
- There should not be any regulatory restrictions on entry, operation and exit for business.

2. Improve standard of living through manufacturing growth

Manufacturing growth should result in overall welfare and improved standard of living for people in India. To achieve this,

- The growth should be based on improved utilisation of India's labour resources
- Workers should be enabled to graduate from lower value added to higher value added jobs.
- SSIs and cottage industries should be encouraged to grow and become competitive, even in the de-reserved environment.
- Education should focus on fostering a culture that encourages innovation and manufacturing, so that people are trained for alternate avenues of employment.

3. Develop India into a strong player in the global market

India should emerge as a significant player in the global market, and a global hub for products in key industries like automotive, pharmaceuticals, electronics and textiles. For this,

- Trade barriers should be further reduced progressively.
- There should not be any barrier to free market competition, except in areas of national security.
- India should look at developing bilateral trade agreements with key countries.
- FDI should be encouraged actively.
- The business climate in India should be made comparatively more attractive than neighbouring countries.
- Indian manufacturers should be encouraged to export and compete in the world markets.

The policies we recommend to address the above goals are elaborated in the following chapter. It should be noted that policies at the Centre should cascade down to the states and be applicable across states, for comprehensive implementation.

Above all, Indian manufacturers, without waiting for the support offered by the government's policies, need to look at their individual operations, and work towards improving process efficiencies and meeting global benchmarks.

6

CHAPTER SIX

Manufacturing Policy Recommendations

IN this chapter, we outline and explain our policy recommendations in key areas. Each sub section deals with recommendations in one area, and is structured to include:

- Goal of the policy
- Imperative for the policy
- Policy recommendation
- Case studies

Our recommendations are:

1. Identify specific industries for the 'fast track' and provide them all support for radical growth
2. Remove reservations in the SSI sector
3. Promote collective industry efforts to transition the SSI sectors and leverage India's large labour force
4. Promote industry - academia - government collaboration
5. Rationalise tax structure
6. Aggressively promote FDI inflow
7. Develop bilateral trade relations with ASEAN countries and China
8. Reform bankruptcy and labour laws
9. Reform and privatise the power sector
10. Privatise ports
11. Improve road infrastructure
12. Restructure government departments

6.1 Identify Specific Industries for the 'Fast Track' and Provide them all Support for Radical Growth

6.1.1 Goal of the Policy

To achieve robust overall growth, India needs to identify industries where it has comparative advantage and put them on a 'fast track' with suitable fiscal incentives. By achieving radical growth in these 'fast track' industries, India can grow faster than the normal pace. In addition, policies need to be put in place to facilitate companies move up the value chain and gain a continuous comparative advantage.

6.1.2 Imperatives for the Policy

Over the last decade, even though the Indian economy has been growing at close to 6% per annum, India needs to target faster consistent growth rates of 10% as visualised in the 10th plan. China (now) and countries like Taiwan and Korea (late eighties and early nineties) have achieved growth at such a fast clip. One striking facet of these successful economies has been a focus in certain key segments to lead the overall growth.

6.1.2.1 The Growth Continuum

Growth in value added for different economic activities normally follows a continuum - starting from basic, low value added activities like agriculture, going on to low-end, labour intensive manufacturing, to high-end manufacturing, and services. Different countries take up different positions on this continuum, depending on their comparative advantage and state of development. (Exhibit. 6.1.1)

EXHIBIT. 6.1.1

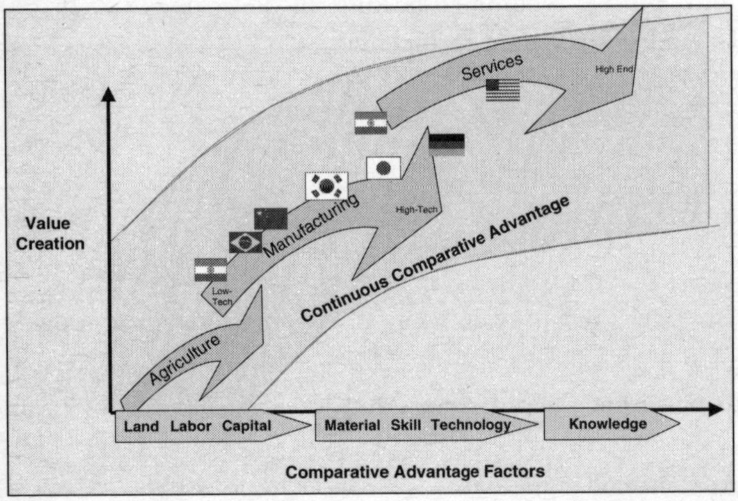

As countries progress up the continuum, they tend to outsource the activities at the lower end. For example, a developed economy like US will be composed predominantly of high-end manufacturing and service, and will export these products, while importing agricultural and low-end manufacturing products from countries lower down the continuum. In addition, they focus on high value added activities of the manufacturing value chain such as concept development, research, and design while outsourcing the rest of the activities such as manufacturing to countries that have a comparative advantage. India is currently at a labour intensive, low-end manufacturing position.

Its comparative advantage also lies primarily in availability of raw material and low cost of labour. India has a large pool of engineers that can be effectively leveraged globally for research and design activities, which will also enable it to move up the value chain. Countries such as China and South Korea have significantly moved up the value chain and are not just a challenge for India but in due course are likely to give a stiff competition to the developed countries as well.

6.1.2.2 The Need to Leapfrog

To achieve radical or discontinuous growth, there is a need to leapfrog over the normal evolution cycles, to get directly to high-end manufacturing and related services. Given India's relatively low level of development, and the fact that the bulk of India's workforce is engaged in low-end manufacturing, it will be a challenge for the entire manufacturing industry to make this jump. Hence India needs to identify focus segments, push for discontinuous growth in these segments and let the overall economy ride on the success of these "fast track industries".

6.1.3 Policy Recommendation

We recommend:

- Identify a few fast track industries and develop a clear action plan for discontinuous growth in these industries.
- Provide an integrated support to these industries through co-ordinated fiscal incentives, institutional support for fast clearance and low regulatory intervention, policy supporting easy technology transfer and adoption, incubation infrastructure like R&D facility and venture funding and focussed industrial parks with quality infrastructure.

The need of the hour is not only creating new facilities, but also co-ordinating the fiscal policy, giving a thrust to activities of research institutions like Centre for Scientific and Industrial Research (CSIR) and promoting linkages with the industry.

6..1.3.1 Identifying Fast Track Industries

All segments that can provide India a comparative advantage through either availability of raw material (like cotton), or process capability (like in the Pharmaceuticals, auto components, castings), or large local demand (like two wheelers) or local product development capability (like satellites, space launch vehicles) or skills (design and engineering) or a combination of these factors are potential segments to focus on for growth.

EXHIBIT 6.1.2

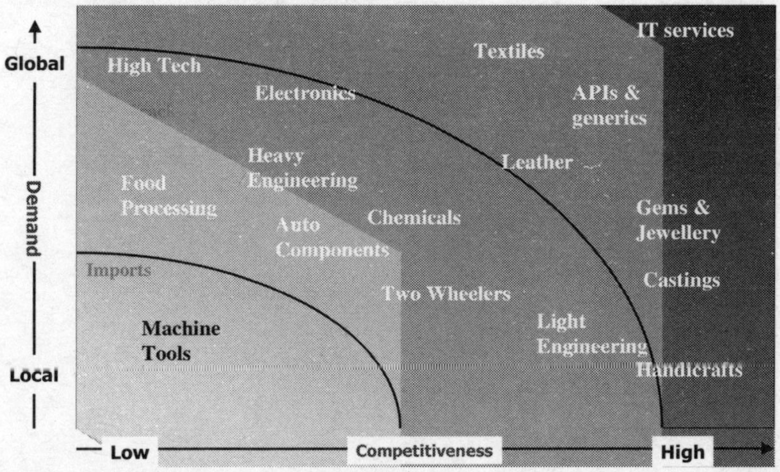

These areas of comparative advantage need to be then evaluated against the extent of global demand. Areas where India's comparative advantage is medium to high with simultaneously medium to high global demand, are areas which have potential to be chosen as fast track industries. Exhibit. 6.1.2 illustrates a sample analysis for identifying such fast track industrial segments.

6.1.4 Case Study

India has several pockets of excellence in different industry segments, where individual companies have achieved global impact through discontinuous growth. India today boasts of companies which are the largest manufacturer in their segments. Some of them are Ajanta Clocks, Hero Honda and several pharmaceutical companies with global scale of operations.

EXHIBIT 6.1.3

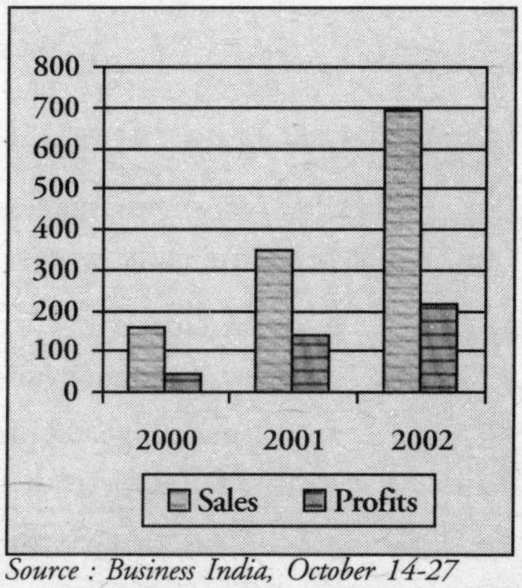

Source : Business India, October 14-27

Moser Baer India Limited

Moser Baer is a good example of an Indian company achieving global impact. Today, it is the third largest manufacturer in the world of CDs and DVDs. This company has had dramatic growth in sales from close to Rs 160 cr. in 1999 to close to Rs. 700 cr. in 2002 (Exhibit 6.1.3).

It is the lowest cost manufacturer of CDs with its cost lower by 26% compared to its Taiwanese competitors. The key reasons for success of Moser Baer are:

- Investment in R&D and Process development leading to low material costs.
- Flexible manufacturing process leading to quick changeover between CD and DVD.
- Own development of capital equipment or out-sourced development leading to 15%-20% cost advantage.
- Competitive labour costs of India.
- Global scale capacity leading to economies of scale.

Moser Baer started with technology assistance from a foreign company, but has successfully adopted and improved on it through in-house R&D to achieve global excellence.

Taiwan's Success in Electronics Industry

Taiwan's economy has grown dramatically riding on the export growth (Exhibit 6.1.4). Manufacturing is the largest segment of Taiwan's export and within manufacturing, electronics was the largest segment, propelling the dramatic growth (Exhibit 6.1.5).

EXHIBIT 6.1.4

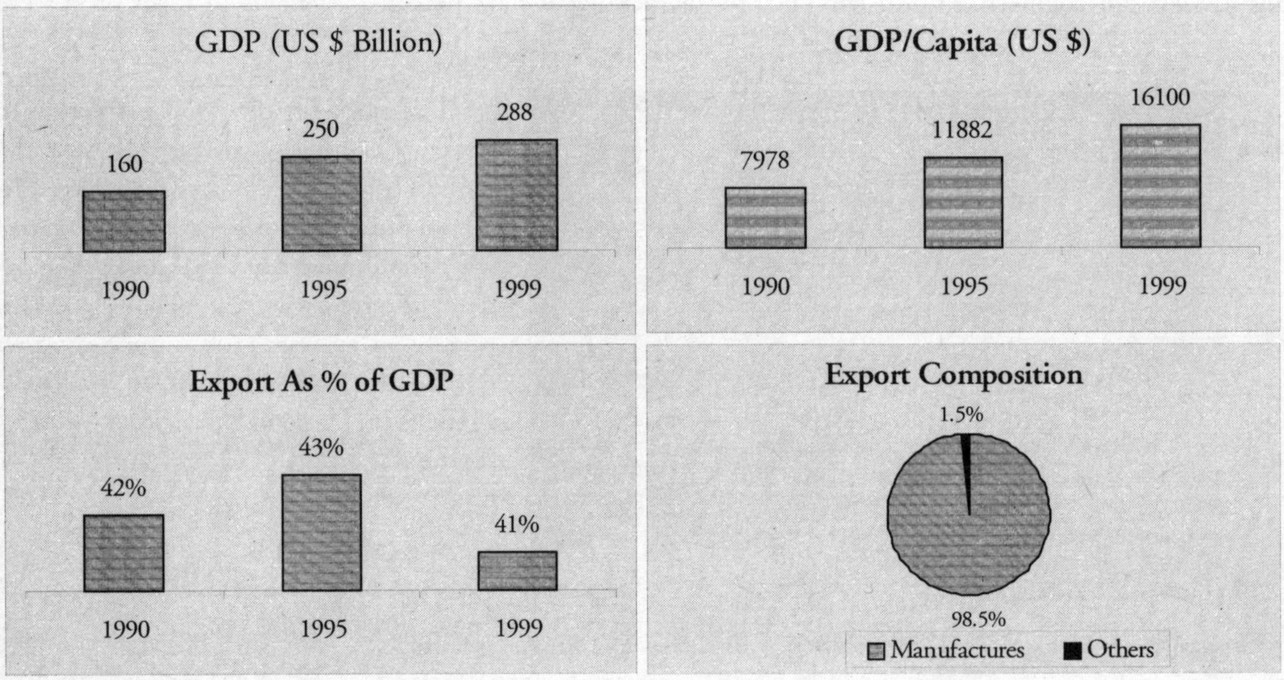

Source : Andersen Analysis, WTEC reports, World Bank.

The key reason for success of the Taiwan's electronics industry is the government's 7 point agenda to promote the segment. These are:

- Provide R&D and plant investment assistance for selected products - promoting strong industry, academia and government interaction.
- Encourage technology transfer to ensure self-reliance in key technologies - facilitated by easy technology transfer and FDI guidelines, duty exemptions and tax incentives.
- Create a favorable climate for High Value Added and low cost R&D, production and marketing

- strong government supported R&D through institutions like ITRI (a government sponsored non profit organisation).

EXHIBIT 6.1.5

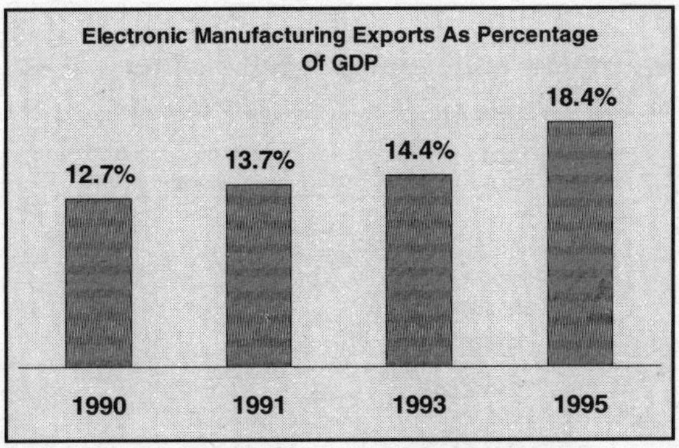

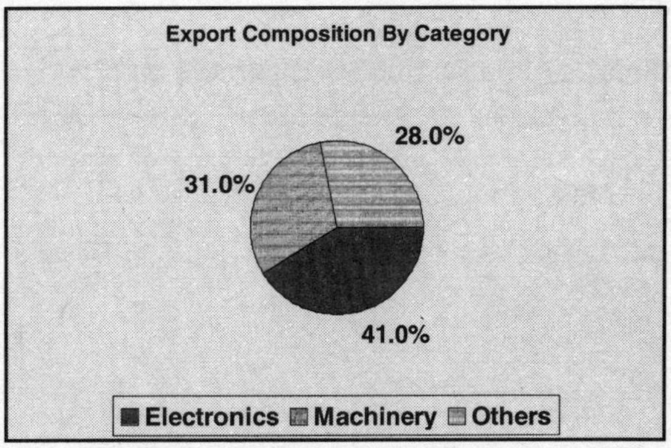

Source : Andersen WTEC reports World Bank UNIDO org

- Strengthen upstream industries to support middle and downstream development systems and vertical industry integration.
- Utilise government for purchasing projects to obtain foreign expertise and technology
- Encourage fair competition in the domestic market with updated laws and regulations - characterised by ease of entry, exit and quick bankruptcy.
- Co-ordinate agencies to establish development plans for top key industries.

6.2 REMOVE RESERVATIONS IN THE SSI SECTOR

6.2.1 Goals of the Policy

Reservations, subsidies and incentives for the SSI sector have resulted in most of the firms remaining small, inefficient and unproductive. There is a need to remove reservations in this sector, which employs the bulk of India's labour force, so that it becomes more competitive. The reservations can continue for items where mass manufacturing is not possible or items such as handicrafts.

6.2.2 Imperative for the Policy

6.2.2.1 The SSI Policy - No Longer Relevant

The small-scale industries sector contributes 40% of the gross industrial value added in the Indian economy. It provides 80% of private Industrial employment (Exhibit 6.2.1) and contributes over 45-50% (Exhibit 6.2.2) of Indian exports. The SSI reservation policy had two objectives:

- Expand employment opportunities through setting up small scale industries.
- Ensure increased production of consumer goods in the small scale sector.

With increasing global integration through WTO, opening of the Indian economy and formation of free trade zones, the SSI reservation policy, designed to promote self sufficiency and protect local employment, has lost its relevance. Reforms in the SSI sector are crucial for India to emerge as a competitive manufacturing base.

The SSI sector in India grew from 0.87 million units in 1980-81 to over 3.2 million units in 2000. SSI exports showed a growth rate of 10.8% from 1994-95 onwards and the sector growth rates have been higher than the Industry as a whole. Though these figures may sound impressive, they hide a deeper malaise:

- Increasing incidence of sickness: At present over 10% - 40% are estimated[1] (Exhibit 6.2.1) to be sick and only 15000 sick SSIs have been found to be viable for production.

EXHIBIT 6.2.1

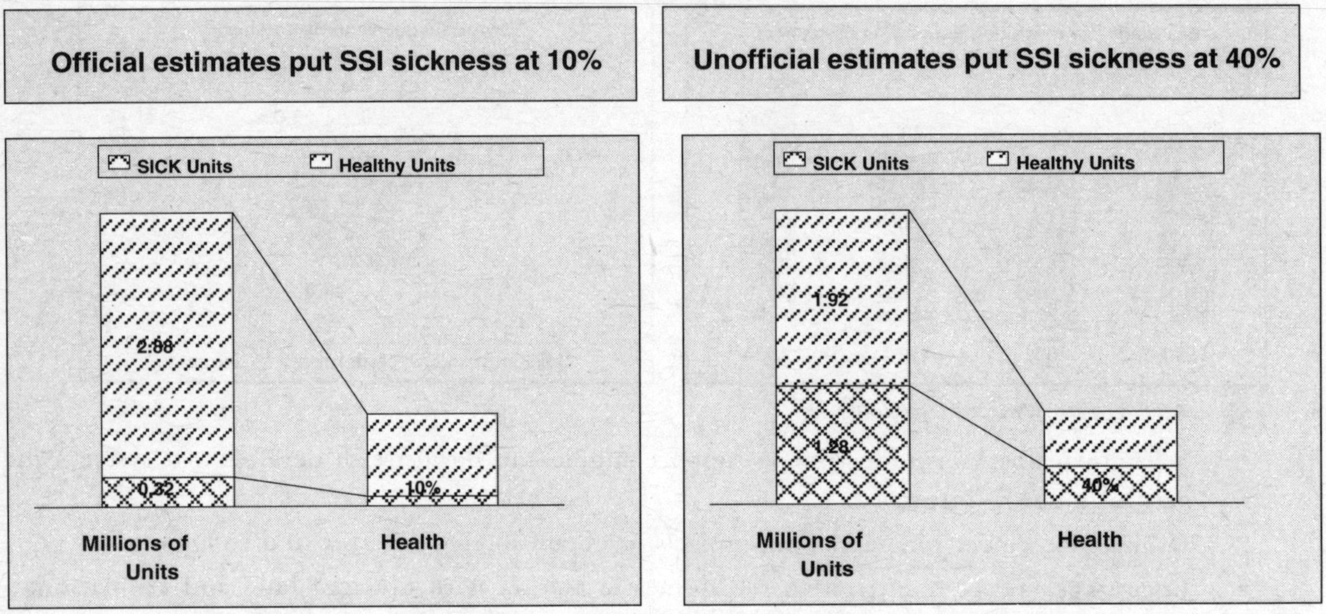

- Lack of capital, technology and productive human capital - which makes these units perform at a fraction of global benchmarks.
- Inability to compete with increasing competition, especially from China, both domestically and internationally.
- Threat to protected quota markets abroad, which account for a large fraction of SSI exports, after the phasing out of quotas under the WTO. India has removed import restrictions over the last decade due to WTO obligations, by allowing items to be imported on the open general license (OGL), but continues to provide protection to a majority of SSIs by reserving 63 items that constitute over 80% of the SSI output (Exhibit 6.2.2).
- Protection of these units from competition through product reservation, excise waivers, etc., under the guise of employment protection.

The need is to make our SSI sector competitive and dynamic. The policy changes recommended would make the SSI sector poised for growth, by attracting increasing capital, professional management and development of productive human capital.

EXHIBIT 6.2.2

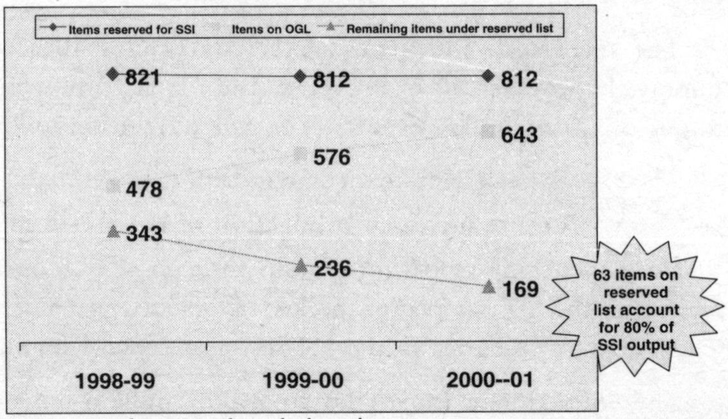

Source : Andersen Analysis, laghu-udyog.com
Note : OGL(Open General License)

6.2.2.2 SSI Policy is Having a Detrimental Effect on Indian Manufacturing

1. *Quasi-SSI structures:* Existing policy promotes a quasi-SSI structure for firms that want to expand. For

[1] *Official estimates put the figure at 10%*

example in the apparel industry, one of the largest firms has over 6,000 personnel, but they are spread over 20 factories of 300 people each. This organisation structure had been maintained for two reasons: the fiscal benefits of SSI still accrue to a large player, and it avoids the threat of organised labour by keeping the workforce artificially small.

2. *No incentive to expand:* Many successful SSI firms do not want to expand as they would loose direct fiscal incentives and face labour legislation if they become too large. They limit their production and firm size intentionally to remain under the SSI sector.
3. *Lack of best practice:* By keeping the organised sector out of reserved categories, these sectors face an acute lack of professional management, best practices and investment in human capital (training etc) as the capital base of these firms is small. Productivity and quality levels of these firms are way below global standards. An interesting case in point is the Textile industry - India's largest exporter. At present exports to the US and EU is based on a quota system for each country. Both India and China have guaranteed markets in the Quota countries - but whereas China has a market share of 38 % in the top 10 Non-Quota countries, India's share languishes at 1.6% (Exhibit 6.2.3). With productivity and quality levels high above Indian standards, China stands to gain a lot after the phasing out of quotas in 2005. This is a direct threat to the Indian textile industry. A snapshot of the Chinese export basket shows that labour intensive exports constituted close to $ 61 Billion in 1999 (Exhibit 6.2.4), which is more than the total Indian exports of $ 43.7 Billion in 2001-02.

EXHIBIT 6.2.3

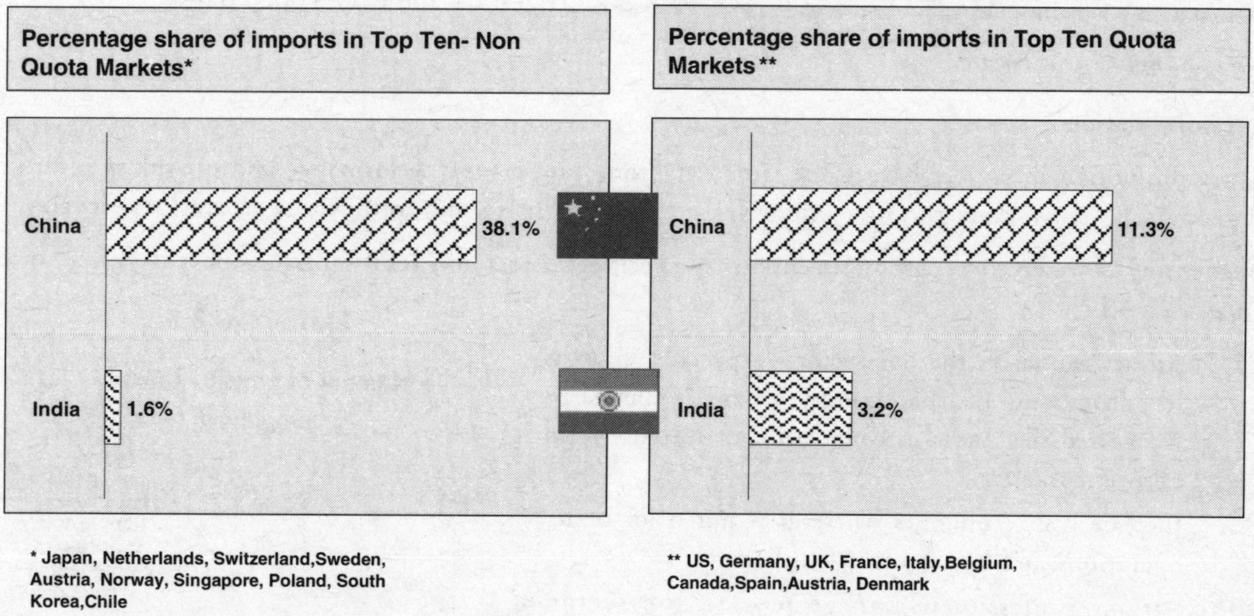

4. *Inefficient capital utilisation*: The SSI policy directly affects capital investment decisions in the Indian manufacturing sector. At present over 85% of Indian factories have less than Rs.1 Crore ($ 0.2 Million) of capital employed (Exhibit 6.2.5). This prevents the benefits of scale economies being accrued to Indian domestic consumers wherever applicable.
5. *Degradation of jobs in the formal sector*: The SSI sector promotes the movement of jobs from the organised sector to the SSI sector- primarily because of the fiscal incentives available and the

lack of labour laws on informal workers. Studies have observed this shift in industries like pharmaceuticals, beedis and textiles. The end result is a degradation of jobs in the formal sector.

6.2.3 Policy Recommendation

We recommend that the government take a 3-step approach towards the SSI sector:

- **De-reservation**

The government must eliminate all reservations in the SSI sector, starting with the 63 items, which constitute over 80% of the total output of the SSI sector. The reservations can, however, continue for items where mass manufacturing is not possible or items such as handicrafts.

EXHIBIT 6.2.4

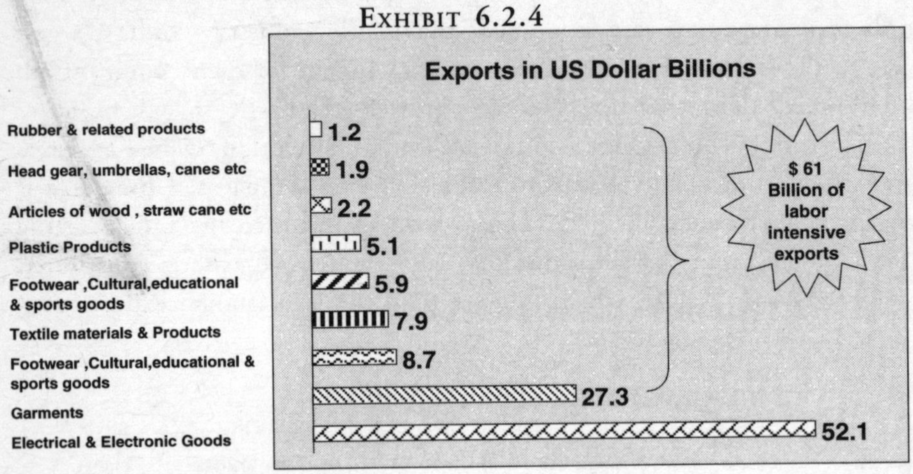

Source : Andersen Analysis, Various Reports

- **Cluster Formation**

The state governments and industry bodies have to take a lead to identify SSI clusters, promote co-operation between businesses & local authorities for cluster development, and formulate policies that attract investment to these clusters. (This recommendation is covered in detail in collective industry Policy 6.3)

- **Policy reformulation**

Industry promotion incentives have to be aimed at promoting export performance and providing indirect support like training to workers through initiatives between Industry and authorities at the local level.

Implementing de-reservation, promoting industry specific clusters and reformulating the SSI policy on a performance basis will:

1. Attract capital to the SSI sector - especially in areas like Toys and Leather goods- where it has been recognised that India can play a major role in world export markets.
2. Increase scale economies wherever required for better operations and better capital utilisation
3. Attract professional management to the sector, bringing with it best practices that will assist in improving productivity and quality levels.
4. Better training of the SSI workers due to more capital attraction, influx of the organised sector and the need to increase quality for export markets.

EXHIBIT 6.2.5

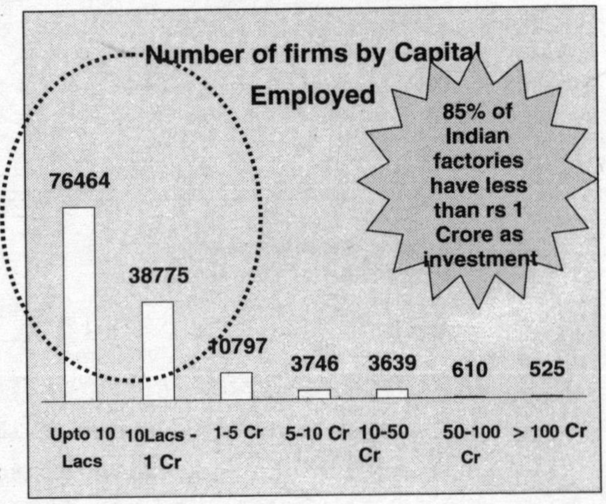

6.2.4 Case Study - China's Town & Village Enterprises

China has effectively used its 'Town & Village Enterprises' (TVEs) as an engine for manufacturing growth and promoting employment. Synonymous with rural industrial firms, TVEs accounted for 32.6% of China's exports, provided employment to over 120 Million people, accounting for close to 20% of the national employment. Over 26% of the rural workforce in China were involved in TVE activity.[2]

Unlike India, the Chinese government has allowed market forces to dictate production, pricing and labour. The Chinese policy has been to provide fiscal incentives only for an initial period of time in terms of tax holidays for initial 1-3 years and waivers or reduction of taxes for unprofitable TVEs for 2-3 years.

Incentives are provided based on export performance and no sectors are reserved. There are no rules preventing exit from TVEs and no incentives are given on firm size. The Indian policies are exactly opposite in many ways.

This has resulted in free flow of capital and labour in the TVE sector in China, generating rural prosperity and rural industrialisation. The surprising factor of this success is that the community owns the TVE, and success has been achieved without clear ownership of the firms.

6.2.4.1 Comparison of TVEs & SSIs

Chinese TVEs far outperform the Indian SSIs in all aspects. This is remarkable, as the Chinese TVE policy started only in 1984, compared to the Indian SSI policy which had started much earlier.

Though both TVEs and SSIs had a comparable share of exports in 1994, in value terms the TVEs exported 4 times as much as the Indian SSIs, employed 7.5 times as many people and used up less than half the percentage of national credit than Indian SSIs (Exhibit 6.2.6).

We believe, the Chinese success is based on the following principles:

1. No market distorting policies favouring TVEs.
2. Easy entry and exit.
3. Letting the market decide the success of the firm.

6.3 Collective Industry Efforts to Transition SSI and Leverage Large Labour Force

6.3.1 Goals of the Policy

Given the highly fragmented structure of small industries and lack of capabilities in individual firms, co-ordinated efforts through collective industry bodies are required to improve competitiveness in the domestic and global markets. This policy aims to spur manufacturing growth by leveraging the vast labour pool that exists in the small industries, sub-urban and rural areas of the country.

6.3.2 Need for Collective Bodies

As the Government implements the SSI de-reservation policy, existing SSIs and the large pool of labour (constituting 80% of the total labour force) need to find ways of transitioning out of the SSI structure. Clearly in the context of globalisation, any removal of existing fiscal incentives available to SSIs will invariably result in elimination of small industries, since they would have to face superior competition in terms of quality and marketing from global players.

---

[2] *All figures mentioned for 1994*

EXHIBIT 6.2.6

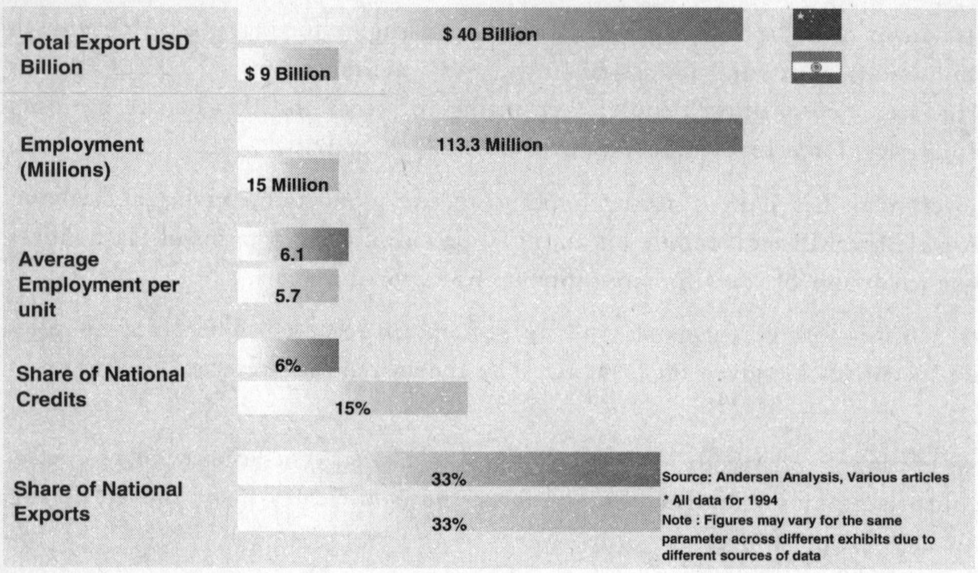

One of the best ways of ensuring survival of SSIs in the post de-reservation scenario is to leverage collective power of the currently fragmented units. World-wide it has been observed that small scale units have survived wherever they were part of a cluster, where all the units worked towards using common infrastructure, procurement and marketing. India too has several cases of success in leveraging collective power for success.

6.3.3 Policy Recommendation

This recommendation is directed at the industry. We recommend:

- The industry associations should come together to promote industry wise collective bodies to work towards alleviating industry standards, by sharing infrastructure, leveraging collective bargaining power for procurement, marketing and distribution

6.3.3.1 Setting up Industry Wise Collective Bodies in India

We believe that the small and tiny scale industries can learn from national and international experiences and set up industry level collective bodies to achieve global competitiveness.

Small enterprises in the country face several key issues, due to lack of scale advantage. They are beset by high cost of funding, high difficulty in sourcing funds from banks which demand large collateral from the units, inability to adopt quality processes leading to poor output quality, inability to invest in product and process improvement research, inability to successfully access markets in India and abroad due to lack of marketing set-up and high raw material cost due to fragmented procurement.

We suggest that industry level collective bodies be set up in different industries to overcome these issues that they face today.

6.3.3.2 How Would the Collective Body Work?

If such a collective body is established in the garments industry, this body would front end the marketing for the garment industry cluster and obtain bulk orders from overseas clients. These orders would then be split amongst the different small units. The body will also negotiate for the bulk procurement of all necessary raw materials achieving, economies of scale. During execution of the order by different units, the body would provide quality supervision and expert assistance to ensure all customer requirements are met. The body would ensure that the delivery schedule is also appropriately adhered to. This body would ship all the finished products with quality assured.

FIGURE 6.3.1

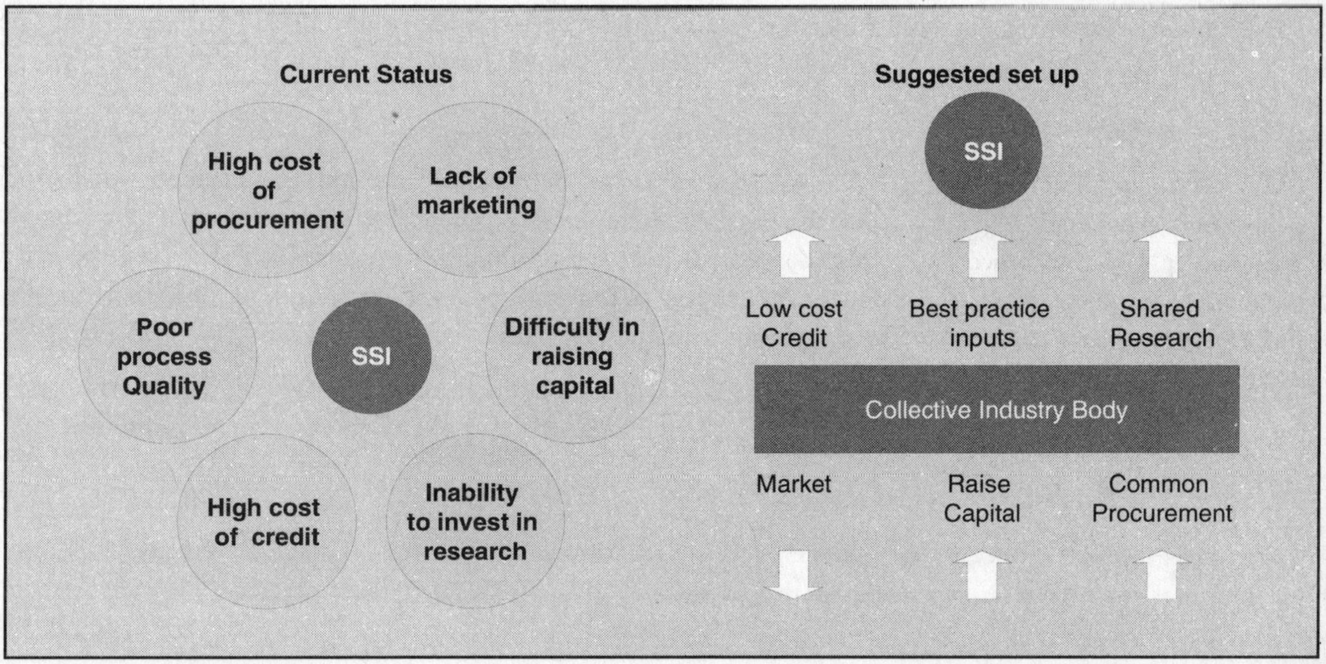

Such a body eases the pressure on the part of the outsourcer by providing a single window ensuring large volume of quality sourcing. In fact, a substantial portion of China's garment exports happen through entities in Taiwan and Hong Kong which obtain bulk orders and route them to multiple manufacturing units in China. These entities have also pumped in substantial FDI in by taking equity stake in these Chinese garment makers.

In a similar manner, we can visualise collective bodies functioning in various key sectors like knitwear, gems and jewellery, pharmaceuticals, leather goods, sports goods, handicrafts and other areas.

The recent promotion of India brand jointly by the Tea industry association and the successful promotion of India as a sourcing destination for software and IT enabled services by NASSCOM are illustrations of how collective power of fragmented players can impact in raising the overall level of the industry. What we are recommending is extending the same concept to larger areas of co-operation amongst all the industry players.

6.3.3.3 Role of the Collective Bodies

The collective bodies would be a channelling agency for each industry helping all the players:

- Do group procurement by consolidating the raw material requirement.
- Carry out common research in product and process development on behalf of industry and share the best practices amongst all units.
- Provide an integrated marketing face to customers allowing them to interact with a single point to source all their requirements and provide quality and delivery assurance.
- Build the brand image of the country as a whole in the international arena.
- Negotiate and get competitive interest rates and facilitate in easy finance availability. In addition, the co-operative would assist financial institutions in easy recovery of their dues by securing the sales realisations and paying the units after set-off of dues.
- Facilitate in getting necessary national and international certification/accreditation.
- Facilitate training and skill development of the masses in sub-urban and rural areas.

6.3.3.4 Role of the Units Participating in the Collective Bodies

The units participating in these collective bodies would be bound to route all the purchases and sales through the co-operative. Making these bodies a collectively owned entity can ensure this discipline.

6.3.3.5 Nature of Co-operation

The nature and extent of co-operation would depend on the nature of the industry itself. A wider extent of co-operation extending to all areas would be possible in commodity product markets where there are common raw materials, production processes etc. As the sophistication/ specialisation of the product increases, the extent of co-operation would gradually come down, culminating in just joint country branding kind of alliance in niche product categories, since there would be a divergence in the raw materials used, production processes etc.

The extent and nature of these bodies would also depend on cultural aspects in the industry and availability of dynamic leadership that can bring together all participants in the industry under a single umbrella.

6.3.4 Case Study: International Experience

The most interesting case study for rural industrialisation is the Town & Village Enterprises (TVEs) in China. The most surprising aspect of this entire situation is that TVEs were never part of a planned strategy[3].

TVEs

Firms in rural areas consist of two ownership types: TVEs that are community public firms and private firms. The crucial feature of TVEs is the community (i.e., township or village) government control of firms, in contrast with private or central government control. The community government plays a critical role in protecting TVEs in an environment lacking secure property rights.

In transition and developing economies, capital is one of the scarcest resources, and new entry firms have great difficulty in obtaining it. TVEs with community government control have several advantages in contrast to private firms:

- Community governments make use of their political connections to get loans from state banks.
- On economic grounds the community government is able to reduce the risks borne by the banks through cross-subsidisation among its many diversified enterprises or it can use accumulated collective assets as collateral or as co-investment funds to reduce potential hazards in the lending borrowing relationship.

Reasons for TVEs Rise

Regional decentralisation policy adopted by the Chinese government has played a key role in the rise of TVEs. Under this scheme, governments at the township and village levels are empowered with comprehensive authority for local economic development and they are also provided with incentives to do so since they can keep the revenues generated.

6.3.4.1 Learnings from TVEs

The success of TVEs in China is indeed remarkable. The TVEs in China started in 1984, and have surpassed the Indian SSI sector in all aspects of performance - Exports, Share of National Credit, Employment etc.

---

3 *Deng Xiaoping said on June 12, 1987 : " The greatest achievement that was totally out of our expectation is that rural enterprises [TVEs & private enterprises] have developed' (Economic Daily, June 13, 1993*

The TVE concept is a unique result of a number of factors - primarily China's political system, its economic policies and its domestic policies. India being a very different country will have to search for its own model for rural industrialisation.

6.3.5 Case Study : Indian Success Stories

India has the home grown success model like "Amul', led by the Gujarat Co-Operative Milk Marketing Federation (GCMMF). The world's largest co-operative, Amul has over 2.23 million producer members, over 10,852 village societies and a yearly milk collection of 1.67 Billion litres a year in 2001-2002. It has made India the largest producer of milk in the world.

Another successful small enterprise from India is Lijjat Papad. A collective enterprise of 40,000 women owners, Lijjat is about Rs. 350 crores in size and growing.

The government has successfully developed the tobacco crop in the country by setting up the Tobacco Board and Central Tobacco Research Institute (CTRI) that provide regulatory and research support to a fragmented farmer base in the country. Tobacco board advances finances to the farmer and sets off the dues against the sales realisation of the farmers, in the auctions facilitated by the board. CTRI develops better yielding seeds and farming best practices that improve the farm productivity.

There are many more of these examples if we look carefully around our own country. These enterprises provide examples of different models of leveraging the collective power of fragmented stakeholders under a common framework, without compromising the individual ownership of assets and sharing of profits, and enhancing overall prosperity.

6.4 Promote Industry - Academia - Government Collaboration

6.4.1 Goals of the Policy

The dazzling growth of services in India has turned the focus away from manufacturing. India needs to develop a culture of providing thrust on innovation and developing a passion for manufacturing. There is a need to create and maintain a large pool of skilled manpower through primary, secondary and tertiary education. To achieve this, collaborated effort spanning the government, industry, research institutions and academia is required.

6.4.2 Imperative for the Policy

6.4.2.1 Low Level of Interaction Between Academia, Industry and Government

India has a vast science and technology infrastructure with more than 400 National Laboratories, over 200 Universities including institutes of technology, and about 1300 in-house R&D units in the industrial sector. However, the current level of interaction between academia and industry is relatively low. The interactions occur mainly between premier institutions / universities and the industry on a limited basis. The main forms of interaction between these universities and industry are:

- Use of university faculty as consultants in industry
- Research projects in university funded by industry
- Students hired by industry
- Visiting professorship filled by industry scientists

- Advisory committee membership to university faculty
- Universities providing special courses to industrial personnel

Translating innovation into new products / services / processes, often requires new organisational structures, innovative financing systems and supportive policies of the government. To promote innovation and research, the Government of India set up the National Science and Technology Entrepreneurship Development Board (NSTEDB) and initiated the Science and Technology Entrepreneurs Park (STEP) programme in 1984. However, the levels of interaction still continue to be low.

6.4.2.2 Inadequate Funding For Research Projects

The current allocation of investments for R&D in India are low. Against a proposed spending of 2% of GDP on science and technology, the current levels of spends are only 0.66%. However in developed countries the level of investment in R&D is considerably higher but the outflows are decreasing. R&D investment as a percentage of GDP for USA decreased from 2.8% to 2.4% from 1991 to 1995. A similar trend was observed for Japan and Germany where investments decreased from 2.9% to 2.7% and 2.9% to 2.4% respectively.

6.4.3 Academia and Industry Collaboration – Key for Innovation and Competitiveness

6.4.3.1 Innovation

Innovation lies at the heart of economic growth. Governments need to recognise that the market will not invest sufficiently in blue-sky research and therefore assume responsibility for sustaining and enhancing a good quality science and research base.

Government has a key role in ensuring that business is able to exploit results of our science base effectively. Knowledge transfer activities can overcome information failures between business and the science base. The government also has an important role in fostering innovation, in encouraging R&D collaboration and knowledge sharing thus enabling individual companies to capture knowledge spin-off from each other's research, and collectively to enjoy the benefits of economies of scale and scope in innovation.

Universities have a vital role to play in the innovation process. The cornerstone of this is the role of universities in educating highly qualified manpower. This not only provides business with the manpower to undertake R&D and innovation (in its widest sense) but is also the basis of networks, which exist between universities and industry.

Manufacturing industry needs to work more closely with academia to spot commercial opportunities from scientific research and ensure that our outstanding record of scientific discovery can be turned into new products, services and processes. In the United Kingdom 38% of innovating businesses cited universities or other research institutes as a source of new technological or useful scientific knowledge.

6.4.3.2 Competitiveness

In a world where globalisation increasingly drives company strategies and where national boundaries are no longer seen as limiting factors in company development, it is becoming clear that competitive success depends on localised concentrations of skiled people and technology. No nation and no state can be competitive in every field.

Increasingly, countries are recognising that a critical mass of individuals, firms, educational and other institutions are needed to give a competitive advantage to a region. Within this grouping, knowledge is readily transferred by formal and informal networks. Such networks enable small to medium enterprises to link together on a large scale to compete successfully with the best large enterprises. Many countries are taking active policy decisions to develop such concentrations, in contrast to previous passive decisions to allow such concentrations to happen, for example, the Co-operative research centres in Australia, the Industrial Technology Research Institute in Taiwan etc. Measures such as technology parks associated with universities or advanced technical institutes or with national research facilities are being increasingly used, together with co-operative centres linking industry, government and academic R&D activities in a particular area of science and technology.

6.4.4 Policy Recommendation

We recommend:

- Focus on increased collaboration between academia, industry and government.
- Increase investments in research and development.
- Increased awareness for manufacturing and crafts by including activities like carpentry etc in school curriculum.

6.4.4.1 Focus on Increased Collaboration Between Academia, Industry and Government

Proposed Model of Government-Industry-Academia interaction

The proposed model for Government-Industry-Academia interaction is shown in Figure 6.4.1

FIGURE 6.4.1

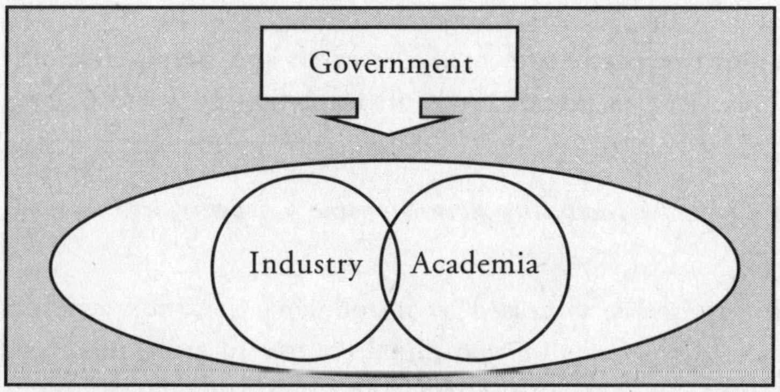

The model proposes increased collaboration between industry and research institutes / universities. The interaction would assist the development of research institutes / universities in the form of funding assistance, commercialisation of research etc. Industry would gain access to vast amounts of research that would be economically infeasible for them to undertake. This interaction would flourish within the framework of suitable policies formed by the government.

The first dimension of the model is internal transformation in each of the spheres, such as the development of lateral ties among companies through strategic alliances.

The second is the influence of one sphere upon another, for example, the role of the central government in instituting an educational policy that has an impact on academia or an environmental policy that affects industry.

The third dimension is the creation of a new overlay of tri-lateral networks that emerge from the interaction of the three spheres, formed for the purpose of coming up with new ideas and formats for innovation. The model denotes the academia-industry-government relationship as one of relatively equal, yet inter-dependent spheres that overlap and take the role of the other.[4]

Bilateral relations between government and academia, academia and industry and government and industry have expanded into triadic relationships among the spheres, especially at the regional level. Older economic

[4] *The Triple Helix of University - Industry-Government Implication for Policy and Evaluation by Henary Etzkowitz*

development strategies, whether based primarily on the industrial sector as in the US or the governmental sector as in Latin America, are being supplemented, if not replaced, by knowledge-based economic development strategies, drawing upon resources from the three spheres.

Universities and industry, till now relatively separate and distinct institutional spheres, are assuming tasks that were formerly largely the province of the other. Thus universities are looking to commercialise research while industry is looking to provide training and to contract out its research.

The cumulative result of these three paradigm shifts is that there is a complex interaction between government, industry and universities driven by the need for wealth creation, the need to maintain competitiveness on a global scale and the need to ensure optimum linkages and outcomes between research and commercialisation.

The onus and benefits of making this collaboration work primarily rests with the industry and academia. The role of the Government would be to facilitate and encourage this participation by providing fiscal incentives to both academia and industries.

6.4.4.2 Increase Investments in Research and Development

The government has to make a conscious effort to provide increased support to R&D institutions. In addition, investments made by the private sector in R&D centres can be exempt from taxation or taxed at lower rates to promote such investments.

Increased investments by the private sector for R&D activities can also take place through sponsored research programs e.g. in the case study for formation of the National Supercomputing Grid by IIT Delhi; interested pharmaceutical and biotechnology companies can participate in the research and bear part of the principal cost of the project. Sponsored research will also lead to greater levels of interaction between industry and the academic sector.

6.4.4.3 Increased Awareness for Manufacturing and Crafts by Including Activities like Carpentry etc in School Curriculum

An awareness of the various facets of manufacturing can be increased by introducing workshop activities like carpentry etc in the curriculum at schools. An increased focus on vocational courses in semi-urban and rural areas, that would prepare students for specific jobs in manufacturing, trade and industry, would increase not only the skill level of the populace but also their employability.

If these skills were viewed as avenues to gain suitable employment an increased number of people would pursue them. Industries could then employ these persons with basic skills in areas of relatively low value added, labour intensive activities leading to higher employment levels for the country.

6.4.5 Case Study 1: Taiwan Electronics Industry

The Gross Domestic Product (GDP) of Taiwan has grown from $160 billion in 1990 to $288 billion in 1999, a growth of 80%, riding on a wave of strong manufacturing exports by the electronics industry. Exports as a percentage of GDP was over 40% for all the years with manufactured exports accounting for over 98% of the total exports. The electronics industry displayed a similar trend with exports of electronics products as a percentage of GDP rising from 12.7% to 18.4%.

The electronics industry has grown due to synergies between government support and incentives, cutting-edge research and its dissemination and world class manufacturing processes. The interaction model:

FIGURE 6.4.2

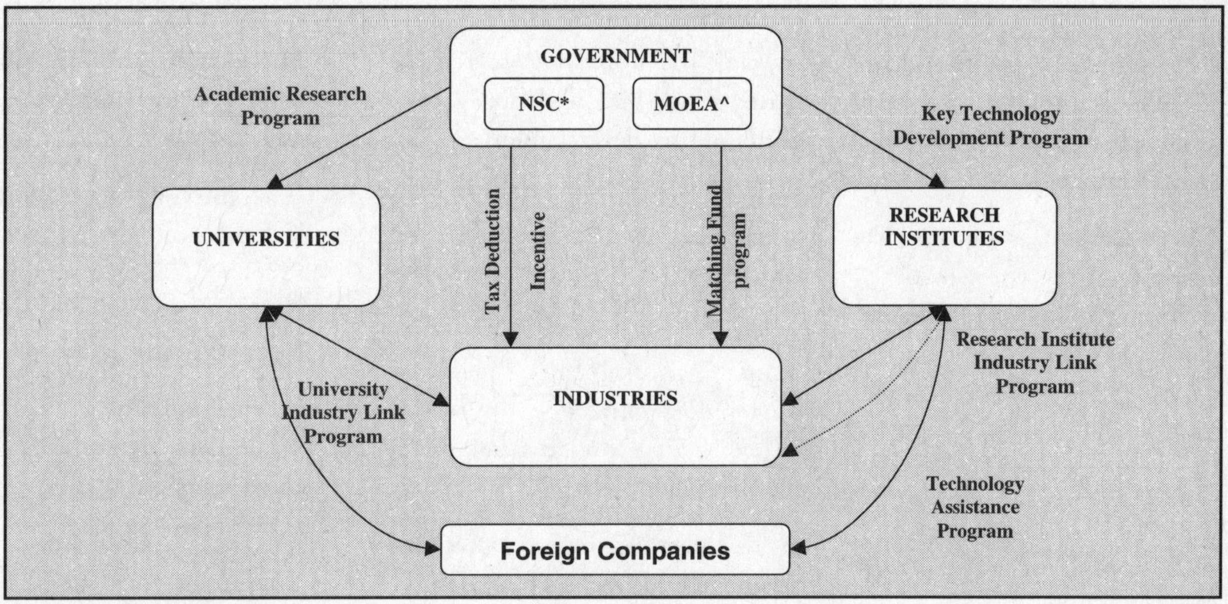

NSC : National Science Council*
MOEA^ : Ministry of Economic Affairs

- **Role of the Government**

The government has spearheaded the emergence of Taiwan to a high technology manufacturing base. The government has identified ten high growth industries that would drive growth and then provided support and incentives to these industries. The laws and tax exemptions that were provided to these industries are:

  - One stop permits and special incentives
  - 5 Year corporate income tax exemption
  - Tax credits for investments in R&D, Factory automation & HRD support
  - Duty exemption on imported components, equipment and materials

- **Role of Universities / Research Institutes**

Industrial Technology Research Institute (ITRI), a government sponsored non-profit organisation has fuelled the high technology drive in Taiwan. ITRI is the sole source of R&D for most companies as most of them are too small to afford R&D facilities.

- **Government University/Research Institute collaboration**

The government and ITRI have together identified critical components for technology development in Taiwan. These are classified into three groups:

  - Group A – Research Institutes like ITRI to develop technology
  - Group B – MOEA to employ government assistance for qualified manufacturers to conduct development plan
  - Group C – Government to provide investment incentives to manufacturers

- **University / Research Institute Industry Interaction**

The collaboration between industry and ITRI is shown in the schematic below:

Identify	*Invest, Inclubate or Transfer*	*Diffuse*
◆ Key high-tech industries are identified and list of critical products and critical components prepared. ◆ These industries become the priority– as they will be driving growth in the future.	Group A Technology: Research institutes like ITRI will apply own technology in development. Group B Technology: MOEA to provide assistance to qualified manufacturers to develop technology. Group C Technology: Help in transfer of technology to Taiwan by encouraging foreign firms to set up etc. Provide incentives to firms that take lead in technology development.	◆ Diffuse technology to the private sector / market as soon as possible. ◆ If technology development is by research institutes like ITRI or ESRO- programs are spun off after 4 years, when technical base is adequate. ◆ License technology from global leaders for use in Taiwan– to develop local industry and generate revenues. ◆ Encourage local firms to apply for patents. ◆ Negotiate flat royalty rate with global leaders in technology to reduce product costs and risk of law suits. ◆ Spin off research teams, transfer talent to industry- Over 11,000 people have been transferred to private firms from ITRI alone.

- **Role of Industry**

The industry uses research from ITRI to produce world class products. The research is also used to continually streamline and improve the processes used in the manufacture of electronics products.

The growth of industries was also fuelled by the development of industrial parks where all industries could avail of tax exemptions listed above. Close proximity to airports and ports helped reduce the costs and time needed to export the products.

Reasons for the successful Government, University and Industry collaboration are:

- Focussed support for the identified industries by the government
- Strong Government, Industry, Research Institution & Academia inter-linkages
- Favourable market friendly Government policies
- Strong human capital accumulation
- Macro-economic stability

6.4.6 Case Study 1: Formation of a National Supercomputing Grid by IIT, Delhi

In 2001, Indian Institute of Technology (IIT, Delhi) proposed the idea to augment drug development in

India through the formation of a National Supercomputing Grid that would cut down the time and costs involved in developing new drugs. The grid would link up the Indian Institute of Science, Bangalore; bio-technology centres in Chennai, Pune and the Kolkata universities; IIT, Delhi; AIIMS and the Jawaharlal Nehru University's biotechnology centre. Each institute would work on a speciality area.

Using specialised software and supercomputing power to the tune of 1 terraflop (1000 billion operations per second) the institutes' could simulate the reactions between various chemical molecules and human genes. The proposed cost of the project would be in the range of Rs.200 crores.

However, the Information technology ministry rejected the proposal stating that the costs were unrealistic. As a result IIT, Delhi had to work on a scaled down version of the project, wherein the computing power of the grid would be reduced to 150 gigaflops from the proposed 1 terraflop. This would ensure that the project costs would be limited to Rs.7.5 crores.

The major problem with the proposed approach is limited computing power. A protein has thousands of atoms and forces on every atom need to be evaluated which requires millions of calculations that are best handled by super-computers. The current architecture can be used to solve less difficult problems and once proof of concept is established additional funds could be procured.

6.5 Rationalise Tax Structure

6.5.1 *Goals of the Policy*

High direct and indirect taxes increase the cost of doing business in India, and also suppress the demand potential. There is a need to bring down taxes to levels found in other developing economies, widen the tax base to ensure that revenues are not affected and simplify the tax administration procedures

6.5.2 *Imperative for the Policy*

6.5.2.1 The Existing Taxation Regime in the Country is Complex and Tax Levels are High

India has a very complex tax regime with several taxes, direct and indirect, being levied on manufacture of goods.

Direct taxes: India has one of the highest direct taxes on corporates and individuals amongst the comparative countries as illustrated in the figure below.

Exhibit 6.5.1

Marginal tax rates for corporate, 1998

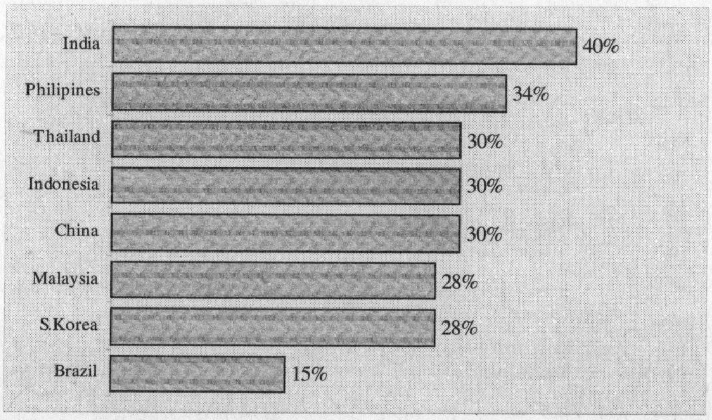

Exhibit 6.5.2

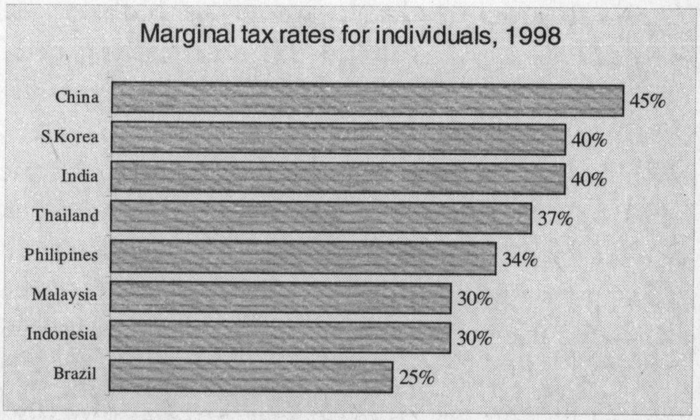

Also the taxation regime is complex with minimum alternate tax provisions, several exemptions and deductions.

Indirect taxes: There is a central tax on manufacturing and state level tax on sales. In addition, there is Octroi levied by local municipalities/ corporations. Indirect tax rates too are high in comparison, as illustrated in the following figure:

EXHIBIT 6.5.3

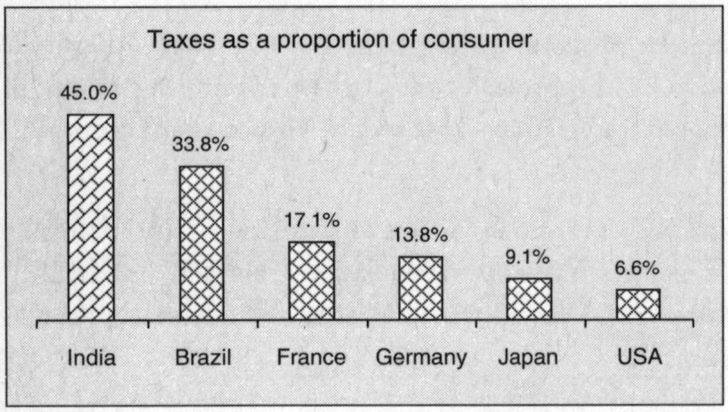

The SSI policy in India gives indirect tax exemptions to small-scale units in the country. This has led to the proliferation of small units even in segments where scale advantages are necessary. The impact of these policies are felt today as the quality of our exports compare poorly against that of competing countries like China.

The indirect tax administration is also very complex - for example, there are 20 slabs of customs duty ranging from 0% to 182% and in addition to basic customs duty there are 7 additional levies (like SCD, CVD, SAD etc.). There is central and state level sales tax. Each state follows a different sales tax structure, with different rates and exemptions schemes. Predominantly the sales tax regime is a single point tax. However, some states have partial VAT implemented as well.

6.5.2.2 High Taxes are Counter Productive

There are several negatives of the existing tax regime in India, as elaborated below:

- **High Tax Rates Lead to Costly Products:**

Very clearly, high tax is one of the key reasons why products are much costlier in India. For example, a Honda City cost about Rs.9,00,000 on the road in India. A comparable car in the US is available at less than Rs.7,50,000. High cost has a direct impact on the demand.

- **High Individual Tax Rates Result in Lower Demand:**

High individual tax rates lead to lesser disposable income in the hands of the end consumers, leading to lower personal consumption expenditure. This has a direct impact in lowering demand.

- **Tax Exemptions Lead to Scale Inefficient Units**

SSI units in India are given indirect tax exemptions. This results in SSI units having price advantage over larger scale players, leading to many industries in India being fragmented with small players. Such industry fragmentation leads to loss of scale advantage, resulting in Indian products being much costlier than competitors in the global market, and much poorer in quality levels. This point has been elaborated in our SSI policy related recommendation.

EXHIBIT 6.5.4

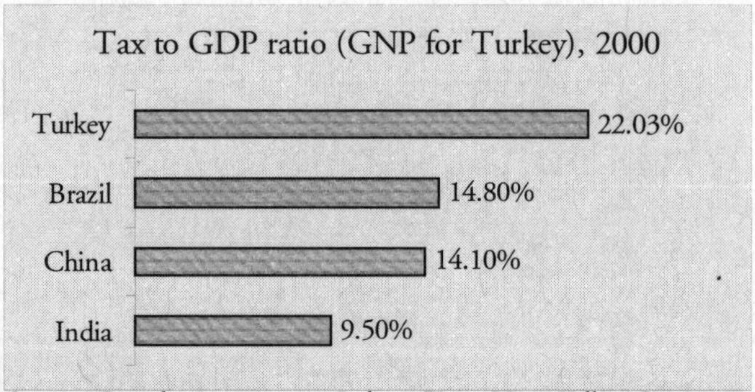

Source: Consultation paper on Indirect Taxes, Vijay Kelkar Committee, 2002

- **High Taxes and a Complex Tax Structure Encourage Low Compliance Resulting in Low Tax Collections**

The following figure illustrates the tax to GDP ratio of India and some comparative countries.

The low tax to GDP ratio of India is an outcome of a complex tax structure and high tax rates, in addition to the absence of service tax and tax on agricultural income. These provide the motivation for tax evasion and enough means to creatively account for taxes.

- *Complex Tax Regime Leads to High Transaction Costs in the Economy*

  Complexity of the tax regime leads to a lot of time being spent by executives in managing the tax compliance activity and increases scope for corruption. These lead to time delays and speed money usage thereby increasing the cost of doing business in India
- *The Tax Structure in the Country is Leading to Supply Chain Inefficiencies*

  Due to the existing complex tax regime, there are multiple layers in the distribution chains of companies. To avoid the central sales tax issue, most companies go in for local stock points in each state, in addition to one or more aggregating godowns near factories. This is one of the key reasons for low inventory turns in the country.

6.5.2.3 High Taxes is the Key Obstacle for Firms Doing Business in India

In the recent Business Environment survey conducted by the World Bank amongst firms operating in the country, high rates of taxation have been highlighted as the leading obstacle for carrying out business in India, as illustrated below:

Tax administration also figures as the fourth major business impediment in doing business in the country.

6.5.3 Policy Recommendation

We recommend that:

- tax distortions based on SSI classification be removed.
- direct and indirect tax levels in the country be reduced in a calibrated manner.
- taxation regime be simplified by reducing the number of slabs and minimising/ removing deductions and exemptions.
- taxation net be expanded to include services and agriculture.
- implementation of a national VAT regime.

Several committees and task forces that have been constituted by the Government have made similar recommendations in the past.

6.5.3.1 Tax Restructuring

Clearly India needs to overhaul its tax administration. Based on our analysis of the existing situation in the country and the experiences of other similar economies we recommend the Government do the following in the taxation regime:

EXHIBIT 6.5.5

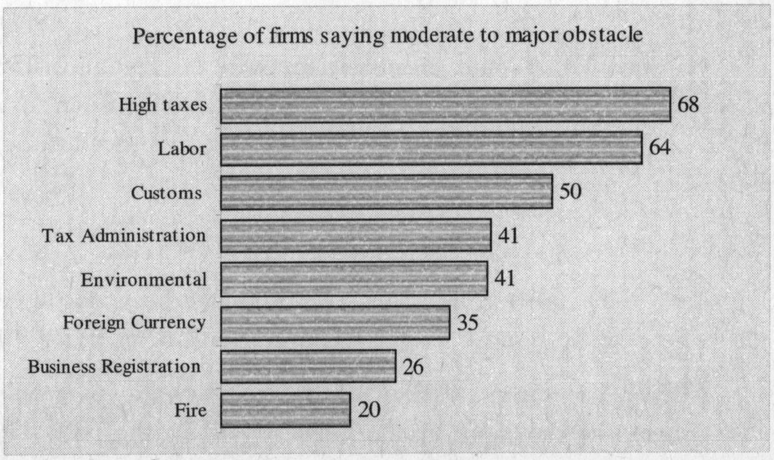

6.5.3.2 Reduce the Absolute Levels of Taxes Across Different Categories:

A key concern has been the impact of lower taxes on the total receipts of the Government. The experience in several countries has illustrated that lower taxes have led to demand surges that more than compensate for the drop in the tax rates and protect or improve the overall tax receipts.

Brazil's auto industry presents a good example of impact of taxation on the total volume of sales of vehicles. The exhibit below indicates the changes in the vehicle sales volumes with respect to tax rate changes in the country. When the taxes were cut in 1991 there was a surge in volumes and when the taxes were raised again in 1997 there was a sharp fall in volumes.

EXHIBIT 6.5.6

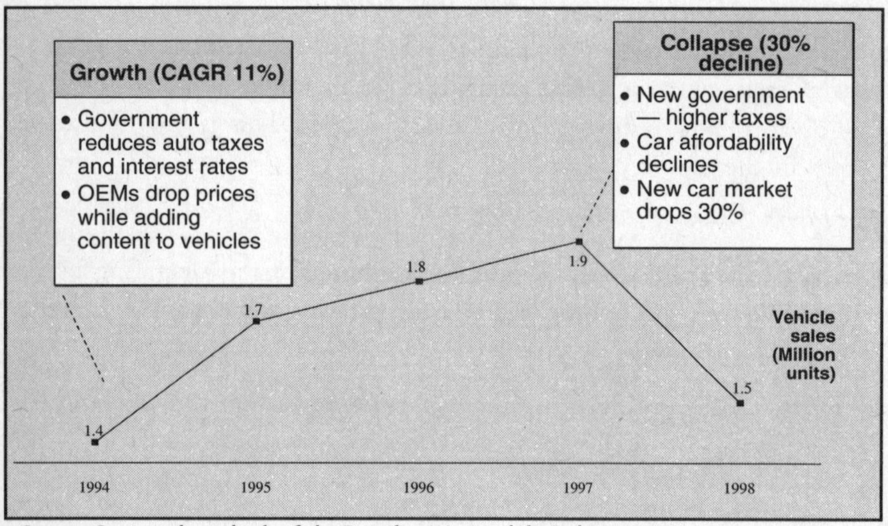

Source: Statistical yearbook of the Brazilian Automobile Industry, 1999

In fact, India's own experience in the past has seen compliance levels going up after reduction in the direct taxes area.

6.5.3.3 Move to a Simple Tax Structure

The multiplicity of slab rates in the existing tax regime make tax computations open to interpretation leading to unnecessary delays and a lot of time spent by corporates in handling tax related issues. This can be alleviated by moving to a regime with just one or at best two slabs of taxes. In the transition period, the tax slabs can be based on revenue neutrality.

6.5.3.4 Expand the Tax Net to Include Services and Agriculture

Study of comparable countries like Brazil, which had similar tax to GDP ratio as India in the early 90's and are much better off than India today clearly indicate that inclusion of services and agriculture is a key step in improving the overall tax collections. In India's context, services contribute to close to 50% of the GDP today and hence is a key segment that needs to be included in the tax net.

6.5.3.5 Move to National VAT

National VAT would streamline the sales tax regime in the country and would also result in establishment of efficient supply chain, leading to substantial reductions in working capital costs for the industry. VAT will also result in simplification of the existing set of procedures and lead to better compliance, leading to better tax realisations.

6.5.3.6 Remove Tax Exemptions Based on Scale

Tax exemptions/ concessions for SSIs have been detrimental to the overall development of the industry. These need to be removed. Our recommendation on SSI policy deals with this recommendation in detail.

All these recommendations are straightforward in terms of their rationale and have also been made by several committees set up by the Government in the past. The need to revamp the taxation structure in the country is well recognised by the Government. This is reflected in the gradual reduction in tax rates and rationalisation of the number of slabs of tax, both of which have been taking place in the country in the past few years. Initiatives to move from single point sales tax regime to a multi-point VAT regime is also on the cards, with a few state governments even coming out with draft VAT legislation.

However a key impediment is the slow speed of implementation of the necessary changes. The recurring delay in passing implementation of VAT is a clear case in point. Also moving to a national VAT is much better than implementing state level VAT as is being proposed currently.

Several committees have commented on tax reforms in the past and highlighted most of what has been listed here. The government needs to accelerate the pace of the tax reforms.

6.6 Aggressively Promote FDI Inflow

6.6.1 Goal of the Policy

India needs to get a greater share of foreign direct investment flowing into Asia. For this, restrictions in inflow of FDI need to be removed, so that India becomes as attractive for FDI as other Asian economies like China, Thailand and Korea.

6.6.2 Imperative for the Policy Change

India's share in total FDI to developing countries declined from two per cent in 1997 to one per cent in 1999. This is despite the fact that the inflow of FDI to developing countries has been progressively increasing throughout the decade of 1990s. China, on the other hand, continues to be the largest absorber of these flows, while South Korea experienced rapid jump in inflows in 1998 and 1999.

Exhibit 6.6.1

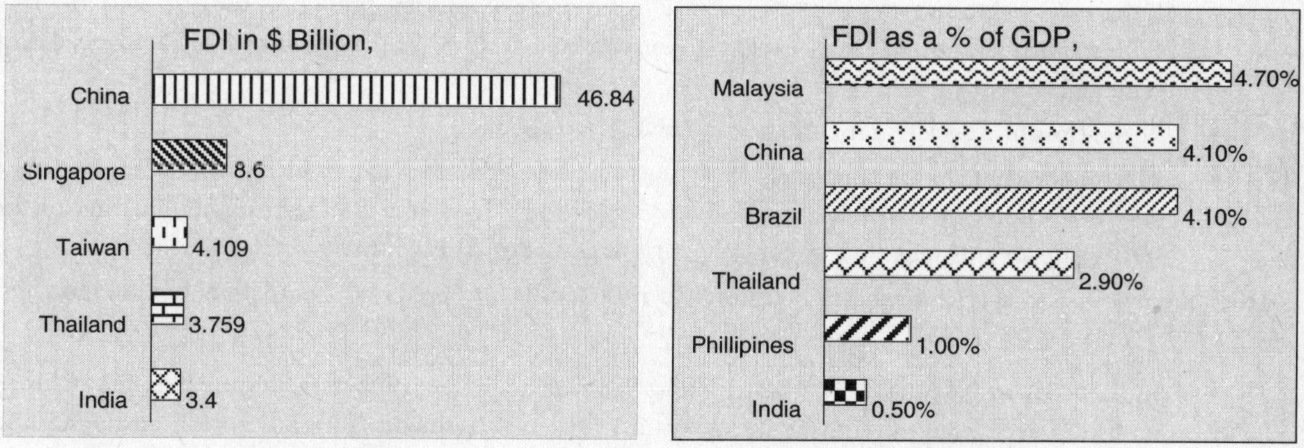

Source : World Development Report, 2001

There are several reasons for this dismal performance by India:

- Setting up a business in India is very difficult. A CEO in our survey said that his collaborator wanted to set up manufacturing base in India but was discouraged by the slew of clearances that were needed and the number of departments that they needed to visit to get their project going. Instead of the government proactively marketing India as a FDI destination, the investors need to run around to get their projects started.

- Even though labour cost is low in India, the productivity level is low as well partially neutralising this advantage. In addition, the regulations governing labour create huge exit barriers for prospective investors, thus discouraging investment, especially in comparison to other countries.
- The cost of doing business in India is also higher due to poor infrastructure. Poor roads, slow customs clearances, poor quality and high cost of power, and time spent in managing regulatory interactions lead to the net cost of manufacturing being very high compared to other competing countries.
- Reservations give unfair advantages to SSI players, discouraging investment flowing in sectors reserved for SSIs. In addition, in sectors reserved for SSI, any foreign participation beyond 24% needs to go through a licensing process with the government, implying procedural delays. A key area where China has been attracting FDI is in garment making which is reserved for SSIs in India.
- Privatisation delays are another reason for poor FDI. In the latter half of the 1990s, a large proportion of global FDI flows were related to mergers and acquisitions (M&As), with little going to green field investments (in the last few years there has been an overall reduction of FDI flow through cross border M&A deals, though). In India, from 1999 to 2001 (up to November), acquisition of equity accounted for a mere 12 per cent of the FDI.
- Slow pace of reforms in sectors like power is another contributory factor for poor FDI scene in the country.
- Countries like Mexico have done well on the FDI front due to their participation in NAFTA. Lack of association with such powerful trade blocs is another reason for poor FDI.

6.6.3 Policy Recommendation

We recommend that:

- India takes aggressive steps to attract FDI in different sectors by providing fiscal incentives. 100% FDI should be allowed in all except a few strategic sectors. FDI restrictions in retail need to be removed.
- Formation of Special economic zones (SEZs) is expedited
- FDI policy changes are supported by actions in associated areas like granting tax benefits, enabling ease of technology transfer, easing labour regulations, removing SSI restrictions, facilitating easy setting up of business and enhancing infrastructure in the country

The action points to accelerate the FDI inflows to make India the preferred destination for foreign capital are clear.

- India needs to be equal if not better in comparison to other competing destinations like China in terms of the different incentives given to attract foreign investment. Currently several of the incentives given by China are not available for foreign investments in India (refer Case Study - China Experience). This calls for adoption of a much more proactive incentive policy towards FDI.
- On the lines similar to China, India has announced the formation of Special Economic Zones. These need to be expedited. And like in China, these economic zones should not be governed by retrograde regulations and should be managed as separate enclaves with progressive regulations. For example, labour flexibility should be ensured in such economic zones for all firm sizes.
- 100% FDI needs to be allowed in the retail sector. This would result in the entry of MNCs in one of the largest sectors and have spin off benefits in terms of improvement in supply chain

efficiencies. For example, Wal-Mart, one of the world's largest retailer, has been a principal contributor to the enhancement of overall productivity of the US economy.

- These changes need to be supported by:
  - De-reservation of SSI.
  - Enhancement of infrastructure by accelerating privatisation initiatives across key sectors, especially Power.
  - Enhancing ease of setting up and running manufacturing units in the country - Only a few states have single window system of clearance of projects. This needs to be made a standard feature in all the states in the country.

6.6.4 Case Study - China Experience

China has managed to attract the highest share of FDI flowing in the world by adopting an FDI friendly policy regime. The success of China's FDI policy is illustrated in 47% of their exports being from units set up with FDI.

China is today the third largest — behind the US and Japan — producer of computers and peripherals. It is also the world's largest manufacturer of many consumer durables such as colour TVs, washing machines and refrigerators. China's dominance in the hardware sector has been mainly because of large FDI.

Some of the reasons for China's FDI success are:

- Several tax incentives have been offered to lure foreign investors
  - Equipment (and related technology), components and spares brought into China by the foreign invested Research and Development (R&D) centres are exempt from import duty
  - Foreign enterprises transferring advanced technology to China are exempt from both business and income taxes.
  - Foreign invested firms that increase their technology funding by more than 10 per cent over the previous year are eligible to deduct 50 per cent of the funds actually spent on technological development from their income-tax dues.
  - Further, the technology, equipment and components imported by foreign investors for upgrading enterprises considered high-priority by the state are exempt from import duty.
- The setting up in the 1990s of Special Economic Zones (SEZs), where business is regulated by an independent authority.
- High priority to improve its power infrastructure, which was in shambles a decade ago, also helped. At present, China's power generating capacity is thrice India's, and its power tariff, only half as much.

6.7 Develop Bilateral Trade Relations with ASEAN Countries and China

6.7.1 Goal of the Policy

To take up a position of strength in the world market, India needs to have strong trading partners. It also needs to leverage its strategic geographical location, to increase trade with other South East Asian countries. For this, India needs to begin by working out bilateral agreements with individual countries, and aim to be a full-fledged member of key trading blocs in the long-term. Given the increase in trade from South East Asia, India cannot afford not to be a part of this.

6.7.2 Imperatives for the Policy

6.7.2.1 ASEAN cannot be Ignored

ASEAN consists of a regional grouping of 10 countries in South East Asia with a cumulative population of half a billion, a per-capita GNP ranging from $ 260- $ 30,170 and a total trade of $ 619 Billion in goods alone, per year.

The countries in ASEAN are at varying levels of economic development. Singapore is ASEAN's leading economy. Other countries like Malaysia, Thailand and Indonesia comprised the South East Asian Tigers before the Asian financial crisis. Though Malaysia and Thailand have managed to recover from the crisis, Indonesia continues to be troubled along with Philippines. Vietnam, Laos, Cambodia & Myanmar have been included in ASEAN only since 1995 and economies are small but growing.

6.7.2.2 Imperative to Become a Full Member of ASEAN

India at present is not part of any regional free trade grouping except SAARC[5], which is ineffective due to India-Pakistan rivalries. India is currently a full dialogue partner with ASEAN with an observer status. This deprives it of any preferential treatment that can be accrued by being a full member. The following points outline the immediate need for Indian engagement of ASEAN members and China:

- *ASEAN is too large a market to be ignored*: With over half a billion people, and the third largest economy in the world, India has to get access to its markets to continue its growth trends in the years to come.
- *ASEAN is India's leading trade partner in Asia*: The two way trade between India & ASEAN at US $ 8.02[6] - $10[7] Billion consisted of 10.05% of India's total trade in 1998, but only 1.28% of ASEAN's total trade in the same year. Indian trade with ASEAN is more than its trade with either China or Japan. Most of Indian exports to ASEAN are traditional low value items.
- *China is about to steal a march ahead of India in ASEAN as well* : China & ASEAN have agreed to establish a free trade area within ten years. Since such a free trade area will have a policy of reducing mutual trade barriers among member countries and reducing non-tariff barriers as a sequel, non-members will be uncompetitive in this market. India cannot afford to be left out of one of the world's most dynamic free trade areas of the future.
- *China will become the world's largest market in 2025* : (by estimates of current growth rates.) India needs business from China and access to its domestic markets. Clearly, we cannot ignore the dragon in our own backyard. It is important that we learn from China and become competitive or risk having a neighbour that is over ten times richer than us by the end of this decade.

6.7.3 Policy Recommendation

6.7.3.1 Engage ASEAN Members Individually

Political considerations have prevented India from having a trade and investment pact with ASEAN over the past decade. Though the government must continue to push for the same, India must parallely engage ASEAN[8] members individually. This is because some members are more inclined towards building a

[5] *South Asian Association For Regional Co-operation*
[6] *1998 Figures*
[7] *1999 Figures*
[8] *Association Of South East Asian Nations*

relationship with India (Singapore, Thailand, Vietnam) than others (Indonesia, Brunei, Malaysia) due to political considerations. We suggest that the Indian government have free trade agreements with ASEAN members individually, on the lines of a customs union.

6.7.3.2 Engage China

Despite the apprehension of Chinese goods flooding Indian markets, China's domestic market cannot be ignored. Moreover it is likely to be the driver of growth in Asia for the next decade. A policy of engaging China on trade and perhaps even having a free trade zone needs to be considered.

6.7.3.3 Have Sector Specific Focus

This strategy can be used to identify sector specific free trade arrangements with countries. For example in the case of China– the worlds largest mobile telephony market, import of Chinese-made mobiles could be free of import duties. This could result in giving a fillip to phone penetration rates in India. We could bargain for similar concessions with the Chinese for access to their markets.

6.8 Reform Bankruptcy and Labour Laws

6.8.1 Goals of the Policy

Ease of entry and exit for business in India is much lower compared to other countries. This results in companies continuing to operate inefficiently, when there is a need to downsize or close down. This creates a poor environment for fostering excellence. There is a need to amend the bankruptcy and labour laws to make it easier for sick units to close down.

6.8.2 Imperative for the Policy

6.8.2.1 Current Bankruptcy and Industrial Dispute Laws are Counter Productive

The Indian government had formulated the Sick Industrial Companies Act (SICA) (1985) with a view to assist in the rehabilitation of sick units and recommend closure wherever required. The Board for Industrial & Financial Reconstruction (BIFR) was the quasi-judicial body that decided on the plan for the firms which had filed under SICA.

SICA

The concept of Industrial sickness in India is defined on the basis of net worth erosion (greater than 50% erosion can classify an industrial undertaking as sick). Firms filing under SICA get complete protection from lawsuits and recovery proceedings from creditors. The BIFR decides on the recommendation for closure, sanction and monitoring of a rehabilitation scheme and eventual discharge from BIFR after revival.

Industrial Disputes Act (IDA), 1947 [Sections 25(M), 25(N), 25(O)]

Under the above act, an industrial undertaking employing 100 or more personnel needs prior permission of the government for closure of operations or layoffs. Government often turns down these applications, which leads to the employer appealing to high courts and tribunals (BIFR). The procedure of closure thus stretches anywhere between 3-15 years.

Effect of SICA & IDA 25[M,N,O] On Indian Manufacturing

SICA & IDA 25 [M,N,O] have been detrimental policies that continue to exist in India, which systematically undermine free market forces, stifle innovation and starve the industry of venture capital. A dynamic market needs fast entry and exit from business, so that capital and labour can flow to the opportunities that give maximum return. These regulations have done exactly the opposite due to the following :

1. *Direct Government Intervention*: The BIFR was a quasi-judicial body, which made recommendation for closure or rehabilitation. A rehabilitation plan was not only sanctioned, but monitored as well by the BIFR. Direct government intervention in running firms only added to the already large bureaucratic network that Indian business had to grapple with. By undertaking what should essentially be a creditor-lender decision, the pace of resolving problems lengthened considerably. The mean delay for solving a case varied from 3-4 years
2. *Definition of Sickness*: By defining industrial sickness on the basis of net worth rather than the world-wide norm of debt default, the SICA regulation created two anomalies. One, even though default would have started early, it could not consider a firm sick as the net worth was high - thus delaying insolvency. It also prodded many firms to fudge their accounts and seek protection under SICA from creditors[Exhibit 6.8.1]. Secondly, there is a difference in NPA classification. For a financial institution, the asset is classified as an NPA[9] once default has happened for 180 days, but for a firm under SICA, it is when the net worth is eroded by a defined level (50%).
3. *Creditor Unfriendly Environment:* Long procedural framework once a case has been accepted by the BIFR, automatic suspension of all lawsuits against debtors, and intentional filing under SICA by unscrupulous firms to avoid debtors have all created a scenario in which financial institutions do not have a risk appetite for investing in the Industrial sector. This can be gauged from the fact that banks have not implemented the P.R. Nayak committee recommendations (1992) on minimum working capital financing to SSI units at 20% of the output value. Worse, according to estimates of the Kapur Committee only 15-20% of SSI units can access bank credit. By having a creditor unfriendly environment, we systematically starve both entrepreneurs and existing units of capital.
4. *Gross Misuse of SICA* : It has been found that many firms file under SICA intentionally when they do not want to payback their creditors. With delays for upto 3-4 years for a BIFR recommendation on accepting the firm under SICA and more time for resolution, the promoters of such firms are shielded from creditors for all practical purposes.

Companies (Amendment) Bill, 2001

The Companies (Amendment) Bill, 2001 has been introduced by the Government and is pending in the Lok Sabha. On the surface the government realises the shortcomings of SICA, BIFR and existing mechanisms pertaining to bankruptcy in India. Though some aspects of the new bill are definitely going to reduce the lead-time in deciding closure or reorganisation of sick units, we feel that the government must reformulate the bill to reflect ground realities.

EXHIBIT 6.8.1

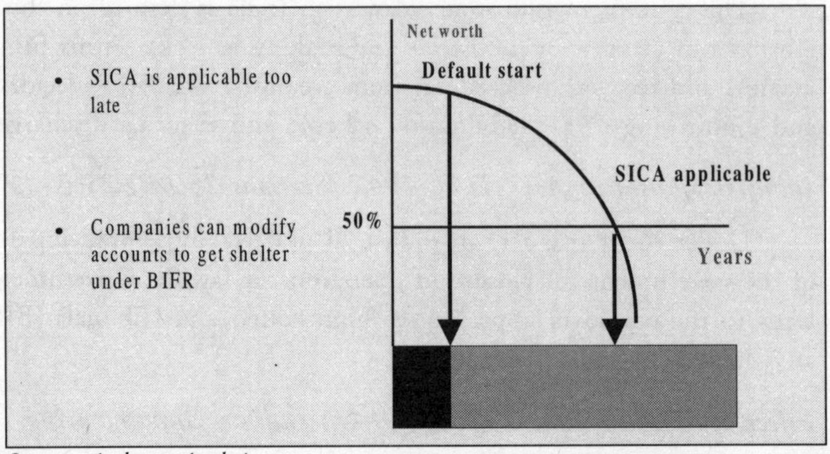

Source : Andersen Analysis

[9] *Non Performing Asset*

We welcome the government's decision to abolish BIFR and the company law board (CLB) and to replace it with a ' National Company Law Tribunal'. By transferring the powers and jurisdiction currently being exercised by these judicial and quasi-judicial bodies to the Tribunal, the multiplicity of litigations before various bodies for revival , rehabilitation , merger , amalgamation or winding up of companies will be avoided leading to a reduction in the lead time for closing cases.

However the regulation has a number of retrograde provisions:

1. The government has long postponed the abolishment of SICA and the current amendment only lifts the SICA provision by transposing them into the Companies' Act. The entire scheme of appointing an operating agency and preparation of rehabilitation schemes for financial reconstruction will now rest with the NCLT. The government is not devolving the responsibility of closure or rehabilitation to the market.
2. Levying a cess on firms to contribute to the 'Rehabilitation and Revival Fund' to rehabilitate, revive and protect the assets of a sick industrial company must be repealed. We recommend this on three counts:
    - The government must not take up the burden for rehabilitation of industrial units. This has to be left to market forces.
    - The government already gets money from companies through collections under 'Investors Education and Protection Fund' under section 205C of the companies act. This money can be used for the same rather than levying an extra cess.
    - The entire bill excludes the service sector that contributes over 48% to the Indian economy. One could argue that services sector should also be taxed for the same- however services firms don't come under SICA provisions. Moreover service sector firms in India function under free market dynamics and status-quo should be maintained.

6.8.2.2 Key Effects on Indian Manufacturing Due to SICA & IDA 25[M,N,O]

1. Financial institutions are reluctant to provide credit to Industrial units. One of the key reasons for SSI sickness in India is lack of credit.
2. High NPAs for banks leads to higher interest rates to credit seekers. This reflects directly in the cost structure of many industrial units, which find it difficult to compete with low cost imports.
3. Large amounts of capital are stuck in unproductive firms that must be allowed to exit. This prevents free flow of capital and redeployment to opportunities that offer higher returns.
4. Entrepreneurship in India is directly affected, as it is very difficult not only to start firms, but also to exit them.

6.8.3 Policy Recommendation

The government must reform Bankruptcy laws in India, by repealing SICA (1985) and altering the Companies Amendment Bill, 2001 and then passing it. The following are the key points of our recommendations:

- Continue with formation of the 'National Company Law Tribunal' under the Companies Amendment Bill, 2001.
- Redefine sickness on the basis of debt default and not on net worth.
- Remove rehabilitation as a function of the government. This is a creditor- lender decision to be determined by market forces.

- Repeal the proposal in the proposed Companies Amendment Bill, 2001, to levy cess on companies to contribute to rehabilitation fund.
- Have time bound negotiations between debtors and creditors.
- Provide safety net to affected workers through policies like severance payments based on liquidation proceeds and re-skilling through training programmes.

Exit policy for businesses must be made easier by amending the Industrial Disputes Act, 1947 ,[sections 25(M), 25(N), 25(O)] to allow industrial units to close down/ restructure their operations through layoffs without permission from the government. To prevent the misuse of this amendment by employees and employers alike, the act can specify modalities like taking majority concurrence from employees before closure and punitive fines on employers for malafide closures, if proved.

6.8.4 Implementation Results

Our recommendations will assist in making the manufacturing sector dynamic in the following ways:

1. Creditor friendly policies will mean better access to credit for entrepreneurs and industrial units.
2. Faster liquidation will allow for faster capital redeployment.
3. Lower interest rates to firms will help in making them more cost competitive in the market.
4. Ease of exit will attract more investors in the manufacturing sectors.

6.9 Reform and Privatise the Power Sector

6.9.1 Goal of the Policy

Availability of good quality and cheap power is a critical success factor for industrial growth. Indian power sector is in a poor state with poor reliability and high cost of supply. The policy aims to unshackle the power sector in the country through unbundling and privatisation.

6.9.2 Imperative for the Policy

6.9.2.1 High Transmission and Distribution Losses

The Transmission and Distribution (T&D) losses for the various Indian SEBs are amongst the highest in the world. A sample of the T&D losses across states reveals an increasing trend.

Delhi Vidyut Board has the dubious distinction of recording the highest T&D losses. These losses rose from 23.56% in 1992-93 to 43.71% in 1998-99 and to 47% in 1999-00, dipping marginally to 45% in 2000-01. Other states that have been saddled with high T&D losses include Andhra Pradesh, Jammu and Kashmir, Karnataka, West Bengal, Assam and Mizoram. In Andhra Pradesh, the losses rose from 19.88% in 1992-93 to 34.09% in 1998-99, and went down marginally to 31.4% and 31% in 1999 – 00 and 2000 – 01. In Karnataka, the losses increased from 19.55% in 1992-93 to 30.45% in 1998-99 and levelled off at 30% in 1999-00 and 2000-01[10].

The exhibit below[11], shows the fluctuations of the T&D losses across some states that have decided to undertake power sector reforms. The T&D losses are frequently restated with the difference between historical and revised T&D losses nearly as high as 100%.

---

[10] *http://www.hinduonnet.com/thehindu/2002/08/31/stories/2002083102281600.htm*
[11] *HT Energy Audit - The crucial starting point for curbing revenue loss*

EXHIBIT 6.9.1

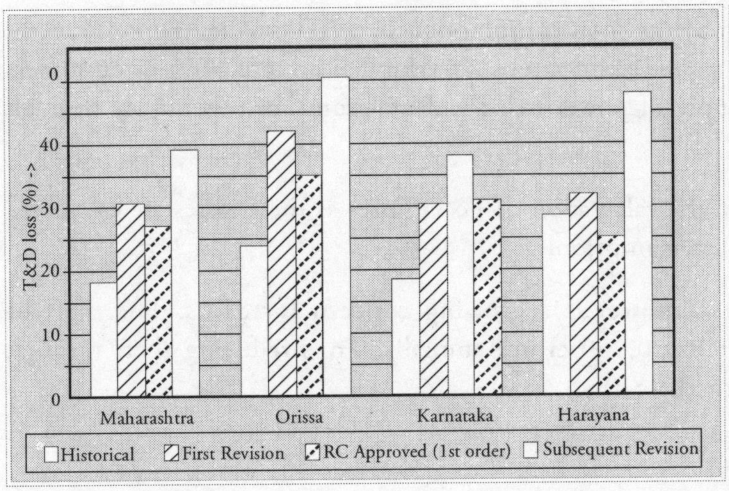

TABLE 6.9.1

Country	Transmission & Distribution Losses (%)
Philippines	16
Vietnam	16
Indonesia	12
Thailand	9
China	7
USA	7
Japan	3

In 1998, percentage T&D losses for some countries in the Asia Pacific region, Japan and the USA are given in Table 6.9.1.

Source: 2001 World Development Indicators (The World Bank).

6.9.2.2 Inequitable Tariff Structure and Burden on Industrial Sector

The tariff structure existing in 1999 – 2000 according to consumer categories along with the 5-year CAGR is shown in Table 6.9.2.

TABLE 6.9.2.

	RS/KWH	5-year CAGR
Domestic	1.49	9.9%
Commercial	3.54	11.2%
Agriculture	0.25	5.6%
Industrial	3.50	9.6%
Railway	4.11	9.5%
Export	1.21	1.6%
Average	2.08	10.2%

From the tariff structure shown above we see that tariffs for agricultural consumption is almost zero and the tariff hike was the lowest in all customer segments. Also in some states like Andhra Pradesh, electricity is supplied free of cost to agricultural consumers, which has resulted in an increase in the agricultural consumption from 2,569 million units (MU) in 1985-86 to 9,866 MU in 1998-99. With tariff for this sector not likely to rise due to political considerations, consumption in the agricultural sector needs to be cross-subsidised by the industrial and commercial segments. Thus the industrial and commercial segments have to pay a higher cost.

As a result there has been a large proliferation of captive power plants as the cost of generation for these plants is significantly lower than the cost of purchasing from the SEBs. This compounds problems for the SEBs as they are losing high value customers while consumption in the low value categories is increasing, thereby swelling their power purchase cost.

6.9.2.3 Percentage of Metered Consumers to Total Consumers is Low

In India, the percentage of metered consumption to the total consumption is quite low. In Andhra Pradesh, only 43% of the energy is used by consumers with metered connections. In Mumbai around 90% of customers have metered connections, against 41% in Delhi. Supply of power to agricultural consumers is mostly through unmetered connections.

The low level of metering has an adverse effect on the calculation of T&D losses as these losses are calculated based on estimated consumption rather than actual consumption.

In addition to unmetered consumption, theft of power is a growing concern for SEBs. The theft of electricity across various states accounts for a loss of Rs.20,000 crores annually. Thus reducing theft through efficient metering can provide considerable savings to the SEBs.

6.9.2.4 Poor Quality of Power in Most Indian States

The quality of power in most Indian states (except for some pockets) is very poor. In order to overcome the problem associated with poor quality of power an increasing number of firms / commercial institutions are using their own generator sets. The percentage of firms having their own generator sets in Indian states is shown in Exhibit 6.9.2.

6.9.3 Policy Recommendation

We recommend:

- Un-bundle State Electricity Boards (SEBs) into generation, transmission and distribution companies.
- Improve operating efficiencies.
- Rationalise the tariff structure to reduce cross subsidisation.

6.9.3.1 Un-bundle the SEBs into Generation, Transmission and Distribution Companies

SEBs that currently have a monopoly in power generation, transmission and distribution need to be unbundled into separate companies for generation, transmission and distribution. The policy should also allow private sector players and other SEBs to maintain a stake in these companies leading to increased competition amongst the various players. The unbundling of monolithic SEBs also brings in more accountability for each of the entities.

EXHIBIT 6.9.2

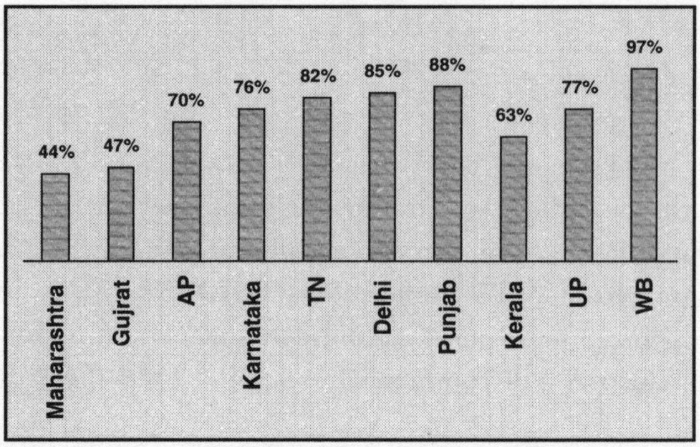

Thus, transmission companies would be able to purchase power from the most efficient and lowest cost producer leading to a reduction in their power purchase costs. The transmission companies would be driven to reduce their losses due to increased price pressure by the distribution companies.

The distribution companies would be able to focus on areas that have traditionally been neglected like increasing the number of metered connections, increased collection drives, and lowering pilferage, etc. as these factors now directly affect the revenue and profitability.

6.9.3.2 Improve the Operating Efficiencies of the Generation, Transmission and Distribution Companies

The Generation, Transmission and Distribution companies need to focus on measures to improve their operational efficiency. The regulators need to set minimum benchmarks for operational parameters that these companies need to comply with. The operational parameters that need to be monitored and tracked are:

- Timely billing for all customers.
- Collections both for the current billing cycles as well as collection against arrears.
- Disconnection drives for defaulters.
- Efficient exception handling for cases like burnt / stuck meters, door locked cases etc.
- Timely and effective customer management including customer complaints handling, new connection requests etc.
- Efficient handling of breakdowns for lines and transformers.
- Regularisation of the metering process.

The consumer base of an SEB can be categorised as industrial and commercials paying a tariff higher than the average cost of supply, residential & agriculture paying a subsidised tariff, and unauthorised users.

Currently the percentage of metered consumers for the SEBs is very low (42%). Coupled with high T&D losses that most of the SEBs incur, the cost of electricity paid by non-subsidised metered consumers is around 2 – 3 times the cost of generation. The primary reason for this anomaly is the fact that a small number of consumers pay for electricity usage by a much larger base.

By increasing the number of metered consumers the distribution companies can reduce the number of consumers in the non-metered agricultural sector and the unauthorised users. This allows distribution companies to transfer the same electricity purchase and distribution cost to a much larger consumer base, effectively reducing the cost of electricity for the industrial, commercial and residential segments.

The installation of new high quality meters would also ensure that energy consumption is measured and billed more accurately.

6.9.3.3 Rationalise the Tariff Structure to Reduce Cross-Subsidisation

The aim of rationalisation of the tariff structure should be such that cross-subsidisation across categories is reduced. The structure and levels of tariff should be designed in a manner to provide incentives to the service provider to improve efficiencies and level of customer service.

The tariff hikes should be such that industrial consumers (HT consumers) have to bear the least hike in tariffs while the residential and agricultural sectors have to bear a steeper increase of tariff. Such increases in tariff will lead to lower levels of cross subsidisation and higher revenues for the SEBs.

The tariff structure followed by SEBs is complex as tariff categories and slabs were introduced to meet specific demands and requests. The process of tariff rationalisation should also include a simplification of the tariff structure. A reduction in the number of Low Tension (LT) and High-Tension (HT) categories through abolition or merging of certain categories as well as a reduction in the number of slabs for each category.

6.9.4 Implication of Power Sector Reforms for the Manufacturing Industry

6.9.4.1 High Quality Power at Lower Cost

With the implementation of structural reforms and control of theft, T&D losses of the electricity companies will reduce. Also with increased competition, the generation companies would be pressurised to improve their power load factor from the current levels of 60% to over 90% leading to higher operating efficiencies. This will ensure that the quality of power supplied to manufacturing units increases. With improved power quality, industry will not need to invest heavily in backup systems / generators thereby lowering their operating costs.

6.9.4.2 Reduction in Power Purchase Costs

With the rationalisation of the tariff structure and reduction in cross subsidisation the per-unit cost of electricity for manufacturing units would reduce. This would have a significant impact on the costs and hence the profitability of these units.

6.10 Privatise Ports

6.10.1 Goals of the Policy

Poor port infrastructure in India has resulted in long clearance times of goods, leading to high product costs and poor service levels. This has adversely affected manufacturing growth. Global manufacturing excellence requires the presence of world class port infrastructure, with easy and quick transit of imports and exports. Privatising port operations is a key step to achieve this.

6.10.2 Imperative for the Policy

6.10.2.1 The Regulatory Framework Governing Indian Ports is Archaic and Cumbersome

1. Major Ports:

In India, the major ports are placed under the Union list of the Indian Constitution, and are administered under the Indian Ports Act, 1908 and the Major Port Trust Act, 1963 by the Government of India. Under this Act, a Board of Trustees appointed by the Government governs each major port. The powers of the trustees are limited and they are bound by policy directions and orders from the Government of India.[12]

2. Minor Ports:

Minor ports are placed in the Concurrent list of the Constitution and are administered under the Indian Ports Act, 1908. The Act defines the jurisdiction of Central and State government over ports. It lays down rules for safety of shipping and conservation of ports and regulates matters pertaining to the administration of port dues, pilotage fees and other charges.[13]

Both these acts were modelled after the erstwhile British practice in managing ports and carry a lot of baggage from acts of the 19th century. The Board of Trustees who are appointed by the Government of India to administer ports represent government departments involved with port operations, labour and service providers such as stevedores, shipping agents etc. Their interest lies more in protecting their turf and not in promoting the commercial well-being of the ports.

6.10.2.2 The Productivity of Indian Ports is Poor as Compared with the other Ports in the Region

1. High Vessel Turn-around Time:

The average vessel turnaround time for Indian ports varies from 3.3 days to 8.3 days, as compared to 15 to 35 hours in major European ports and less than a day in Singapore.

---

[12] *Port restructuring in India - S Sundar TERI, New Delhi*

[13] *ibid.*

2. Low Equipment Utilisation

The average availability of equipment in Indian ports is around 70% as compared to 85%-90% for other Asian ports. Jawaharlal Nehru Port Trust (JNPT), which has India's most modern container terminal, handles just 10 containers per ship per hour as compared to 30 in Colombo and 69 in Singapore. While productivity and efficiency of Indian ports have improved since 1995, they are still below international standards.

3. Low Labour Productivity

The average output per worker in 1999-2000, for all 12 major ports put together, was about 4,546 tons, based on the data published by the Indian Port Association. The best performance average was of nearly 10,874 tons per annum by the Kandla Port. In comparison, Singapore port averages an output of 2303 TEUs, (nearly 34,545 tons per worker).

EXHIBIT 6.10.1

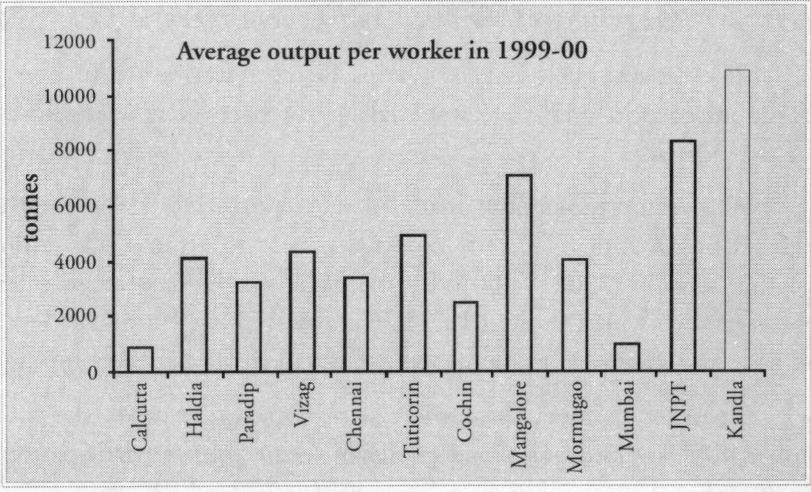

Source: maritime consultancy Ports in 2002

One of the major problems of the Indian port sector is excess labour and a strong union. For the Mumbai Port Trust, out of the total revenue of Rs.5 billion, 80% goes towards the wage bills. Further, the presence of a strong labour union implies frequent incidence of strikes that further reduces the productivity of ports.

6.10.2.3 High Vessel Handling Expenses is a Deterrent to the Growth of Ports

The vessel handling charges for Indian ports are substantially higher than other ports within the region. This is a cause for concern as it prevents mainline vessels from calling at Indian shores, resulting in loss of valuable revenue.

A comparison of the port-wise vessel handling charges is shown in Table 6.10.1:

TABLE 6.10.2.

Port-wise Comparison of Vessel-Related Charges (in $)

Vessel Size GRT	*Chennai*	*Singapore*	*Colombo*	*Jebel Ali (Dubai)*	*Port Klang (Malaysia)*	*Salalah (Oman)*
32,534 tons/ 2,950 TEUs	21,978	9,292	7,239	3,946	617	3,630
31,356 tons/ 2,500 TEUs	20,974	9,238	7,004	3,901	631	3,591
21,963 tons/ 2,000 TEUs	12,101	8,419	5,137	3,249	554	3,009

Source : Urgent need for Chennai Port to examine vessel-related charges www.exim-india.com

6.10.2.4 Regulated Tariff Structure at the Major Ports

As the private sector was allowed entry into the major ports to provide services in competition with the Port Trusts, an independent regulator, Tariff Authority for Major Ports (TAMP) was set up in March 1997 through an amendment of the Major Port Trust Act 1963, to fix port tariffs. This was done to ensure that there was no unfair competition between the private sector players and the Port Trust.

The 1996 guidelines provided for the establishment of the Tariff Authority for Major Ports to fix and revise port tariffs. All powers for fixing tariffs in major ports lies with TAMP, but it has no jurisdiction over minor ports or private ports. TAMP is only an authority for fixing tariffs and has no other regulatory functions or powers.

6.10.2.5 Recognition of the Need and Actions taken so far

The Government is taking steps for corporatisation of ports. It started with the Ennore Port as the first corporatised port and the Jawaharlal Nehru Port; New Mangalore Port, Mormugao Port and Tuticorin Port are in the pipeline.

As per the projections made in the report "Draft Perspective Plan for Indian Port Sector - vision 2020", the traffic on Indian ports is expected to rise to around 537 million tons by the same time. If the traffic rises to this level by 2006 - 2007, Indian ports would require an investment of around Rs.200 billion in the next few years, of which around Rs.35 billion has already been made and another Rs.30 billion is committed. Much of the investment is presently being diverted towards developing new facilities.

In order to meet the projected traffic requirements the Government of India has decided to adopt the concept of landlord ports, and gradually secure private participation in the provision of port services. Accordingly, in October 1996 policy guidelines were issued which provided for private sector participation/investment in the following areas:

1. Leasing out existing assets of the port.
2. Construction/creation of additional assets, such as:
   - Construction and operation of container terminals.
   - Construction and operation of bulk, break bulk, multipurpose and specialised cargo berths.
   - Warehousing, container freight stations, storage facilities and tank farms.
   - Cranage/handling equipment.
   - Setting up of captive power plants.
   - Dry-docking and ship repair facilities.
3. Leasing of equipment for port handling and leasing of floating crafts from the private sector.
4. Pilotage.
5. Captive facilities for port based industries.

However, the landlord model of port privatisation that India follows is very different from the way it works the world over. The port authorities in other countries only own common infrastructure and provide marine services. Nowhere do they operate the container berths, which are all leased out to private firms. India is perhaps the only country, where the port authorities are landlords-cum-competitors, which skews the competition in their favour.

Rules on tariff also weigh down the port trusts. Shipping major, Maersk India, for example, switched its business to Nhava Sheva International Container Terminal (NSICT) because it offered more competitive rates. But, major ports in India are impeded by the Major Port Trusts Act, which insists that tariff should be fixed by the Tariff Authority for Major Ports (TAMP). On the other hand, minor ports are completely free of tariff control, allowing them to woo customers through attractive packages like neutralising the increased transport cost by offering lower cargo handling charges.

Looking at the shortcomings in the ports sector we suggest the following recommendations for efficient operations of the various major and minor ports in India.

6.10.3 Policy Recommendation

We recommend:

- Operate the ports using the classical "Landlord model"
- Rationalise the port tariff structure.
- Undertake labour reforms.
- Invest in infrastructure so as to set up world class operating infrastructure.

6.10.3.1 Operate Ports on the Classical "Landlord Model"

The Port Authority owns basic infrastructure only, (land, access and assets) and leases them out to operators on a long-term concession basis while retaining all regulatory functions. Examples of landlord ports are Rotterdam (Netherlands), Buenos Aires (Argentina), or Laem Chabang (Thailand). Each port would have multiple private operators to induce competition. This would also reduce the risk of formation of private monopolies. There would not be any restriction on the set-up and operation of competing facilities by service providers. The proposed model of operation for ports is as shown in Table 6.10.2.

TABLE 6.10.2.

PortType	*Port Authority Responsibilities*		
	Infrastructure	*Superstructure*	*Stevedoring*
Landlord	Yes	No	No
Tool	Yes	Yes	No
Service	Yes	Yes	Yes

Source : Restructuring & strategies of sea pots & companies: The United Kingdom, Peter Turnbull

All operations such as stevedoring would be licensed to private operators. By withdrawing from stevedoring operations, port authorities would be able to divest themselves of many attendant risks associated with cargo handling, as stevedoring is inherently irregular. By entering into joint ventures with other companies (risk sharing) or licensing new independent operators to undertake stevedoring (risk displacement), port authorities can be in a position where most of their revenue is guaranteed (from rent, licenses, equipment hire, and port charges), but they no longer carry the cost of employing a large number of registered dock workers.

6.10.3.2 Rationalise the Port Tariff Structure

The port tariff structure should not be governed by TAMP but decided by competing market forces depending on the demand-supply conditions. This structure is currently followed by minor ports and has been successful, as the share of these ports has increased to around 20% in 1999-00. The cargo handled by minor ports has increased by about 80% in 1999-00 from 1998-99. The tariff structure should be such that the costs incurred by the major shipping lines for availing services at ports in India are comparable with those of other major ports in the region.

The tariff structure followed needs to be simple. This can be achieved by reducing the number of tariff heads and the number of variables within each tariff head.

Port tariffs should include only the following four heads:

- Navigation charges (port dues, tug services etc.)
- Berth charges (berth hire, wharfage etc.)
- Cargo operational charges (stevedorage, packing/unpacking, storage etc.)
- Licensing (real estate etc.)

6.10.3.3 Undertake Labour Reforms

The Dock Workers Act 1948 that governs port workers provides a remarkable degree of protection to dock workers. This has led to various mal-practices such as speed money, over-manning, low productivity, idle time, etc. The present policy does not provide for an exit policy for labour and therefore, acts as a dis-incentive for private investment. The Mumbai Port Trust (MbPT) handled 27.03 million tons of cargo in 2001-02 and had a workforce of around 31000 whereas the Jawaharlal Nehru Port Trust, (which was privatised in 1998) handled 18.58 million tons with only 1500 workers.

In order to overcome the problems associated with over manning we suggest that:

- Modifications to the existing Dock Workers Act 1948 be carried out such that there are provisions which give the various port trusts an exit policy for additional labour.
- The various port trusts introduce voluntary retirement schemes that reduces the problem of over-manning.
- Reduction of retirement of the dock workers from the current level of 60 years.
- A temporary freeze on recruitment of dock workers till the work force is right-sized.
- Training dock workers on new equipment to improve productivity.

6.10.3.4 Invest in Infrastructure so as to set up World Class Operating Infrastructure

The government has already announced guidelines for private/ foreign participation that permit formation of joint ventures between major ports and foreign ports, major ports and minor ports and major ports and companies in addition to investment policy. The Investment Policy followed for participation of private players is as follows:

- Foreign investment up to 100% is permitted in construction and maintenance of ports and harbours, and in projects providing support services to water transport, such as operation and maintenance of piers, loading and discharging of vehicles without requirement of an approval.
- Open tenders will be invited for private sector participation on Build-Operate-Transfer (BOT) basis.

The Major Port Trusts (Amendment) Bill, 2001, which enables government to vest control of major ports to private companies is pending before the parliament. This bill needs to be passed expeditiously.

The focus of private investment for port infrastructure should be on building containerised terminals, as this is the fastest growing segment. Containerised cargo accounts for half the exports in Europe and the Americas whereas in China it currently accounts for one third of the exports and is growing at a rate of 20% per annum.

Another area of investment in infrastructure is the development of Special Economic Zones (SEZ) around port areas. This has a tremendous impact on reduction of time and cost and improves the efficiency of companies located in the zones. The ports have also benefited in terms of increased volumes.

6.10.4 Implications of the Port Reforms on the Manufacturing Sector

Ports form a major gateway for exports from India. Some implications of the suggested port reforms on the manufacturing industry are as follows:

6.10.4.1 Lower Product Costs

After undertaking port reforms the port and freight handling charges at the major ports would be based on market forces leading to a substantial decrease in these charges. With improved efficiencies in cargo handling and a rise in productivity would reduce the operating costs for these ports and improve service levels. These saving could then be transferred to the export oriented manufacturing units thereby reducing their overall product costs; making them more competitive in the international markets.

6.10.4.2 Lower Inventory Levels and Vetter Service Levels

With world class infrastructure present at Indian ports, the time required for freight handling would reduce from the current levels of 4–5 days to 15–30 hours. Improved berth and freight handling procedures time slots would be allotted for freight loading/unloading. This would help manufacturers produce goods when required rather than building up inventories. Thus the manufacturers would benefit from maintaining lower inventory levels and increased competitiveness due to lower response times.

6.11 IMPROVE ROAD INFRASTRUCTURE

6.11.1 Goal of the Policy

Good roads facilitate free movement of goods resulting in efficient supply chains and low cost of operations. Indian road infrastructure needs improvement and requires huge investment to make it world class. Private participation needs to be accelerated to quickly improve the roads. This policy addresses this.

6.11.2 Imperative for the Policy

6.11.2.1 Inadequate Growth of Roads Infrastructure

India has over 3 million kilometres of roads, making it one of the most extensive road networks in the world. However, the conditions of Indian roads continue to remain poor and a vast majority of them still remain unpaved. A comparison of the percentage of paved roads for representative Asian countries in 1997 is shown below:

Over the past 10 years, vehicle population has increased over 100% while the road length network has increased only by a factor of seven. The annual average rate of traffic growth stands at around 8-10%.

National Highways	57,737 km
State Highways	165,704 km
District Roads	2,788,000 km

The National Highways constitute only 2% of the total road length in India but carry over 40% of the total traffic. In order to spur growth the network of national and state highways need to be strengthened.

6.11.2.2 High Costs Involved in Road Development Projects

The National Highways Development Project (NHDP) undertaken by the National Highways Authority of India (NHAI) is proposed to cost Rs.54,000 crores based on 1999 prices. Of this cost only Rs.4,000 crores or 7.4% is proposed to be raised from private sources whereas the balance is to be raised from cess on petrol and diesel, external assistance and market borrowing.

The costs of building one km of national highways is around Rs.4 – 4.5 crores whereas a six lane expressway costs around 8 – 10 crores. As the costs related to the construction of new roads or improving existing roads is high financing becomes a key issue. Thus the government has to look at alternative sources of financing for road development projects to make them financially viable.

6.11.2.3 Delays Involved in Road Development Projects

1. Financing Delays

The financing nature of these projects is such that they require a large investment upfront with long gestation periods. In India, cess is the one of the major components of financing for road projects. A "Central Road Fund" has been set up for distribution of the collected cess for various activities like

- Construction and maintenance of state highways by state governments.
- Development of rural roads by state governments.
- Construction of rail over- bridges by Indian Railways.
- Construction and maintenance of national highways by NHDP and Ministry of Road Transport & Highways.

However the process of fund disbursement is entangled in red tape as the Ministry of Surface Transport (MOST) disburses the funds. Also, formation of the Central Road Fund was delayed due to internal wrangling between MOST (which wanted a dedicated fund) and the Finance Ministry (which wanted the money to remain in a consolidated fund).

6.11.2.4 Shortfalls in Targets

Physical targets/achievements for the first three years of the ninth plan are as shown in the Table 6.11.1:

TABLE 6.11.1

S No	*Scheme*	*Unit*	**Ninth Plan Target*	*1997-98*		*1998-99*		*1999-2000***	*Achievement (1997 98+1998-99) and Target (1999-2000 as % age of 9th Plan Targets*
				Target	*Achieve ment*	*Target*	*Achieve ment*		
1	Widening to two lane	Km	1194	159	164	958	217	218	50.17
2.	Widening to four lanes	Km	202	65	49	56	127	29	101.49
3.	Strengthening weak 2 lanes	Km	2908	443	340	2213	471	584	47.97
4.	Bypasses	No.	20	7	0	9	1	4	25.00
5.	Major Bridges	No.	40	19	6	23	8	7	52.50
6.	Minor Bridges	No.	226	94	18	263	49	44	49.12

***(1999-2000) Targets as per Ninth Plan document of Planning Commission*
**Ninth Plan Targets, as per Ninth Plan document of Planning Commission*

Thus we see a serious shortfall in all the proposed activities except for widening highways to four lanes. With the Prime Minister reducing the deadline for completion of the Golden Quadrilateral by one year to December 2003, the agencies involved would be hard pressed to complete the project on time.

6.11.3 Policy Recommendation

We recommend:

- Enter into Public-Private Partnerships (PPP) for large scale road development projects.
- Eliminate multiple intermediary agencies with one regulatory authority for all major road development and maintenance projects.
- Simplify the procedures for private participation.

6.11.3.1 Enter into Public Private Partnerships (PPP) for Large-scale Road Development Projects

Public Private Partnerships ('PPPs') are partnerships between the public sector and the private sector for designing, planning, financing, constructing and/or operating projects which would be regarded traditionally as falling within the purview of the public sector. A key objective of a PPP is to allocate risk to the person best placed to manage and deal with the particular risk. Certain risks may be more effectively managed by the private sector rather than the public sector.

The regulatory authority should invite private players to participate more actively in the development of roads. The selection of private players would be done through a bidding process where the capabilities of the interested parties would be evaluated both on technical and commercial basis. The interested parties could be Indian or International companies or consortiums. The regulator would need to do away with all price preferences for Indian companies that exist currently so that the bidding is undertaken based only on market considerations.

The private sector players would execute the project on a Build, Own, Operate and Transfer (BOOT) basis wherein the private sector players arrange the required finance themselves from financial institutions. In return concessions are given to the private sector to operate a toll facility for a specified period (anywhere from 15 to 45 years).

Another method of obtaining finance is through the issue of infrastructure bonds for the purpose of investing the proceeds from the bonds for development of roads. The use of capital raised through infrastructure bonds eases the financial burden on the government. Repayment of the bond capital can be made through tolls and levies collected during the concession period.

In order to ensure that the private players do not cut corners to maximise their profits stringent quality control measures need to be put in place. Quality control inspectors / consultants need to be appointed to ensure that world class standards are followed for construction and maintenance of the contracted roads.

The government limits its role to setting the general framework and financing road segments which are unlikely to be a commercial success, but which are important for social and developmental reasons.

6.11.3.2 Eliminate Multiple Intermediary Agencies with one Regulatory Authority for all Major Road Development and Maintenance Projects

The various road development and implementation authorities currently operating in India are:

- Ministry of Road Transport and Highways (MRTH) is responsible for all policy matters relating to national highways.
- The National Highways Authority of India (NHAI) is the implementing agency for select National Highway stretches entrusted to them by MRTH and the National Highway Development Project.

- Implementing Agency (IA) carries out the policy implementation and regulatory functions. The IA ensures that the highway facility is available to all and that there are no unfair or restrictive practices.
- In addition to the agencies mentioned above the development and maintenance of state highways and district roads is under the purview of the State Public Works Departments.

According to the rules for FDI investments in roads foreign equity up to 100 per cent (with total foreign equity up to 1500 crores) is permitted for the construction and maintenance of roads and highways. The construction and maintenance would be contracted out using the bidding process. Therefore it is essential that there is only one agency that can float the tenders and evaluate the interested parties.

Currently the development and maintenance of state highways and district roads falls under the jurisdiction of the local public works department. Setting up of a single regulator and implementation agency for all highway (national and state) construction and maintenance related activities would ensure the benefits like projects would be prioritised based on an all India basis and finances would be awarded to projects based on priority rather than being governed by local agenda.

This agency would form a single point of contact for private investors and would ensure single window clearances for all approvals. The agency would operate with the following objectives:

- Simplified private participation procedures that would be easily available to all interested parties on the agency's web-site. All the necessary forms and documents needed for participation would also be available on-line.
- All the required licenses and clearances would be obtained from this agency without the need to interact with other government departments.
- Act as a single point for resolution of issues and ensure that all issues are cleared within a specified time frame.

This would expedite the process of tendering and awarding these tenders to interested parties.

6.11.3.3 Simplify the Procedures for Private Participation

The procedures and guidelines for participation of private players using the PPP arrangement should be clear and simple to follow. Some indicative guidelines are given below:

- There should be fixed time-lines for the clearance of each proposal.
- There should be single point of contact for all project proposals and a single window clearances for the proposals.
- Invite foreign bidders for concessions and contracts on par with Indian franchisees and contractors.
- Litigation should be settled through internationally accepted procedures or clearly defined local laws. There should be set time-frames within which all litigation proceedings should be settled.

6.11.4 Implications of Reforms in Development of Roads

6.11.4.1 Expansion of the Serviced Markets

Better roads infrastructure leads to an expansion of serviced markets. This is because manufacturing units can now service a larger group of customers and not be limited by geographical considerations like proximity to market. Improved transport infrastructure also permits a more efficient use of resources as it allows markets to become larger and better integrated.

6.11.4.2 Increased Savings in Terms of Transportation Costs and Lower Service Times

With improved roads infrastructure the time taken to transport goods to customers reduces leading to considerable savings in terms of fuel and maintenance cost for the logistics providers; who could then transfer these savings to manufacturers which may lead to a reduction in product costs. Savings in transportation time and costs would also allow manufacturers to procure materials from sources that were previously inaccessible to them, thereby reducing their procurement costs.

6.11.4.3 Lower Levels of Inventory

Reduction in the time taken to deliver goods to customers allows manufacturers to maintain lower levels of inventory without sacrificing responsiveness. This also allows the manufacturers to be more flexible in terms of their production schedules thereby maximising their throughput.

6.11.5 Case Study New Mexico State Highway & Transportation Department to Mesa Development Corporation

The New Mexico State Highway & Transportation Department to Mesa Development Corporation entered into a public-private partnership for the development of the NM Corridor 44 road. The objectives of the partnership were to ensure accelerated development while ensuring that the finance for the project was raised without any government aid or loans.

The development of the 121-mile corridor from project conception to contract execution was completed in a record time of 15 months (a similar project in India would take an elapsed time of 26 months). The construction contract was awarded based on competitive bidding with the bidding process also managed by the partners. The finance was raised through the issue of bonds to the public amounting to $105 million.

6.12 Restructure the Government Departments

6.12.1 Goals of the Policy

There are 40 ministries directly reporting to the PM, out of which at least 14 are concerned with manufacturing, one way or the other. This wide span increases bureaucracy which leads to delays in decision making and implementing policies. Spurring global competitiveness requires focus, co-ordination, commitment and efficient implementation. For this, the government's structure needs to be simplified.

6.12.2 Imperative for the Policy

We sensed the need for re-structuring and downsizing of the Indian government during our interviews with Industry, Educational Institutions, and Government Officials. We also analysed government structures in developed countries and in transition economies.

6.12.2.1 Co-ordination Across Ministries, a Challenge

The structure of the Indian government and its agents of execution have been a legacy from the colonial times and the following socialist years. In the 21st century, India needs administrative reforms more than ever.

The central government has over 40 different ministries reporting directly to the prime minister. Co-ordination across the 40 ministries or even a group of ministries would be a key requirement for efficient implementation of reforms that are needed to put India on a high growth trajectory

The current ministerial portfolio hampers policy implementation. For example- if we consider 'Manufacturing Policy', there will be close to 14 ministries – Commerce & Industry, Food Processing Industries, Power, Science & Technology, Railways, Shipping, Shipping, Steel, Heavy Industries & Public Enterprises, Labour, Road Transport & Highways, SSI & Agro & Rural Industries, Textiles, Finance & Company Affairs and Petroleum & Natural Gas - that need to function in a co-ordinated manner to implement and improve policy.

6.12.2.2 Delayed Decision Making and Bureaucracy

Devolution of authority is minimal in the Indian government. Almost every decision is referred to the top (Ministry) and it passes through an average of 8 layers on its way up. This delays decision-making and affects the investment climate in India. Delays hurt our domestic industry, which loses time and money in execution and also discourage FDI. With countries in our neighbourhood are competing with us for FDI and encouraging local Industry at a fast pace, speed of decision making is necessary to gain competitive advantage.

We came across instances of loss of FDI in manufacturing due to bureaucratic delays and a maze of clearances from multiple departments, in almost every industry and sector that we studied. We found that while well-established MNCs do not face many hurdles from central and state governments, smaller foreign firms wishing to set up facilities in India face the most problems. The most common refrain of the investors has always been the long lead times that our governments and states have due to multiple ministries, multiple clearances and multiple touch points with the governments. It is indeed no surprise that cumulative FDI realised in India from 1991-92 onwards has been only about 50% of commitments. China on the other hand has realisations Vs commitments closer to 80-90%.

6.12.3 Policy Recommendation

The Indian government has to be restructured for effective governance of the country. The restructuring must start from a re-organisation of the Ministries at the centre to achieve better and faster policy implementation. De-layering of the bureaucracy must be given top priority to improve administrative efficiency. These reforms at the centre must be followed through at the state and local levels. We provide below only guidelines that the government needs to follow:

6.12.3.1 Consolidate Ministries with Related Functions

We recommend that the following new ministries be set up:

- ***Ministry of Energy:*** This will essentially be a consolidation of the existing Ministries of Power, Petroleum & Natural Gas, Non-conventional energy sources and coal.
- ***Ministry of Transportation:*** This will be a consolidation of the Ministries of Railways, Shipping and Roads & Highways.
- ***Ministry of Justice:*** This will involve the consolidation of the Ministry of Law & Justice, Social Justice & Empowerment.
- ***Ministry of Commerce:*** This will essentially have a consolidation of functions of the Ministries of Heavy Industry, Food Processing, Commerce & Industry, SSI & Agro & Rural Industries.
- ***Ministry of Communications:*** This will perform consolidated functions of the existing ministries of Communication & Information Technology, Information & Broadcasting.

6.12.3.2 Devolve Some Central Ministries to State Functions

We make a distinction here between devolving and abolishing. We feel that these functions are needed today, but they will be better executed at the state level. The ministries of Tribal Affairs, Urban Development & Poverty Alleviation, Tourism & Culture, Rural Development, Consumer Affairs, Food & Public Distribution must be devolved to states. We believe that the functions of these departments will be better executed under pure state and local level control.

6.12.4 Implementation Results

The restructuring of the central government's structure as per our recommendation will cut down the number of central ministries from over 40 to 17 (The disinvestment ministry will get closed after divestment is complete). This will have the following benefits:

6.12.4.1 Increase in Administrative Efficiency

With lesser number of ministries there will be a marked improvement in the administrative efficiency of the central government. This will be primarily due to lesser span of control leading to faster decision executions compared to the current system.

6.12.4.2 Down Sized Bureaucracy

A downsized bureaucracy will be the result of redefining Organisation Of Functions & Tasks (OFT). Several redundant functions being abolished will lead to a direct reduction in the bureaucracy.

6.12.4.3 Faster Policy Implementation

Examples from transition economies, especially China, show that the success of a reform Programme is not "Making No Errors", rather it is "Correcting Errors" as fast as possible when they occur. A smaller ministerial set-up at the centre will lead to faster policy implementation due to an increase in administrative efficiency and downsized bureaucracy

However, the true success of the re-organisation will be when the states follow the centres suit and re-organise themselves. We believe that this would be possible through a series of fiscal and non-fiscal incentives that could be provided to the states, which meet criteria set by the centre.

6.12.5 Case Study - China's Restructuring of Bureaucracy

In early 1998, China undertook a major reform for streamlining the government bureaucracy. Most Industrial ministries like Textiles and Machinery were abolished and replaced by much smaller corresponding bureaus which were then absorbed into the State Economic and Trade Commission.

The number of ministries in the Central Government was trimmed from 49 to 25 and similar actions were underway at the local levels.

The number of civil servants was cut from 8 Million to 4 Million. To compensate for the losses, displaced civil servants were sold apartments at discounted prices according to their seniority and were given an option to study for undergraduate and graduate degrees with tuition and stipends paid by the government for 3 years.

Conclusion

Our policy recommendations are aimed primarily at achieving radical growth for Indian manufacturing and developing excellence in manufacturing in India, thereby enabling the Indian workforce to have a better standard of living, and transforming India into a global manufacturing hub. India has a tremendous opportunity to develop its manufacturing sector, and through it, stimulate economic growth. To do this, it needs to first catch up with, and then get ahead of other fast growing Asian economies like China and Korea, who are currently showing better results. We believe the policies we have recommended, if implemented effectively, will definitely help India to achieve this.

Most of the recommendations we have made are not startling or even radically new. However, we believe that any policy is only as good as its implementation. Thus we emphasise that the key to manufacturing growth in India is not just policy formulation, but effective implementation. The government needs to demonstrate commitment in implementation and have a laser focus on achieving results. Industry, institutions and academia need to collaborate with the government in ensuring this.

While government policies set the direction, Indian manufacturers need to introspect and identify areas where they lag behind their global competitors. Benefits of policy may not be felt immediately, whereas competition is intensifying continuously. Manufacturing firms in India need to be proactive and do all it takes to improve their own capabilities to global levels. Only when they do so, can they effectively leverage policy support to be successful in both the domestic and world markets.

"Knowing is not enough - we must apply....Willing is not enough - we must do."

Goethe

This Policy Draft was written by Sanjay Upendram and team of Andersen Business Consulting, which has since become part of KPMG Consulting.

For further details and clarifications, please contact:

Sanjay Upendram
Associate Director
KPMG Consulting
Supendram@in.kpmg.com <mailto:Supendram@in.kpmg.com>
Tel : 044-28473911
Mobile : 9841078091
Fax : 044-28473912

Manufacturing Policy for India

Indian Institute of Management (IIM) - Lucknow

Authors : **Atanu Chaudhuri** (Team Leader), **Deepak Gaur, Sachin Bery, Shaleen Prakash, Shridhar Sivaraman.**
Faculty Advisors : **Prof. R.K. Srivastava** and **Prof. Anadi S. Pande**

Manufacturing Policy for India
(IIM - Lucknow)

Contents

1

CHAPTER ONE

Executive Summary

THE integration of global markets and the increasing importance of international trade in a nations' destiny have brought manufacturing competitiveness on the radar of national policy making. With 70% of the 7 trillion dollar international trade being in processed manufacturers no nation can afford to neglect manufacturing on the road to economic development apart from the positive spin-offs of manufacturing activity.

However several studies have portrayed worrying picture of industrial decline and loss of competitiveness of the Indian manufacturing sector in the past period. The situation is aggravated by the rapid advances made by south East Asian countries and China in particular who have surpassed the competitiveness of India in manufacturing of basic labour intensive products to sophisticated equipment and machinery. The performance is worth concern on all the basic drivers of competitiveness i.e. skills, local technological effort, foreign direct investment and physical infrastructure. The dream of capitalizing on the small scale and informal sector to provide employment, infuse entrepreneurship and make India a manufacturing force is frustrated and the SME's struggle with myriad issues ranging from high costs of production to lack of basic infrastructure.

Cross country comparisons of India's performance of factors like labour productivity, risks vs. return profiles, export performance, regulatory provisions pertaining to manufacturing and infrastructure all position us on the lower end of the spectrum. In an attempt to unravel the mystery of the current state when we started off much better than most of the counter-parts we delved into the evolution of the regulatory regime to explore if the various components of Industrial, trade, foreign direct investment, public sector and small scale sector policy have over time contributed to the erosion of India's manufacturing prowess.

The study revealed that the major controls exercised in the areas of industrial licensing, import controls and those on pricing and foreign investments led to an uncompetitive, high cost monopolistic industrial structure. Thus, manipulation of the administrative set up made healthy competition, while product innovation, technological developments and cost reduction took a back seat. The growth of the organized industry was marked by high cost, low efficiency and low productivity and the development of small scale industries could not meet the objective of decentralized industrialization because of lack of concerted efforts.

We then focused our attention on the policy initiatives taken by several East Asian and Latin American countries like South Korea, China, Japan, Taiwan, Mexico and Brazil to identify the lacunae in our policies as well as identify the best practices for adaptation in Indian context. Most of the countries had used a combination of selective protection and import substitution, selective liberalization, enabling technology absorption in local industry and financial incentives to spur manufacturing growth. A case in point is Japan where the MITI (Ministry of International Trade and Industry) acted as a nodal agency and architect of industrial policy along with facilitating the dissemination of technological ideas through Japanese firms. Considering the importance of China in the manufacturing scenario a detailed comparison of India and China is contrasted on the different areas of Industrial policy.

The broad policy lessons from the countries prepare a background of the extent to which government initiatives can stimulate manufacturing competitiveness. This along with the framework proposed by well known management guru Michael Porter on the " Competitive Advantage of Nations" is used as a building block for the proposed manufacturing policy of India. Porter has identified a nation's position in factors of production, nature of home market demand. presence of supplier and related industries and the manner in which industry actors(firms) are managed and act as the primary determinants of national competitiveness. Thus a manufacturing policy cannot be created in isolation with the industry and has to address national policy as well as organizational issues. Government's proper role is as a catalyst and challenger enabling the firms to move to higher levels of competitiveness. Government should therefore focus more on creation of supporting structures which facilitate the manufacturing organizations rather than direct interventions which can only provide temporary respite.

The proposed manufacturing policy draft thus recommends for government as well as organizational level initiatives. The policy recommendations within the ambit of government address changes in the regulatory system to facilitate FDI along with regulations to enable training and knowledge transfer to local suppliers and easing the process of regulatory approvals needed to set up new manufacturing. A common body enabling single window clearing is recommended.

Labour reforms for provision of employment flexibility for firms and fixation of minimum wages, payment of bonuses, shift of focus from employment security to income security through re-training etc., expansion of the scope of labour redundancy and national renewal fund are the proposed initiatives. A host of policy measures particularly in the power sector are proposed to encourage re-designing the tariff structure to performance based rate determination and incentives to encourage alternative sources of power to reduce the reliance on SEBs.

A critical policy recommendation is to make industrial clusters (as Sectoral and geographical concentration of enterprises to gain advantages of geographical proximity and close inter- firm collaboration like Panipat, Tirupur, Agra etc.) as integral part of the manufacturing policy. Replacement of the current taxation laws with Value Added Tax and rationalization of tax rates are some of the measures to be taken in taxation reforms.

Organizational measures recommended span from softer issues like leadership development to establishing active institute-industry linkages for dissemination of latest manufacturing know how and best practices across nation. Establishment of a Technology Resource Center (TRC) as a nodal body is proposed to perform these roles. TRC could be created by combining the different industry associations and government bodies with a single mission of promoting manufacturing competitiveness.

A systematic and timely implementation of the proposed initiatives can restore to India the glory it enjoyed in manufacturing. However, it is stressed again that government can only create a supporting environment for the Industry. The burden of rejuvenating the lost competitiveness rests primarily on the corporate and business organizations.

2

CHAPTER TWO

Manufacturing : The Imperative

THE most important imperative for focusing on manufacturing comes from the significance of world trade to economic development. The evidence is overwhelmingly in favor of adopting an outward looking, international trade oriented economy as opposed to an inward looking domestic market focused one. Even supposedly 'large' economies such as the USSR, China, India and Brazil (which probably had a better chance of making a success out of such a strategy than smaller countries) have failed dramatically in creating vibrant economies based on self-sufficiency. The process of globalization only emphasizes the urgency for a strong international trade and manufacturing oriented economy.

In this context, the pattern of international trade (where over 70% of the 7 Trillion Dollar International Trade market is in processed manufactures) points to the necessity of developing global competitiveness in the manufacturing sector (Figure - 1).

From a domestic market perspective, the experience of developed countries and newly industrialized countries (NICs) in the last century categorically points to the significance of the manufacturing sector in overall economic development. The transition from agriculture to manufacturing results in substantial alternate employment possibilities for the agriculture workforce, in the process increasing domestic surpluses. Various studies have pointed towards the positive spin-off effects of increased manufacturing activity - for instance on the overall skill levels (learning by doing), on social issues through drawing women into productive activity and on agricultural productivity itself.

FIGURE - 1

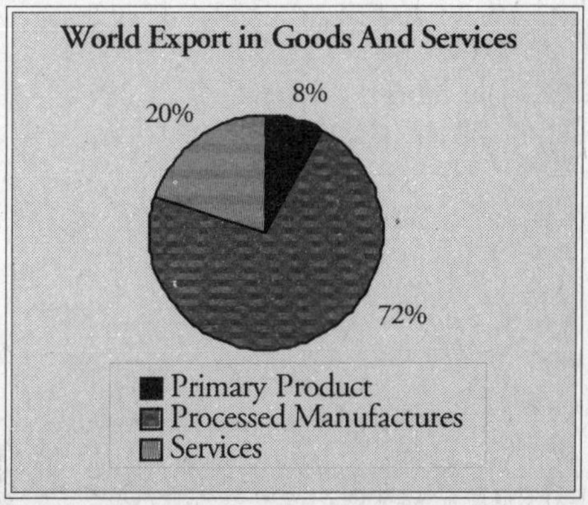

With agriculture providing the primary means of employment to over 65% of India's working population, it is evident that measures taken to address agricultural growth are likely to have the maximum long term impact on the overall economy. However, we believe that such measures will face resistance from interest groups across the country and the common public, and consequently will require substantial political will and consensus to be carried out to any degree of significance. Further, these measures are also likely to have a long lead time before they bear fruit. In the current scenario, where rapid proof of success is necessary to sustain the momentum for continued reform, recommendations for boosting agricultural growth are likely to remain in the 'necessary to do' but 'have not been implemented' list.

At the other end of the spectrum, the services sector has shown strong growth in the 1990s. With sectors such as software showing outstanding results, there is much public speculation on India transforming itself into a 'knowledge economy' without having to transition through a manufacturing-led growth phase. This appears to be a highly improbable scenario.

An analysis of India's demographic profile highlights the stark reality that nearly 75% of India's working age population of over 600 million is educated below middle school level. Services in general, and knowledge based service sectors more so, require a much higher level of education than either agriculture or repetitive labour-intensive manufacturing (Figure - 2).

Hence, we believe that while certain services sectors will continue to grow impressively, their overall contribution to economic growth, employment and incomes will not be adequate in themselves to meet India's economic growth needs. Therefore our choice of focus on the manufacturing sector as the target for India's immediate economic development implicitly recognizes both the importance of this sector to overall economic growth and the likelihood of successfully carrying out specific measures within a reasonable time frame.

Figure - 2

Demographic Profile of Indian Population: 2001

Educational levels	*Urban%*		*Rural %*		*Aggregate*
	Male	*Female*	*Male*	*Female*	*%*
Illiterate	17.9	44.5	43.7	79.0	52.3
Literate but upto Primary	25.4	17.7	29.0	13.7	21.4
Middle	17.5	8.1	13.6	3.9	10.0
Secondary	24.6	15.1	10.9	2.7	10.7
Graduate and above	14.6	14.5	2.8	0.7	5.6
Total	100	100	100	100	100

3

CHAPTER THREE

Attitude Towards Manufacturing
The Missing Link

CII has rightly identified that manufacturing competence is an important ingredient in the strategy for enhancing India Inc's global competitiveness. There is an urgent need for Indian industry to change its mind-set vis-a-vis manufacturing as a corporate function. India has no choice but to strengthen its manufacturing base. Manufacturing should be considered as a function and activity central to a large economy like India.

The UNCTAD Trade and Development Report, 2002 recommends that the larger developing economies should pay particular attention to strengthening their domestic manufacturing capabilities. However, the importance of manufacturing to the economies of large developing countries is yet to be perceived by Indian companies and investors. The Indian economy was believed to have by-passed the traditional path of sectoral movement - primary-to-secondary-to-tertiary; it jumped from the rustic agricultural sector led growth to the sophisticated services sector led growth without having passed through the tough and dirty manufacturing sector led growth phase. While the debate on the economic desirability of this non-traditional transition may go on, the consequences of such a transition on the allocation of human and knowledge resources to the manufacturing sector are fairly visible. The manufacturing sector is a victim of this national belief.

Managing manufacturing has been the main issue in management science. The principles of scientific management enunciated in the early part of twentieth century were in response to making the worker and his work more efficient. The key concern was the productivity of the worker. The practice of scientific management found its place in the factory as industrial engineering. Apart from the productivity of the worker, industrial engineering in its present form deals with production planning, inventory control, plant layouts, and capacity planning.

In fact, the management of the manufacturing system is the concern of industrial engineering or operations management, as it is popularly known. In India, this aspect of management has, somehow, never found a pride of place in most organizations. Rarely one comes across a functional and functioning production planning and control system in any of the medium sized enterprises in the Indian context.

In fact, the use of planning as a tool to manage uncertainty is hardly a pervasive practice. A survey of manufacturing futures done by two academics from the Indian Institute of Management, Ahmedabad, in 1997 revealed that manufacturing is yet to be viewed as a critical function in most Indian companies, although some companies are trying to move towards world class practices in manufacturing. Similar studies in the US indicate that for the American counterparts, manufacturing strategy is an integral part of corporate strategy.

If Indian manufacturing sector has to become a core competence of the country then the practice of manufacturing management has to change rapidly. Indian managers and businessmen have to institutionalize the various systems of manufacturing management in their work context. For this to happen, it is necessary for the top management of companies to realize that manufacturing needs to be as much a part of the strategic intent and content of the company as finance or marketing. In fact, even to be competitive in the IT sector, the rigour and discipline of manufacturing systems will have to be internalized.

And the biggest change that the manufacturing function will have to witness, will be in the human resource system. The prolonged neglect of the function of manufacturing by the Indian industry has resulted in most engineers and management graduates opting for marketing, finance, and software as career options. If India Inc is aspiring to find a place for itself in the global market for manufactured goods then it has to ensure a greater market share for the manufacturing function of the knowledge workers. A drastic change in the prevailing incentive structures would be one of the necessary conditions to make this possible.

A long-term strategic direction for the Indian industry, increasing its quality and cost competitiveness, enhancing its ability to absorb technological advances and adopt innovative management practices and build industry's overall capability to face the pressures of open global market competition still remain among key issues that need to be clinched. Surfeit of developments in the wake of fast-track economic reforms initiated in 1991 have further reinforced the need to deepen the reform process at the enterprise level itself.

Manufacturing and Service

The average annual rate of growth of the manufacturing sector was a little over 5 per cent during the period 1950 to 1980 and increased to over 7 per cent in the 1980s and to over 9 per cent in the 1990s. However, the sharp increase in the first half of the 1990s was partly negated by the slowdown after 1997. Five years of continued slow growth in the manufacturing sector were matched by high rates of growth in the services sector contributing to fancy notions of India having leapfrogged over the industrial stage of development into a 'post-industrial' knowledge and services economy.

Year	Growth
1950-80	5%
1980s	7%
1990s	9%
1997-	< 9%

The service sector includes all manner of activity and a lot of it may be economically wasteful even if it generates demand for goods and services and thereby contributes to estimates of national income. Even if it is the case that India has a global competitive advantage in many areas of the service economy and should invest intelligently to derive its benefits, we cannot afford to ignore the manufacturing sector. That is, we can neither ignore the importance of creating and sustaining both entrepreneurial and labour skills in the manufacturing sector, nor of generating the required domestic demand to sustain a domestic industry with globally viable and competitive scales of production.

Manufacturing and service are complementary to each other. A successful manufacturing industry has benefits beyond the sector.

It can help the flourishing service sector. For example for a factory, producing machine tools, there will be demand for collaboration with designers, marketing, software developers etc. Similarly software project management and design services will be required for many physical products to be produced. Service sector is an enabler, for it to thrive India needs a vibrant manufacturing sector. Large continental economies like the United States and China with a sizeable home market cannot afford to become dependent on imports for vital industrial products. They must foster domestic manufacturing capabilities and capacities in order to create the foundation of skills and the infrastructure of national security and prosperity.

Both the United States and China, where the manufacturing sector has a larger share of national income than in India, have more open economies and much higher shares of world trade than India. The pursuit of trade liberalization, the engagement with globalization, the endeavour to increase our share of world trade must go hand in hand with a concerted effort to expand the domestic manufacturing sector. This is the challenge for India.

Indian manufacturing has been affected by a sustained weak economy and low demand; excess capacity and global competition; unfavorable external environment - ranging from labour laws and the inspector-raj to poor infrastructure; powerful legacies.

No industrial economy in the world has a literacy rate below 80 per cent and India, with its 62 per cent rate, cannot hope to be an exception to a universal rule. These challenges must be addressed both at the macro-level and the micro, that is, enterprise, level. The important thing is to recognize the need to grapple with them. In the hope of riding the "third wave" and leapfrogging into a post-industrial, services economy, we cannot afford to neglect the foundation of a modern economy and of national power, namely, the manufacturing sector.

4

CHAPTER FOUR

Indian Manufacturing : Current Scenario

THE Industrial Development Report 2002-03, published by the United Nations Industrial Development Organization (UNIDO), portrays a worrying picture of industrial decline and loss of competitiveness of the Indian manufacturing sector in the 13 year period 1985-1998.

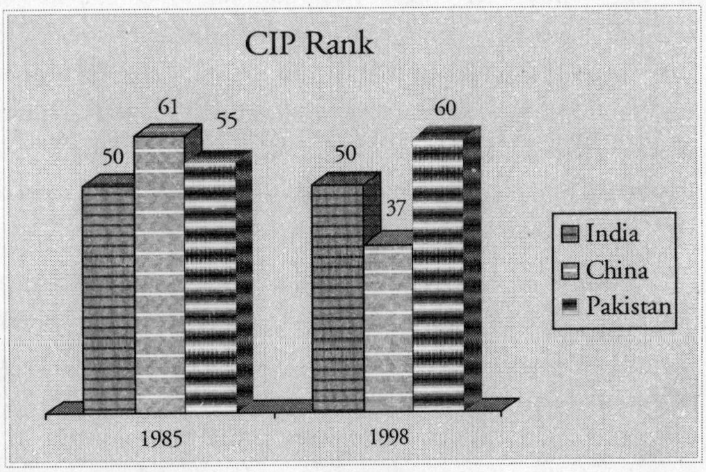

The report states that during the 13 years, 1985-98, not only has India's industrial structure remained static but, more damagingly, "India is falling further behind its competitors". UNIDO has constructed what it calls the 'Competitive Industrial Performance index' comprising four indicators of performance, namely, manufacturing value-added per capita, manufactured exports per capita, share of medium and high-tech activities in manufacturing and the share of medium and high-tech products in manufactured exports.

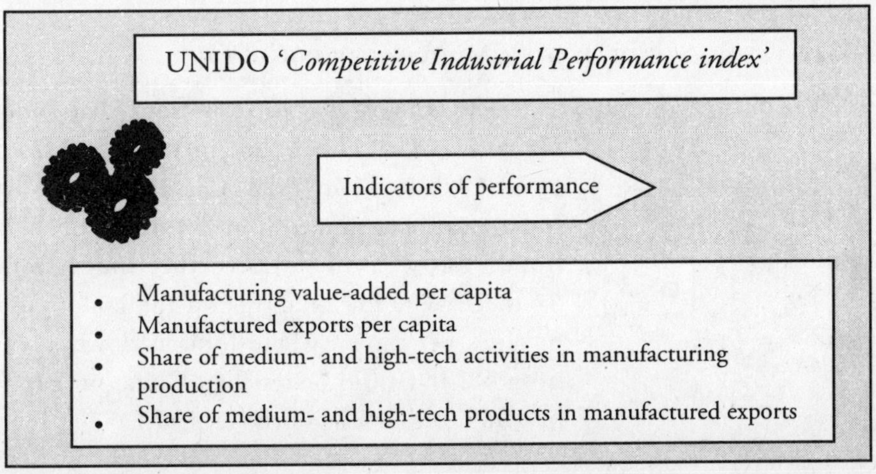

While India's CIP rank remains at 50 between 1985 and 1998, in a list of 87 countries, China's rank has improved from 61 to 37 in this period. Every East and South-East Asian country has improved its CIP rank, and it's a small consolation that Pakistan's rank has slipped from 55 to 60! There has been an improvement in the CIP index value for India, from 0.034 in 1985 to 0.054 in 1998, but this pales in significance when compared to

China's CIP value improvement from 0.021 in 1985 (significantly below India) to 0.126 in 1998 (more than double that of India).

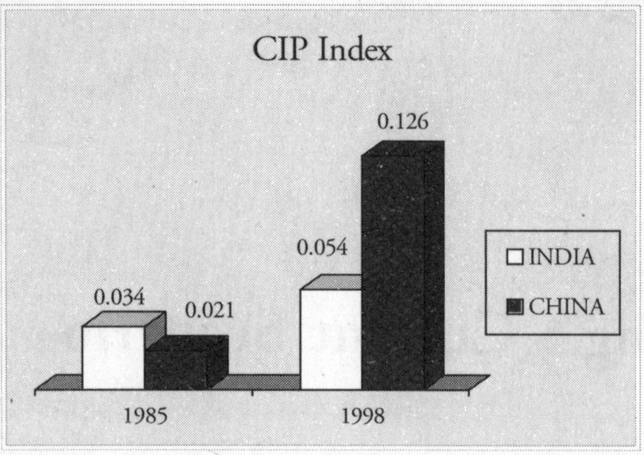

The period 1985-98 is associated with the modernization and globalization of China's manufacturing sector.

The UNIDO report identifies at least five reasons for the poor performance of India on industrial competitiveness. On each of the five "drivers of industrial performance", the factors that have contributed to the improved competitiveness of the newly industrializing economies of Asia and Latin America, India has experienced deterioration in these 13 years. These five drivers are skills, local technological effort (R&D), foreign direct investment, licensing payments abroad (royalties) and physical infrastructure.

India's skill base is seen to be "weak and deteriorating relative to those of competitors"; lags in infrastructure are widespread and here too India is "falling behind its competitors"; "enterprise financed R&D" has remained static between 1985 and 1998 and has declined as a share of national income; while FDI inflows have gone up for India, its competitor countries have seen even higher inflows; moreover, the export-orientation of FDI is minimal, especially compared to China. That last factor, says the report, "puts India at risk of losing out on links with the most dynamic part of industrial export activity and integrated production systems in technology-intensive activities."

FIGURE : THE FIVE DRIVERS OF INDUSTRIAL COMPETITIVENESS

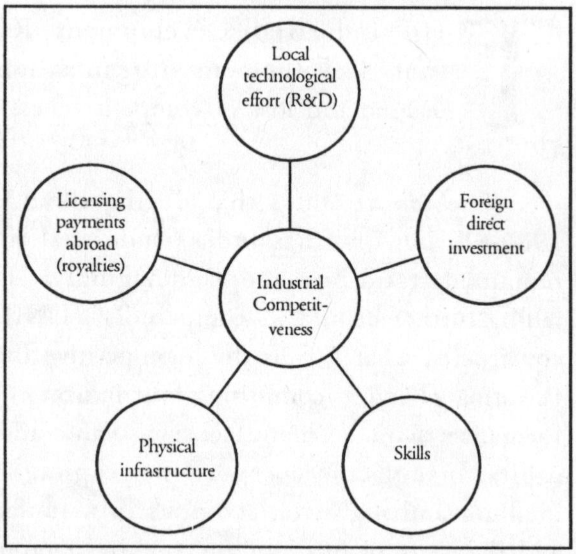

Among the drivers of industrial performance worldwide, says the report, "R&D is statistically the most important, in both 1985 and 1998 and over time. This finding highlights the need for domestic technological effort even at low levels of industrial development. While the causation may run in both directions, theory does suggest that the causation from R&D to industrial performance is likely to be predominant."

FIGURE: RANK CHANGE ON VARIOUS PARAMETERS IN CASE OF INDIAN MANUFACTURING

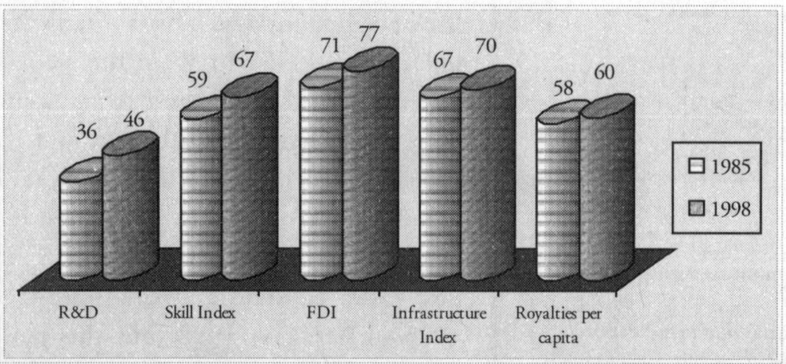

Shockingly, it is on this front that India has done worst. The decline in ranking is from 36 in 1985 to 46 in 1998. On the skills index India's rank has slipped from 59 to 67, on FDI from 71 to 77, on infrastructure index from 67 to 70. The lowest slip is on the ranking on royalties per capita, where India is down at 60 from 58! India did better than China on R&D in 1985 and does poorly in 1998.

The only objection one can have to UNIDO's measures and drivers is the emphasis they place on export-orientation. Given this emphasis, it is not surprising that East and South-East Asian industrial economies come out on top and India fares poorly. To be sure, UNIDO has a reason for focusing on export-orientation because it believes it is only through subjecting itself to the glare of global competition that a country becomes competitive. Competitiveness has to be tested in the market place. UNIDO also views trade as an instrument of technology acquisition and capability building: "If countries fail to build the capabilities to compete internationally, they can become by-standers at the technological feast, stuck with the crumbs - stuck with simple manufacturing activities that do not lead to sustained, diversified growth."

Thus we find that on each 'driver' India has experienced a decline in ranking between 1985 and 1998. It is equally instructive to note that the one positive element in India's case, namely the "depth and complexity of its industrial structure", is the legacy of "its prolonged import substitution strategy emphasizing heavy industry".

In other words, the post-Independence strategy of industrial development did help create a broad-based industrial structure which placed India ahead of most developing countries but it made the industrial sector non-competitive. The non-competitiveness of the industrial sector has become even more pronounced over the 1985-98 period because of the weakness of the five 'drivers' of competitive industrialization, even though policy initiatives have been taken to subject Indian industry to greater international competition.

The Informal Sector and SMEs

The definition of small scale industries has been changing continuously. Initially values of fixed capital and number of employees were combined in defining SSI's. The upper limit of fixed assets was fixed at Rs. 5 lakhs in 1955 and the number of employees at 50, if the unit used power and 100 without power. Subsequently SSI's were defined on the basis of fixed assets alone. Later the ceiling was changed to Rs. 6 million for SSI's and Rs. 7.5 million for ancillary units. According to the current Government definition, small scale industry will include an undertaking with investments in fixed assets in plant and machinery, not exceeding Rs. 10 million and Rs. 2.5 million for tiny units.

The performance of the SSI sector in terms of production, employment and exports have been impressive. The performance data is shown in Table 1.

TABLE 1

Year	*No. of Units (million)*	*Output (Rs. Billion)*	*Employment (million)*	*Export (Rs. Billion)*
1991-92	2.082	1786.88	12.98	13.883
1992-93	2.246	2093.00	13.41	17.785
1993-94	2.381	2416.48	13.94	25.307
1994-95	2.571	2939.90	14.65	29.068
1995-96	2.724	3562.13	15.26	36.47
1996-97	2.857	4126.36	16.00	39.249
1997-98	3.014	4651.71	16.72	44.437

The SSI sector is non-homogenous in structure and includes diverse types of production units ranging from traditional crafts to high-tech industries. The total number of SSI working units in the country is estimated to be around 3 million. In terms of ownership, the vast majority of SSI units are proprietary concerns (80.5%) with only 16.8% functioning as partnerships and private limited companies. The first census of SSIs in the country was undertaken in 1972 and the second in 1987-88. Tamil Nadu occupied the first rank in both years in terms of number of units and employment, followed by Maharashtra. Andhra Pradesh was placed at No. 6 and 7 respectively. By 1993, Punjab emerged at the top. (AIMA, 1997) One important factor to be noted in the statistics on small industries is the mortality rate. While the industries may get registered at the entry point, there is no record of their exit. There are at least 3 lakh units declared as sick and out of production, accounting for 10 per cent of the recorded units.

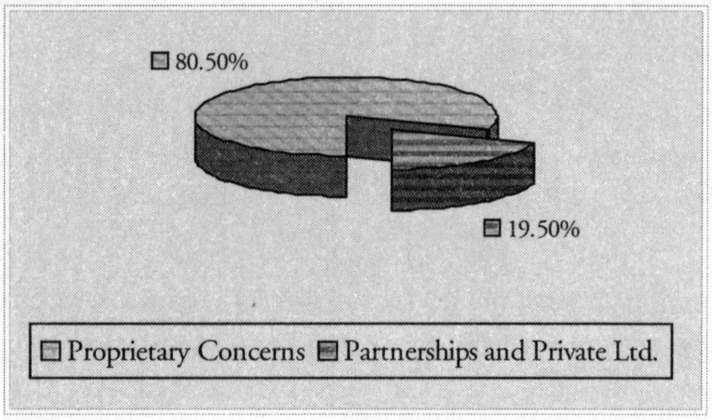

The potential of SSIs to generate employment has remained the strongest argument in their favour. The sector now employs 17 million persons and is the second largest employer of India's workforce after agriculture. The role of SSIs in the economy can be seen from the fact that it now accounts for 95% of all industrial units in the country and 40% of total output. About 7,500 products are manufactured in the small-scale sector. The export share is 35%. The composition of exports shows the largest shares of SSIs are in the industry groups: hosiery and garments (29.0%), food products (21.4%) and, leather products (18%). The industry groups which have recorded high growth rates and a large share in total production of SSIs are: textile products, wood, furniture, etc., paper and printing, and metal products.

FIGURE: THE PERCENTAGE COMPOSITION OF EXPORTS AS PER INDUSTRY SEGMENTS

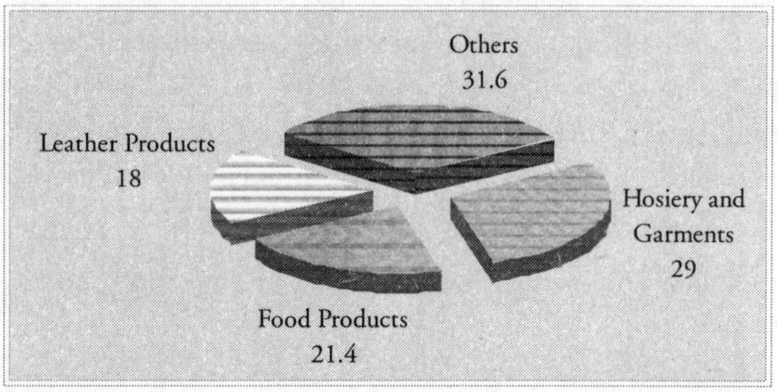

Since independence, the Centre and State governments have devoted considerable attention to the growth and development of small scale industries. The Industrial Policy Resolution, 1956 stressed the State's policy of supporting cottage and small scale industries by restricting the volume of production in large sector, differential taxation, direct subsidies and improving the competitive strength of the small scale producers. The Industrial licensing policy of the Government of India also emphasized the policy of encouraging competent small and medium entrepreneurs.

FIGURE: THE ROLE OF SSIs IN THE ECONOMY

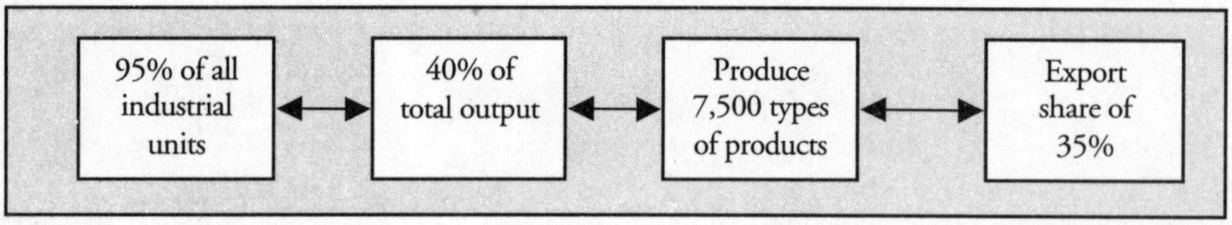

With the deregulation of the Indian economy the Government announced a new package of policy measures for promoting and strengthening the Village and Small Industries sector. The policy measures are aimed at simplifying the regulations and procedures and helping SSI's to improve their competitive strength.

Informal Sector

The informal sector among the SSI's, engaged in manufacturing also play a significant role, which can be seen from the chart below.

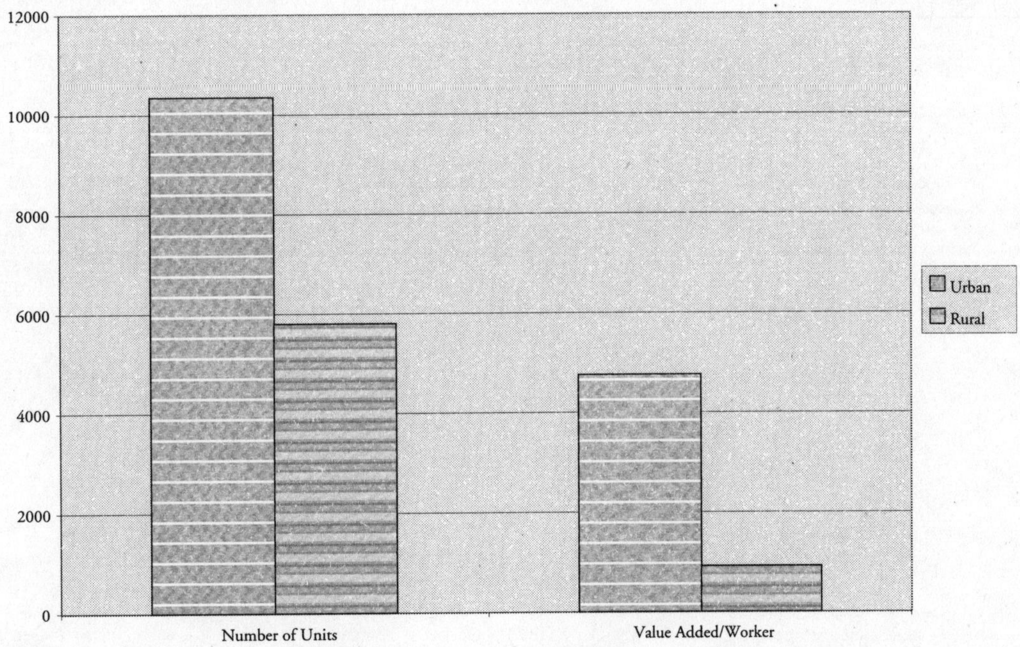

Many of these units operate from the owner's home and have space constraints. This not only hinders their expansion plans but they are also denied registration as SSI and thus the benefits, related to it. These entrepreneurs should be encouraged to set up their units in the land allocated by State Industrial Development Corporations. This will ensure that they get access to power, water, transport and telecommunication facilities.

5

CHAPTER FIVE

Indian Manufacturing : Relative Performance

COMPETITIVE manufacturing demands producing world class high quality products at low costs with high rate of innovations and improvements in product and process technologies. Achieving this goal mandates performance on the following parameters:

- Presence of necessary infrastructure
- Stable political environment
- Favourable regulatory framework
- Visionary business leadership to drive change and performance
- Availability of skilled human resources
- An environment which encourages innovation.
- High proportion of manufactured goods in exports.

To understand the relative standing of India vis a vis its global counterparts in manufacturing we present below some data which highlights the stark differences that exist and need to be bridged:

Labour Productivity

While labour productivity in Indian manufacturing has increased marginally during the last three decades, labour productivity in Taiwanese and Korean manufacturing has increased more rapidly. Since labour productivity is directly linked to economic well-being, the implication of lack of productivity growth on income levels and manufacturing competitiveness is significant (Figure - 3).

Export Performance

Considering that manufacturing exports account for over 75% of total world exports of goods and services, India's export performance also reflects lack of competitiveness in manufacturing. India's share of world exports has grown marginally from a low base (0.5 - 0.7%) during the period 1980 - 2000, while in the same period other developing countries - notably China, Thailand, South Korea and Malaysia have more than doubled their share of world exports (Figure - 4).

FIGURE 3

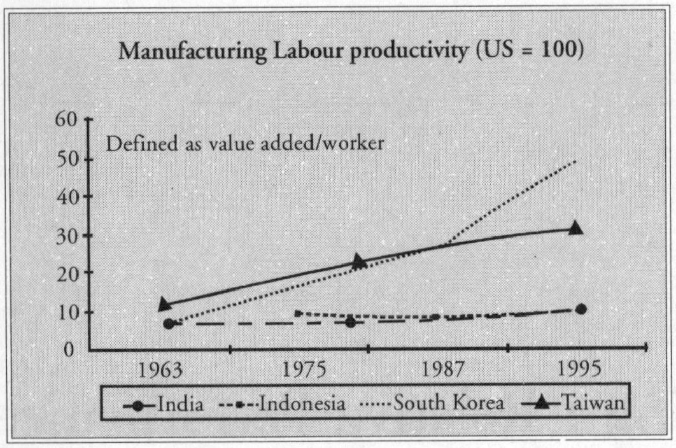

FIGURE 4

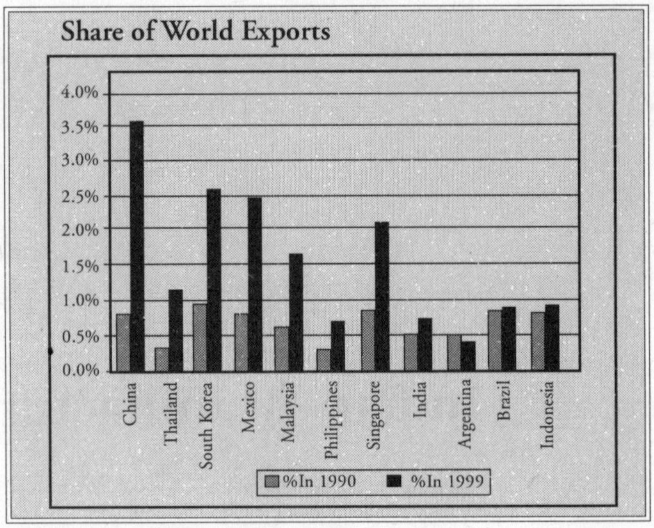

Regulatory and General Environment

The lack of competitiveness is further demonstrated by the fact that despite tariff reductions through the 1990s, India continues to be amongst the world's more protected economies. India has the second-highest average tariff rate in 2000 and continued to have a higher percentage of products covered under non-tariff barriers in 2000 than most other developing countries had in 1990. Finally, when viewed on a sector-wise basis, it is observed that five sectors account for over 75% of exports, suggesting that several sectors are not globally competitive. Hence, cross-country comparisons on direct and surrogate measures of competitiveness point towards a significant lack of competitiveness in Indian manufacturing today. Global competitiveness survey rankings such as those in the WEF's Global Competitiveness Report provide corroborative results, where India has been ranked 37th and 49th (out of 59 countries surveyed) in terms of current and growth competitiveness respectively (Figure - 5).

FIGURE - 5

WEF RANKING ON SELECT INDICATORS OF COMPETITIVENESS (2000)

Parameter	*India*	*China*	*Korea*	*Malaysia*	*Indonesia*	*Taiwan*
Overall quality of Infrastructure	54	46	28	18	42	26
Sophistication of Technology available	38	42	23	27	48	16
Import Fees- combined effect of Import tariffs, license fees and time required for	59	45	32	25	40	24
administrative procedures Average tariff rates, 1999	59	57	40	41	45	13
Base of starting a new business	39	37	33	10	24	6
Local development of product designs	47	35	22	50	51	19
Efficient production processes	42	43	28	29	46	21
Labour flexibility	53	32	18	38	26	13
Pay related to productivity	52	15	23	29	42	2

The result of this unfavourable regulatory environment manifests itself in the FDI inflows into the country (figure 6). What is more alarming is the fact that about a quarter of the FDI approvals do not lead to inflows.(Figure 7). This means that just smoothening of the inflow process without attacking the root causes will have only temporary impact and lead to events like abandonment of projects and flight to better locations. Indications about the poor attractiveness of India can also be gauged from Figure 8 which demonstrates the risk versus

FIGURE 6

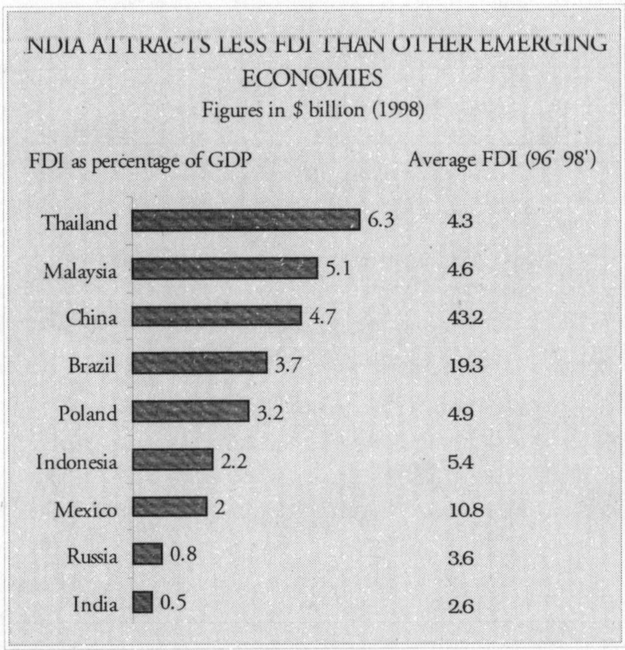

Source: World Investment Report, 1999

FIGURE 7

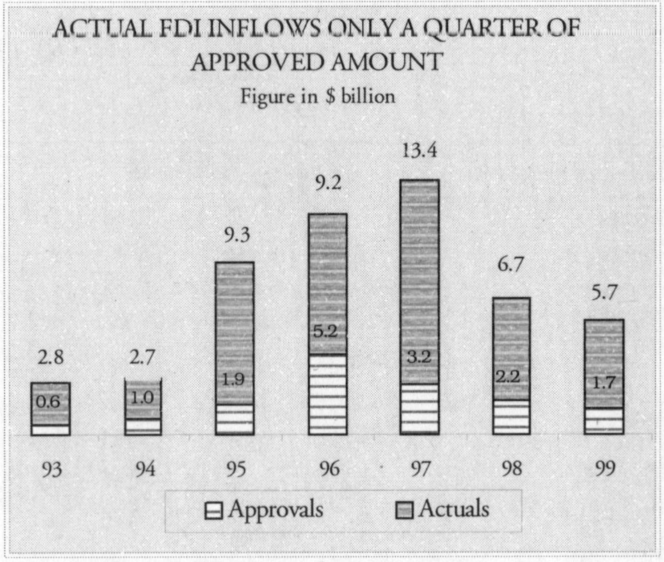

Source: CMIE monthly review of investment projects, June 1999

return profile of Investments in India made by US. Here we see a very glaring anomaly where India offers lower return for a higher level of risk. Similar deductions can be made from Figure 9 where a country risk index for India and other Asian countries is prepared.

FIGURE 8

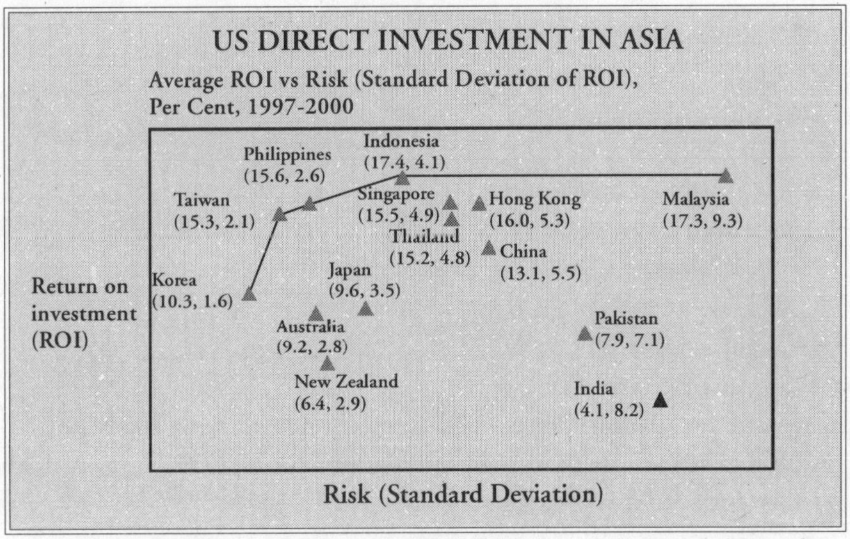

Source : Bureau of Economic Analysis (US Dept. of Commerce), PSI analysis

Infrastructure

Inadequate provision of high quality and reasonably priced infrastructure services represents a major barrier to continued growth of the manufacturing sector because of deeply ingrained dependence on infrastructure for different activities. Inefficiencies and shortfalls in capacity are reflected in congested roads and ports, power failures, and drinking water shortages. A depiction of the current state of key infrastructure variables in India is presented below.

FIGURE 9

COUNTRY RISK INDEX

Country And Market Risk Factors With Weight Across Country

Subtotal weight %	*Country Factors 50*			*Market Factors 50*					*100*
Factor	**1**	**2**	**3**	**4**	**5**	**6**	**7**	**8**	
Weight %	**17**	**16**	**17**	**10**	**15**	**10**	**10**	**5**	**100**
Country	*Political Stability score*	*Transparency score*	*Productive labor practices score*	*Market liquidity & volatility score*	*Investor protection score*	*Capital market openness score*	*Settlement proficiency score*	*Transactions score*	*Wiltshire score*
South Korea	3	3	3	3	1	2	3	2	2.55
Taiwan	3	2	2	3	3	2	3	2	2.52
Philippines	2	3	2	1	2	1	1	2	1.86
India	2	2	1	3	2	1	1	2	1.73
Thailand	2	1	2	1	1	2	3	1	1.64
China	1	1	1	3	1	1	3	2	1.45
Malaysia	1	1	1	2	2	1	2	3	1.45
Pakistan	1	1	1	1	2	2	2	2	1.40
Sri Lanka	1	2	1	1	1	2	1	2	1.31
Indonesia	1	1	1	1	1	2	2	2	1.25

Source : Wilshire Associates, Jan 2002

Roads

India's road network of 3 million kilometers is the third largest in the world. Of this, National Highways form less than 2%, but carry more than 40% of the country's road traffic. Half of roads are unsurfaced, with only 52,010 km of national highways suitable for high-speed traffic.

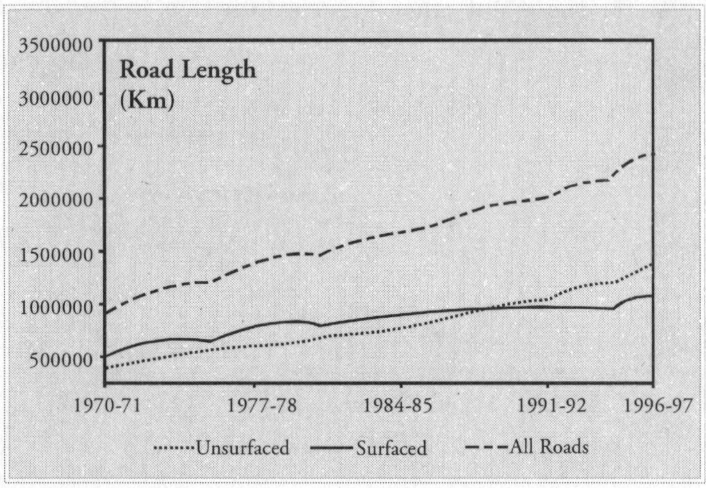

Village Connectivity

Of the 5.89 lakh villages, only 48 per cent were connected with roads by the end of March 1995. While 92 per cent of villages with population above 1500 were connected with roads only 38 per cent villages with population less than 1000 had roads. All states had almost all or a high proportion of their villages having population above 1000 persons connected with roads.

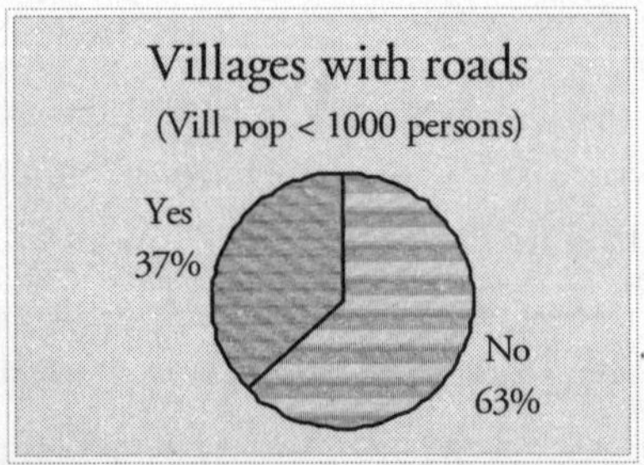

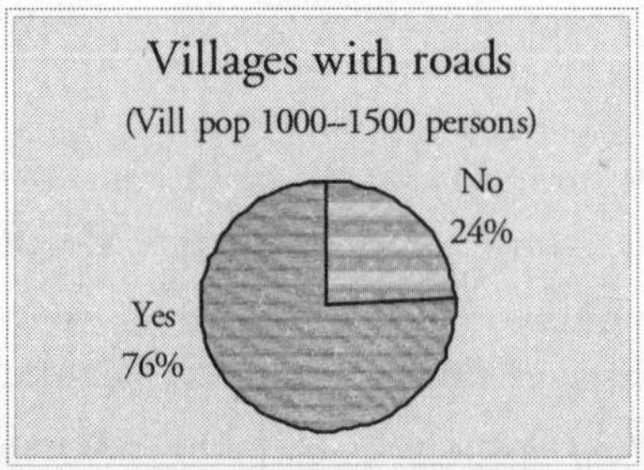

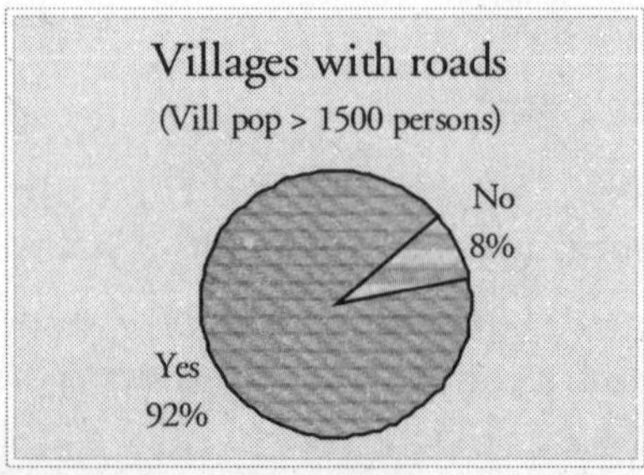

Railways

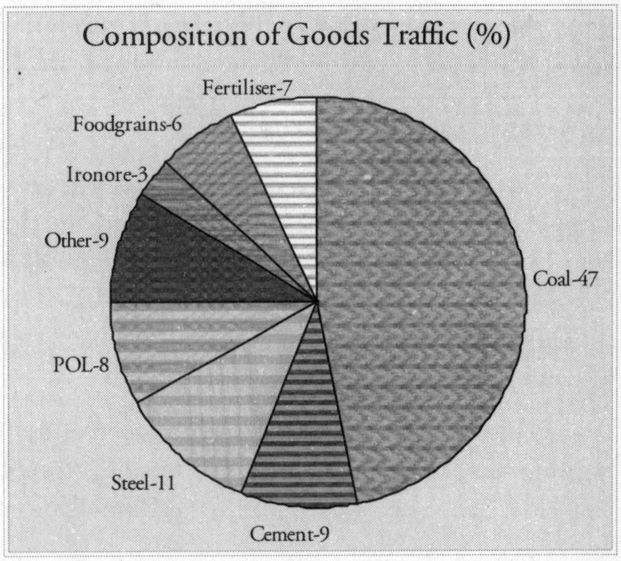

Indian Railway network is one of the largest railroad systems in the world. Indian Railway provides one of the cheapest means of transport in India. Role of railways has been fast transforming from mere means of connecting far-flung areas in India, to more economic means of mass transport system in large cities like Mumbai.

Despite such an extensive network, the Railways are plagued by a lot of inefficiencies. Average speed of goods on broad gauge and metre gauge was just 23.7 km per hr and 18.2 km per hr, respectively. Electric goods train recorded the highest speed of 25 km per hr on the broad gauge. Average wagon turnaround time is about 8 days. All these figures compare poorly as compared to other developing and developed countries. The Railway's Build-Own-Lease-Transfer (BOLT) and Own-Your-Wagon Schemes have not attracted private investors.

Ports

About 90 per cent of Indian imports and exports are transported by sea, but the country's port capacity is inadequate to handle the flows. India's eleven major ports, which account for over 90 per cent of the country's port traffic, handled 272 million tons of cargo in IFY 1999-2000, registering a growth of 8 per cent over last year. Capacity utilization during 1999-00 was 105 per cent. Port traffic has been growing 9-10 per cent annually.

Besides working beyond their optimal capacity, India's ports are very inefficient by world standards, with ship turnaround times averaging 6.9 days. This stems from highly unionized labor, low labor and equipment productivity, and a lack of autonomy for the Port Trusts, which manage the ports under the Ministry of Surface Transport (MOST). To meet the huge gap between demand and available port capacity, the government passed the Major Port Trust Act 2000 enabling port trusts to form joint ventures with the private sector, including foreign ports. Private investment of about Rs. 400 billion is required by 2012 to increase the capacity of major ports. Automatic approval of foreign equity up to 100 per cent is available for the port and harbor construction.

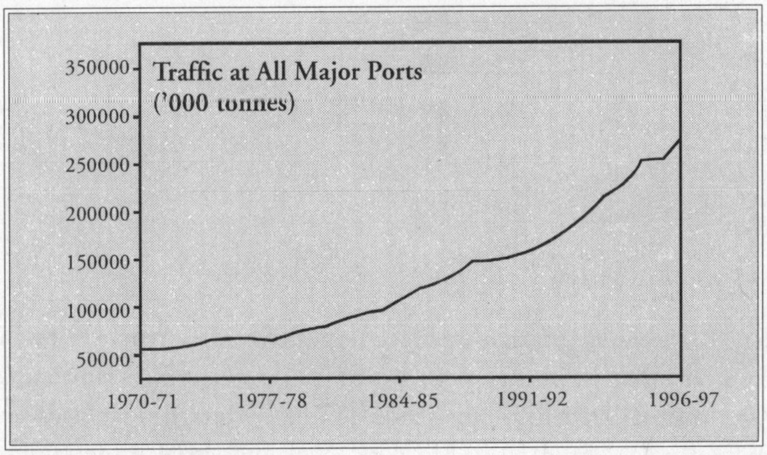

Telecommunications

India operates one of the largest telecommunication networks in Asia comprising over 26,049 exchanges with a capacity of nearly 27 million lines and 23 million working connections as of December 1999. The Domestic long distance service and International Long Distance telephony have been opened to the private sector. These measures are expected to improve the telecom infrastructure quality to international levels. However, the fact remains that the telecom revolution in India has benefited the urban areas more than the rural ones. Rural tele-density still remains at abysmally poor levels.

Information Technology

The software industry has emerged as one of the fastest growing sectors in the economy with compound annual growth rate exceeding 50 per cent in the last five years and total annual receipts of USD 5.7 billion and exports worth USD 3.9 billion in 1999-2000. A separate Ministry of Information Technology (IT) has been established to promote knowledge-based enterprises, internet, e-commerce and IT related education in India. The recently passed IT Act, 2000 in the budget session amended laws to facilitate e-commerce and provide a legal framework for IT in India.

The estimated volume of e-commerce in India was USD 102 million for 1999-2000. Of this, USD 11 million was contributed by retail business-to-consumer transactions and business-to-business dealings on the Internet. However, this figure is very small in comparison with international standards.

Though Internet penetration and usage still lags behind other countries like Japan and China, there has been a sudden splurge in the Internet front, largely owing to the increasing number of private ISP market mushrooming all over the country, offering internet services at more and more competitive prices. The increasing growth of Internet and data traffic is prompting the industry to upgrade infrastructure and introduce high bandwidth or broadband communications to ease data transfer.

FIGURE 8

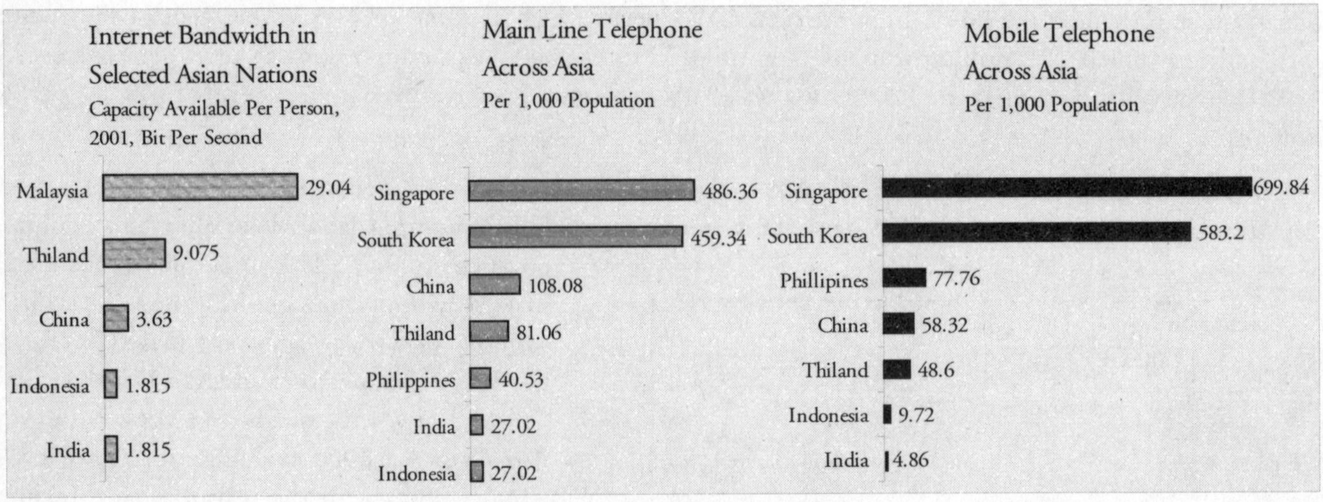

Foreign Trade Zones / Special Economic Zones

Export Processing Zones (EPZs) are designed to provide internationally competitive infrastructure facilities and a duty-free, low-cost environment for exporters. Foreign investment up to 100 percent is permitted in units set up in Export Processing Zones (EPZs), Software Technology Parks (STPs), Electronic Hardware Technology Parks (EHTPs), and to 100 per cent Export Oriented Units (EOUs). New industrial undertakings set up in Free Trade Zones (FTZs) are entitled, subject to various conditions, exemption from income tax on business income. Income of new industrial undertakings set up in FTZs and newly established 100 per cent EOUs are exempt from income tax for a period of 10 years. This exemption has been restricted to units set up prior to March 31, 2000. Such units are allowed to make a specified percentage of their sales in the domestic tariff area. While it lasts, the tax holiday in FTZs replaces all other income tax incentives available to industrial undertakings.

The Government of India requires a minimum value-addition of 20 per cent for most products. There is no need to obtain an industrial license to manufacture small scale industry (SSI) reserved items. EOU/EPZ/SEZ units are obligated to export 66 per cent of their production. The export-import policy announced on April 1, 2000 allowed EPZs, EOUs, SEZs and Software Technology Parks (STPs) to import or obtain goods from the domestic tariff area for exports. The policy authorized new SEZs and the conversion of existing EPZs to SEZs. The SEZs would be regarded as foreign territory for the purpose of duties and taxes to retain their entire foreign exchange earnings as opposed to 70 percent earlier. No sector caps that presently limit FDI in different industries will be applicable to units in SEZs.

EPZs are currently set up in eight designated areas in India: Kandla FTZ, Santa Cruz Electronics EPZ, NOIDA EPZ, Cochin EPZ, Falta EPZ, Vishakapatnam EPZ, Chennai EPZ and Surat EPZ. The rationale for the FTZ scheme envisages no interference by customs authorities in order to get the best from exporters when left without any bureaucratic interference.

Despite these measures, the EPZs, FTZs etc. have not been as successful as they have been in other East Asian countries like China. These zones contribute only about 11% of India's total exports, as against China's 22%, despite Chinese exports being much larger in absolute numbers also. This is because these zones still come under the Labour Act, Industrial Disputes Act, Banking and Foreign Exchange laws, etc.

6

CHAPTER SIX

Evolution of Industrial Regulatory Regime in India

NO policy recommendation can be effective without a proper understanding of the context of the problem. Hence it is imperative that we are aware of the events, policies and milestones that we have covered to arrive at the current situation before identifying the focus of improvement and suggesting a manufacturing policy.

Ingredients of Manufacturing Policy

In this section, we explore the important stages in the evolution of policy in four areas

Industrial policy

Foreign Direct Investment policy

Trade Policy, and

Public Sector policy.

Note that we have not presented the evolution of regulations and the reforms initiated in other important areas such as financial sector (including insurance and banking) policies, capital market (including foreign portfolio investment) policies, policies in various infrastructure sectors including, power, telecom, ports, roads, civil aviation, oil exploration, etc. Also there are detailed policies governing certain industries such as drugs, fertilizers, sugar, hotels and tourism, electronics and computers etc, which are not covered here.

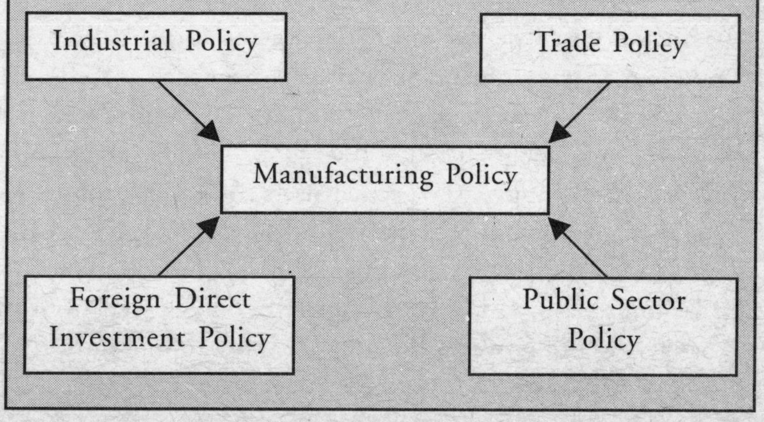

EVOLUTION OF THE INDUSTRIAL REGULATORY REGIME, PRE-1991

The key objective of the economic policy makers after independence in 1947 was to achieve self-reliance in all sectors of the economy. The other main objectives are reflected in the objectives set out for the public sector enterprises and included promoting redistribution of income and wealth, creating employment opportunities, assist in the development of small-scale industries and protect consumers against private sector monopolies. They decided to follow the approach of a planned economy, along other socialist regimes such as the Soviet Union, and designed a regulatory framework that attempted to address these objectives.

Industrial Policy

The key legislation in this field was the Industrial Development Regulation Act, 1951, (IDRA, 1951) which gave powers to the government to regulate industry in a number of ways. The main instruments were the regulation of capacity (and hence output) and power to control prices. The IDRA, 1951 specified a schedule of industries that were subject to licensing. The next important policy statement was the Industrial Policy Resolution, 1956 (IPR, 1956), which reserved certain industries (Schedule A) for public sector monopoly and certain other industries (Schedule B) for public sector dominance. After reviewing the performance of the licensing regime, the government over the years made several modifications to this. The main policies are summarized in the figure below.

Some changes were made on some of the details of the policies over the years. The fixed investment limits for small-scale industries was raised and the list of products reserved for the small-scale sector was expanded. Further, the list of industries where business houses and foreign companies were allowed to operate (Appendix I, IPS, 1973) was expanded in 1978, 1982, 1985 and 1987 (subject to location restrictions or export commitments).

Foreign Direct Investment Policy

The objective of the regulation of foreign direct investment as stated in the Industrial Policy Regulation, 1948 was to ensure as far as possible that majority control was in Indian hands. Consistent with this objective, over the years the government introduced policies that discouraged foreign ownership in most industries. The key policy measures adopted over the years are summarized in a figure on the next page.

Trade Policy

The main objective of the trade policy was to accelerate India's self-reliance, by discouraging imports except in specific products or technologies deemed important for the country's development. The two instruments of policy were import (and export) licensing and tariffs.

In the initial period (up to 1978), the former was used to a greater extent to restrict imports into India. A number of government departments were designated as sponsoring authorities; import licenses would be granted only if these sponsoring authorities certified that the imports were "essential" and gave "indigenous clearance "(i.e. certified that these goods could not be purchased domestically). Over time, import licenses were preferentially given to 'actual users', negatively impacting trading houses. Government owned companies were given a monopoly on the import of certain goods (called 'canalized goods', e.g. steel, oil and fertilizers), where imports enjoyed economies of scale, and domestic users were too small to directly import efficiently. While domestic producers were generally discouraged from importing, in order to alleviate shortage of foreign exchange, exporters were allowed to import inputs duty-free under various schemes.

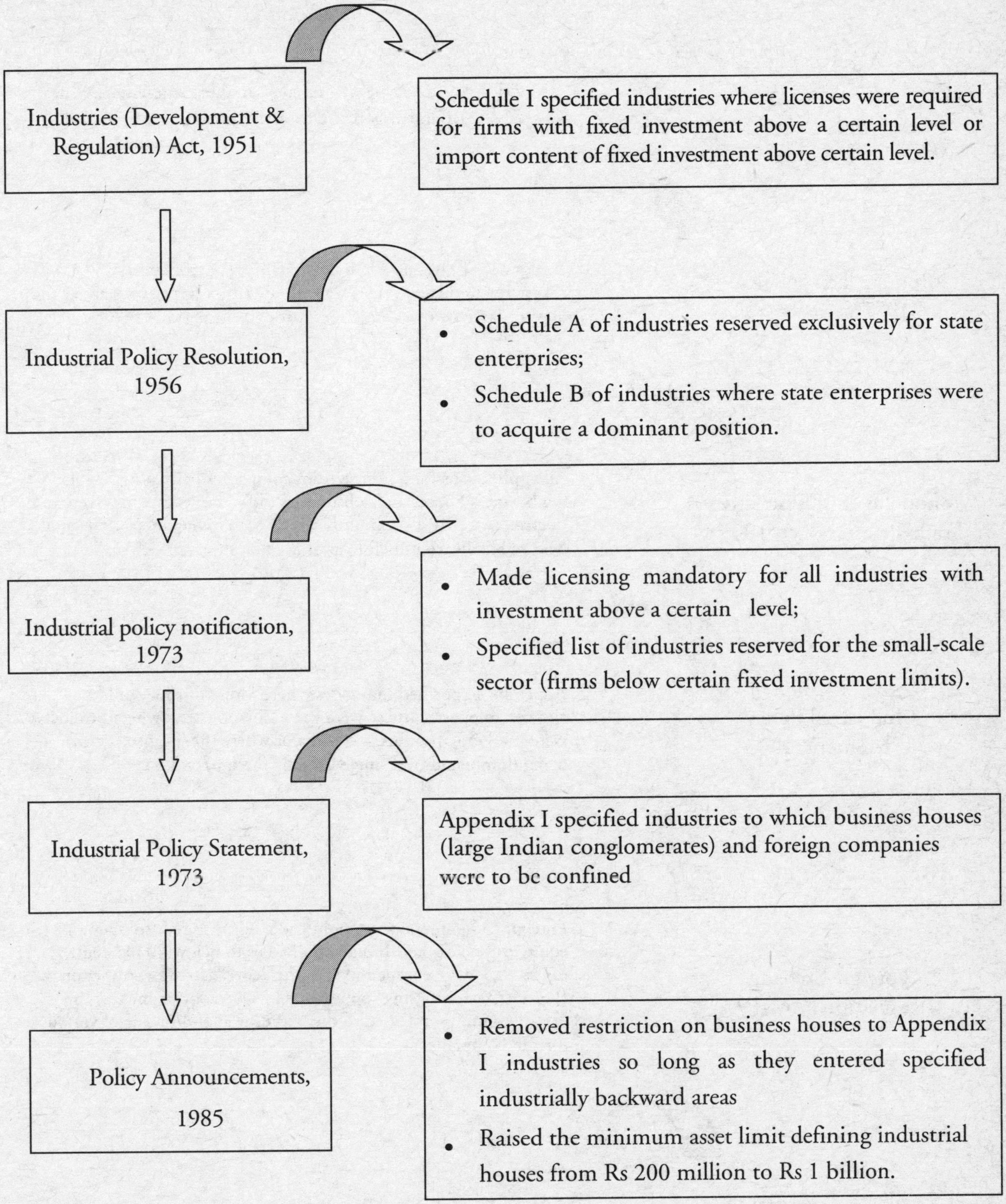
Industries (Development & Regulation) Act, 1951
Schedule I specified industries where licenses were required for firms with fixed investment above a certain level or import content of fixed investment above certain level.
Industrial Policy Resolution, 1956
• Schedule A of industries reserved exclusively for state enterprises;
• Schedule B of industries where state enterprises were to acquire a dominant position.
Industrial policy notification, 1973
• Made licensing mandatory for all industries with investment above a certain level;
• Specified list of industries reserved for the small-scale sector (firms below certain fixed investment limits).
Industrial Policy Statement, 1973
Appendix I specified industries to which business houses (large Indian conglomerates) and foreign companies were to be confined
Policy Announcements, 1985
• Removed restriction on business houses to Appendix I industries so long as they entered specified industrially backward areas
• Raised the minimum asset limit defining industrial houses from Rs 200 million to Rs 1 billion.

Companies Act, 1951	Restrictions on the operations of managing agencies[1], which affected the operations of many British companies in India.
Corporate Tax policies, 1957 to 1991	Corporate tax rates on foreign companies (i.e. companies incorporated outside India) were about 15 to 20 per cent above the rates for large Indian companies through the period 1956 to 1991
Monopolies and Restrictive Trade Practices Act, 1969	All applications for a license from companies belonging to a list of big business houses and subsidiaries of foreign companies were to be referred to a 'MRTP Commission', which invited objections and held public hearings before granting a license.
Industrial Policy Statement, 1973	Appendix 1 specified industries where foreign firms would be allowed to operate; these were generally industries where products were not being produced in India or where the local sector was being dominated by a single (usually foreign)
Foreign Exchange Regulation Act, 1973	Foreign companies operating in India asked to reduce share in equity capital of their Indian companies to below 40 per cent, unless they were engaged in specified 'core' activities (Appendix I, IPS, 1973), were using sophisticated technology or met certain export commitments (led to the withdrawal and/or sale of various foreign companies).
Amendment to MRTP Act, 1985	Set lower limit of Rs 1 billion in assets for referring company to the MRTP Commission, limiting the applicability of the Act.

As the number of domestic producers grew, certifying 'indigenous clearance' became increasingly difficult for government departments. The government set up the Directorate General of Technical Development (DGTD), which tracked capacities and production in Indian industry. They generated data for the central planners to grant industrial licenses and also provided indigenous clearance for licensed companies to import goods. The DGTD tried to ensure that imports were allowed only if domestic producers could not meet the needs of the importer (which they ascertained through various procedures). The government published an import policy annually (sometimes bi-annually) listing all the products that could be imported and including various details on who could import, quantities to be imported, sponsoring authority permissions required, restrictions on sources of import etc.

In 1978, subsequent to an easing of the foreign exchange shortage in the country, the system was reorganized. Instead of announcing a policy every few months for hundreds of goods, they were organized into broadly four lists:

i. Banned goods (later called 'restricted' goods) whose imports were banned,
ii. Restricted goods (later called 'limited permissible' goods) whose imports were allowed only if they were not domestically available,
iii. Open General License (OGL) goods, whose imports were permitted without a license by actual users, and
iv. Canalized goods, whose imports were allowed to be undertaken only by government monopolies.

Over the 80s, the list of OGL goods increased considerably, but so did the list of restricted goods. At the same time, the tariff levels on most goods were raised throughput the 1980s. Import substitution was encouraged also through Phased Manufacturing Programs, where licenses were given to new plants on the condition that they would reduce dependence on imported inputs within a fixed timeframe.

Public Sector Policy

After independence in 1947, India subscribed to a model which placed a strong emphasis on central planning and a big role for government owned Public Sector Enterprises (PSEs) who were established to control the "commanding heights" of the economy. The major objectives of setting up of public enterprises were broadly:

a. to help in rapid economic growth and industrialization of the country and create necessary infrastructure for economic development;
b. to earn a return on investment and thus generate resources for economic development;
c. to promote redistribution of income and wealth;
d. to create employment opportunities;
e. to promote balanced regional development;
f. to assist the development of small scale and ancillary industries; and
g. to promote import substitutions, save and earn foreign exchange for the economy.

With these objectives in mind, the government reserved the 'core' industrial sectors for either public sector monopoly (schedule A of IPR, 1956; refer section on industrial policy above) or for public sector dominance (schedule B of IPR, 1956; refer section on industrial policy above). No significant changes were made to these lists over the years, until July 1991. By reserving these sectors for PSEs, the government was able to effectively regulate output and prices in these sectors.

In 1988-89, the government introduced the concept of 'Memorandum of Understanding', in order to improve the performance of the PSEs. The MOU was an attempt to grant greater autonomy to the PSEs in their day-to-day operations, while increasing accountability by rating the PSEs based on agreed performance measures. The 1988-89 MOUs contained only performance indicators (mostly financial and physical targets) without negotiated 'weights' and 'values' for the different performance measures. In 1989-90, the list of measures was broadened to include qualitative aspects of performance (e.g. 'on-time performance' for Indian Airlines) and measures of "dynamic efficiency" (e.g. 'corporate planning', 'preventive maintenance', etc). Also, weights were attached to these performance measures to arrive at a composite measure of the performance. The focus on a broad range of criterion continued in 1990-91. After the reforms in July 1991, the emphasis shifted to profitability (see discussion below).

IMMEDIATE CAUSE FOR THE 1991-92 REFORMS[1]

The short-term impetus for the reforms announced in July 1991 (which is in the fiscal year 1991-92 and hence referred to as 1992 elsewhere) was the severe balance of payments (BOP) crisis that occurred in this period (refer Figure 1). The immediate cause of the loss of reserves beginning September 1990 was the rise in oil import costs, as a result of the sharp spurt in oil prices after the annexation of Kuwait by Iraq. Indian workers in Kuwait had to be airlifted and inflows from Non-resident Indians (NRIs) in the Middle East were reduced considerably. Further, cessation of exports to Iraq and Kuwait also reduced the inflow of foreign exchange. The payments crisis was worsened by a deterioration of the capital account. Short term credits for imports dipped as creditors were concerned about the government's ability to manage the situation. Amount of foreign currency medium term loans used by financial institutions and PSEs to finance imports declined and net outflow of NRI deposits, which began in October 1990, continued in 1991.

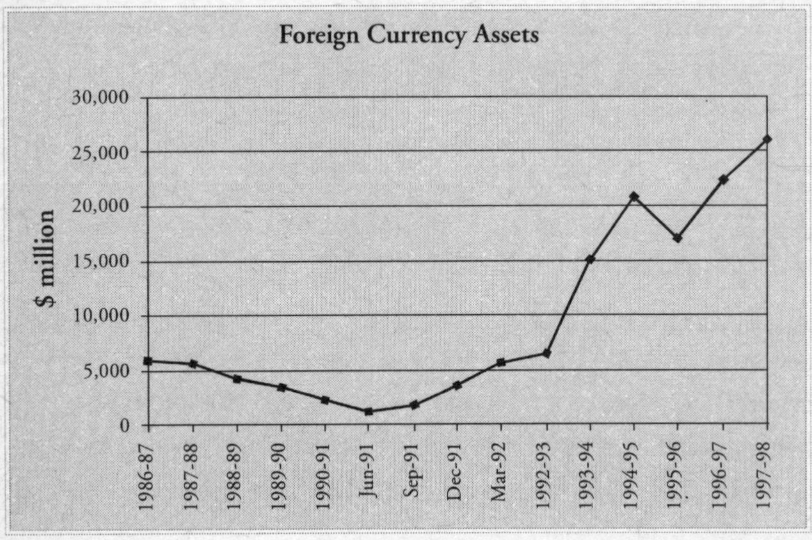

The rapid loss of reserves in the second half of 1990-91 prompted the government to take several short-term measures, restricting imports and raising price of petroleum products. At the same time, there was a significant dip in industrial production, which was negative for most months of 1991-92 (see figure 2). The government presented this as evidence that the import compression policies were counter-productive; it argued that "import compression had reached a stage when it threatened widespread loss of production and employment, and verged on economic chaos." and that "the economy needed substantial reforms if the crisis was to be fully overcome". The government initiated a program of structural reforms of the trade, industrial and public sector policies with the objective of evolving "an industrial and trade policy framework which would promote efficiency, reduce the bias in favour of excessive capital intensity and encourage employment-oriented pattern of industrialization".

Another impetus for the reforms (not explicitly acknowledged in the Economic Survey) came from multilateral aid agencies. Consequent to the BOP crisis, the government had to borrow from the IMF and the World Bank, who negotiated stabilization measures and structural reforms as part of the loan package.

1 *This section draws extensively on the Economic Survey 1991-92*

FIGURE 2: REAL GDP AND MANUFACTURING OUTPUT GROWTH

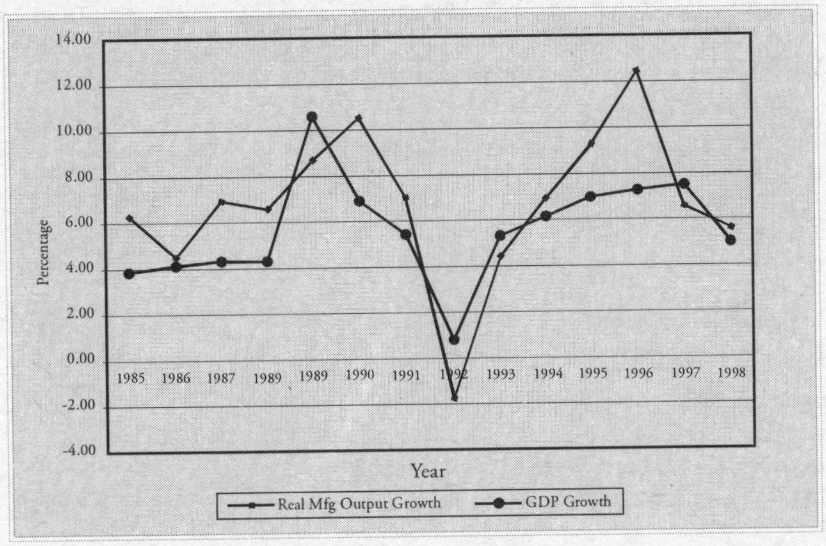

It must be noted that the reforms announced in 1991 were not completely contradictory to the general trend in policy. Concerns about the lack of competitiveness of Indian industry and consequent poor export performance (and poor growth record compared to the east Asian economies that had focused on export-led growth) had led to some liberalization of both the licensing and trade policies in the late 70s and 80s (see discussion in section 1 above). However, the reforms in 1991 did make significant changes in industrial, trade and public sector policies.

REGULATORY REGIME, POST-1991

The significant changes in industrial, foreign investment, trade and public sector policies announced in July 1991 (and thereafter) are summarized below.

Industrial Policy

TABLE 3

Summary of key industrial policy changes, 1991-1998

New Industrial Policy, July 1991	• Abolished licensing for all projects except in 18 industries. • MRTP Act amended to eliminate prior approval for large companies for capacity expansions. • The requirement of Phased Manufacturing Programs (PMP) discontinued for all new projects. • Schedule A of industries reserved exclusively for state enterprises cut down from 17 to 8. • Schedule B of industries where state enterprises were to acquire a dominant position abolished. • Small scale enterprises allowed to offer upto 24 per cent of shareholding to large enterprises.
1992-93	• List of delicensed industries expanded. • Oil exploration and refining removed from list of industries reserved for the public sector. • Power sector opened to foreign and domestic private investment.
1993-94	• Minerals removed from list of industries reserved for Public Sector, bringing the number down to 6. • List of industries where licensing is necessary reduced to 15 (motor car, white goods and leather delicensed). • Readymade garments removed from list of industries reserved for the small-scale sector, subject to certain restrictions.
1994-95	• Licensing for bulk drugs abolished; also added to list of industries with automatic approval of foreign equity. • Basic telecom opened to private participation, including foreign investments. • Reduction in tax rates on foreign and domestic companies. • Major overhaul of excise tax structure.
1995-96	• Number of measures to attract private investment in infrastructure announced. • Policy for drugs and pharmaceuticals liberalized and span of price control reduced.

1996-97	◆ Delicensed consumer electronics, bringing list of licensed industries to 14. ◆ Investment limit for defining small-scale company raised from Rs. 7.5 million to Rs. 30 million; export obligation on non-SSI firm manufacturing products reserved for SSI reduced from 75% to 50%. ◆ Pricing of coal deregulated; coal to be removed from list of industries reserved for public sector. ◆ List of industries in list for automatic approval expanded.
1997-98	◆ Number of industries requiring compulsory licensing reduced from 14 to 9. ◆ 15 items removed from the list of industries reserved for the small scale sector. ◆ Corporate tax rates reduced; tax on dividends removed.

Foreign Direct Investment Policy

TABLE 4

Summary of key foreign direct investment policy changes, 1991-1998

New Industrial Policy, July 1991	◆ Limit on foreign equity holdings raised from 40% to 51% in a wide range of industries; foreign exchange outflow as dividends to be balanced by export earnings. ◆ Procedures for foreign direct investment streamlined by creating a Foreign Investment Promotion Board, which would consider individual applications case-by-case. ◆ Technology imports liberalized by increasing royalty limits.
1992-93	◆ 'Automatic' approval for FDI applications in Appendix III (high priority) industries. ◆ Power sector opened to foreign and domestic private investment.
1993-94	◆ Basic telecom opened to private participation, including foreign investments. ◆ Reduction in tax rates on foreign and domestic companies.
1994-95	◆ Basic telecom opened to private participation, including foreign investments.
1995-96	◆ Pricing norms for raising level of foreign equity liberalized.
1996-97	◆ Automatic approval of FDI up to 74 per cent by the Reserve Bank of India in nine categories of industries ◆ The list of items for automatic approvals of foreign equity by the RBI expanded by adding three industries. ◆ Issued guidelines for expeditious approval of foreign investment in areas not covered under automatic approval.
1997-98	◆ The list of industries eligible for automatic approval expanded. ◆ Taxes on royalties reduced.

Trade Policy

TABLE 5

Summary of key trade policy changes, 1991-1998

1991-1992	◆ Administered licensing of imports replaced by import entitlements linked to export earnings. These entitlements, called exim scrips made freely tradable. ◆ Exim scrips can be used for all OGL items for actual users, limited permissible list, non-sensitive canalized list and non-OGL capital goods not in the restricted list. ◆ Permission to import capital goods without 'indigenous clearance' provided import covered by foreign equity or was 25% of value of plant and machinery (subject to a limit of Rs 200 mn). ◆ Scope of canalization narrowed. ◆ Actual user requirements for import of capital goods, raw materials and components under OGL were removed.

	• Short term import compression measures (interest surcharge on bank credit, cash margin on import payments, etc) were introduced.
1992-93	• Peak tariff reduced to 110%. • General reduction in tariff rates.
1993-94	• Peak tariff reduced to 85% (duties on baggage cut from 255 to 150 per cent). • De-canalized imports of certain petroleum products and fertilizers. • Pruned negative list of imports by removing 146 entries. • Encouragement for export of agricultural products (rice, sunflower seeds and superior quality wheat).
1994-95	• Peak rate reduced from 85 to 65 percent (with a few exceptions); • Major reform of structure of tariffs (including pruning of notifications, end-use exemptions etc) initiated.
1995-96	• Peak customs duty reduced to 50 percent. • List of freely importable goods (especially consumer goods) expanded. • Transferability of import licenses liberalized.
1996-97	• Large number of consumer goods shifted from negative list to the export-linked special import license list. • 488 items were moved from the restricted list to the OGL. • Basic import duties on a large range of goods reduced to 30 or 40 per cent.
1997-98	• 128 items, mainly textiles moved to the OGL list; additional 340 items shifted from restricted to OGL in April 98. • Import of gold and silver liberalized. • Peak customs duties reduced from 50 per cent to 40 per cent.

Public Sector Policy

TABLE 6

Summary of key public sector policy changes, 1991-1998

New Industrial Policy, July 1991	• List of industries reserved for the Public Sector (Schedule A) reduced from 17 to 8. • List of sectors reserved for dominance by public sector (Schedule B) effectively abolished. • Disinvestment in selected public sector enterprises to raise finances for development, bring in greater accountability and help create a new culture in their working which would improve efficiency.
	• Government equity ranging from 5% to 20% in 31 PSEs with 'good track record' disinvested to public sector mutual funds and Financial institutions. • Implementation of the MOU system extended; attempt to attain congruence between profits and MOU scores by attaching higher weights to profits.
1992-93	• Oil exploration and refining removed from list of industries reserved for the public sector. • Around 5 per cent of equity in 16 enterprises (overlap of 14 enterprises with earlier list, bringing total number of PSEs where shares were divested to 32). • Provisions of SICA extended to public sector undertakings.
1993-94	• Minerals removed from list of industries reserved for Public Sector, bringing the number down to 6. • Unfavorable stock market conditions sighted for suspension of disinvestment in 1993-94. • The weight for profit-related criteria set at 50%; two criteria, Gross Margin and Net Profit to have weight of 20% and 30% respectively in every MOU.
1994-95	• Shares in 16 companies sold in total of 3 rounds (some overlap with the earlier list of 32 companies; total number now at 39).
1995-96	• Due to adverse market conditions, only one round of disinvestment in October 1995, in 4 PSEs (all overlap with earlier list of PSEs).

	◆ 75% weight to MOU performance in the Annual Confidence Reports of the MOU signing PSE chief executives.
	◆ 60% weight to financial performance measures
	◆ Decision to extend MOUs to loss making PSEs from 1996-97
1996-97	◆ Disinvestment Commission set up for identifying PSEs (Public Sector Enterprises) for equity disinvestment as well as for working out the modalities of disinvestment.
	◆ Pricing of coal deregulated; coal to be removed from list of industries reserved for public sector.
1997-98	◆ The Government has granted enhanced autonomy to nine selected PSEs referred to as "Navaratnas". These are IOC, IPCL, ONGC, BPCL, HPCL, NTPC, SAIL, VSNL and BHEL. Two more enterprises, GAIL and MTNL, were given same status.
	◆ 696 guidelines, issued for PSEs over the last three decades were withdrawn to provide greater effective autonomy to the PSEs.
	◆ The Disinvestment Commission submitted seven reports on the 41 out of the 50 PSEs referred to it. It recommended varying levels of disinvestment for 12 companies, strategic sale for 21 companies and no disinvestment for 8 companies.
	◆ Dismantling of the Administered Pricing Mechanism (APM) for petroleum products initiated, as recommended by the Disinvestment Commission as pre-condition for sale of equity of companies dealing with petroleum products.

Small Scale Industrial Policy

Some of the salient features of the Government's Small scale industrial policy of 1991 are stated below-

i) A separate package for the promotion of Tiny Enterprises was introduced. This constituted the main thrust of Government's new policy. While the small scale sector (other than 'Tiny Enterprises') were mainly entitled to one-time benefits (like preference in land allocation/power connection, access to facilities for skill/technology upgradation), the 'Tiny' enterprises were also made eligible for additional support on a continuing basis, including easier access to institutional finance, priority in the Government Purchase Programme and relaxation from certain provisions of labour laws.

ii) The scope of the National Equity Fund Scheme was widened to cover projects upto Rs. 1 million for equity support (upto 15 per cent). Single Window Loan Scheme was enlarged to cover projects upto Rs 2 million with working capital margin upto Rs 1 million. Composite loans under Single Window Scheme, which was previously available only through State Financial Corporations (SFCs) and twin function State Small Industries Development Corporation (SSIDCs), were channelised through commercial banks to facilitate access to large number of entrepreneurs.

iii) Emphasis was shifted from subsidised/cheap credit, except for specified target groups, and efforts were made to ensure both adequate flow of credit on a normative basis, and the quality of its delivery, for viable operations of this sector.

iv) To provide access to the capital market and to encourage modernisation and technological upgradation, equity participation by other industrial undertakings in the SSI, not exceeding 24 per cent of the total shareholding was allowed. This was done to give an impetus to ancillarisation and sub-contracting, leading to expansion of employment opportunities.

v) A beginning was made towards solving the problem of delayed payments to small industries by setting up of 'factoring' services through Small Industries Development Bank of India (SIDBI). Network of such services was set up throughout the country and operated through commercial banks.

vi) It was recognized that a large potential of small scale sector in contributing to exports still remained untapped. The SIDO was recognised as the nodal agency to support the small scale industries in export promotion. An Export Development Centre was set up in SIDO to serve the small scale industries through its network of field offices to further augment export activities of this sector.

vii) Industry Associations were encouraged and supported to establish quality counselling and common testing facilities. Technology Information Centres to provide updated knowledge on technology and markets were proposed to be established. It was decided to enforce compulsory quality control, where non-conformity with quality and standards involves risk to human life and public health.

viii) A reoriented programme of modernization and technological upgradation aimed at improving productivity, efficiency and cost effectiveness in the small scale sector was intended to be pursued. Specific industries in large concentrations/clusters were identified for studies in conjunction with SIDBI and other banks. Such studies were supposed to establish commercial viability of modernization prescriptions, and financial support was to be provided for modernization of these industries on a priority basis. Indian Institutes of Technology (IITs) and selected Regional/other Engineering Colleges were chosen as Technological Information, Design and Development Centres in their respective command areas.

ix) A Technology Development Cell (TDC) was set up in the Small Industries Development Organisation (SIDO) to provide technology inputs to improve productivity and competitiveness of the products of the small scale sector. The TDC coordinated the activities of the Tool Rooms, Process-cum-Product Development Centres (PPDCs), existing as well as to be established under SIDO, and would also interact with the other industrial research and development organisations to achieve its objectives.

x) Adequacy and equitable distribution of indigenous and imported raw materials was to be ensured to the small scale sector, particularly the tiny sub-sector. It was decided to give priority to Tiny/Small Scale units in allocation of indigenous raw materials based on the capacity needs.

xi) Need for developing a strong Entrepreneurship Development Programme (EDP) and developing a pool of trainers for EDP was also felt.

7

CHAPTER SEVEN

Critical Evaluation of the Policy Environment in India and Impact on Manufacturing Competitiveness

AFTER a review of the various regulations to which Indian industry and manufacturing has been subjected to, let us examine how these policies have progressively affected the competitiveness of Indian manufacturing and what are the demands on the sector by the current times.

India had an industrial policy instead of a specific manufacturing policy from the early independence period. The industrial policy followed by India was **aimed at attaining self sufficiency in the production of basic and capital goods.** But Indian industry has lacked the dynamism to launch itself on a high growth path. **Initiatives for increasing productivity, technological upgradation and cost reduction measures never really took off.** The performance of Indian industry and its apparent paradoxes need to be analyzed in specific socio-political context of India. India is a country with a heterogeneous society and a complex geo-political configuration.

The country under the leadership of Nehru chose the Soviet model of development in heavy, basic and capital goods industries along with inclusion of Gandhian features and stress on cottage and village industries. The government wanted to have the commanding heights of the economy under its control and the four major areas of control exercised were industrial licensing, import controls and controls on pricing and foreign investments. The industrial licensing policy seriously undermined competition. Import controls ensured that as long as local availability of a product was there, it was protected from foreign competition even if it had high costs of production. These control measures along with restrictions on inflow of foreign capital encouraged import substitution but at the same time led to an uncompetitive, high cost monopolistic industrial structure. Also low purchasing power and small size of the market made large scale production and competition among large number of manufacturers virtually impossible. Business houses very often pre-empted capacity without actually investing. Thus manipulation of the administrative set up made healthy competition, product innovation, technological developments and cost reduction taking a back seat. State became the biggest risk-bearer in the role of incubator for a developing industrial economy. The strategic mind-set that underlay the Nehru-Mahalobnis model of development, distrustful of private sector, made the state the chief instrument of development. The strategy did pay off well by accounting for much of India's initial industrial growth in core segments like power, steel, oil refining, civil aviation, machine tools, heavy engineering, public works and other infrastructure services, where none existed before. The state also engineered the growth of modern services sector institutions such as banking, insurance, institutional credit, scientific and technology research and development and technical education.

However, despite its fairly well-articulated strategic intent, not without its historical merits and achievements, the Nehru-Mahalobnis model soon regressed into what is aptly called the 'license-permit-quota raj". A virtual nexus of disparate politicians of all hues, career bureaucrats and willy-nilly businessmen that emerged subsequently proved to be the lynchpin of Indian economy. Socialistic over-indulgence not only bred its brand of dust-raising populist politics, bereft of sound institutional foundations, but also created a buccaneer business class that took advantage of every possible loophole in the policies of the government to corner production capacities, import licenses, export subsidies and quotas.

There was nothing wrong in the strategy of emphasizing the development of heavy and basic industries, which was expected to have multiplier effects along with small house hold industries to balance the long term objective of capital accumulation and immediate needs of employment, but its implementation was faulty. This occurred because the growth of the organized industry was marked by high cost, low efficiency and low productivity and the development of small scale industries could not meet the objective of decentralized industrialization because of lack of concerted efforts. Thus, the paradox in the Indian situation is simultaneous existence of fairly high degree of technological competence and a substantial industrial base along with low productivity and low technological dynamism.

Relaxations of licensing constraints, endorsing capacity expansions by industrial units for modernization and replacement, quick response to changing demand conditions and fixation of minimum scale of production in many industries over the years have brought about a welcome change. The nature of the developmental process has also led to the growth of a substantial middle class, demanding a variety of industrial goods. This has resulted in the growth in demand and production of consumer durables. Export performance has also improved but the sustainability of the industrial growth is questionable.

The major post-reform transitional trends in Indian industry could be summed up as under:

M&A phenomenon with its inbuilt challenges of restructuring of equity, debt and management controls is now cutting across the entire spectrum of industry and is acting as a curtain raiser on more concrete processes of business process restructuring.

Global quality standards have made improvements in productivity and rebuilding cost-inefficient systems of manufacturing *sine quo non* of survival and growth for the Indian industry, faced with severe import competition in almost all segments.

- **Strategic value drivers** in the industry are now entrenched in the logistics and supply chain pipeline, with focus on just-in time inventories, shorter production turnaround time and smart outsourcing strategies that can cut down distribution and procurement costs.
- **Adapting to information technology** as a key competitive tool in managing the business processes, both within and outside the enterprise, effectively redefining the manufacturing systems and supplier and
- customer-led business processes.

8

CHAPTER EIGHT

Cross Country Comparisons

EAST Asian and Latin American countries were virtually in a similar situation as India around the time of India's independence. However, the pace of industrialization and economic growth has not been the same across these countries. We take a look at what policies were adopted by these countries and their impact on the manufacturing sector in those countries. This serves the purposes of identifying the lacunae in our own policies as well as setting examples and best practices which can be adapted for application in the Indian context.

East Asian Economies

East Asian countries have adopted a wide range of industrial policies, which vary from case to case and over different time periods. What they have shared in common is a commitment to achieving full employment at rising real wage levels. Trade, investment, and industrial policies have been tailored to this objective, with state interventions aimed at optimizing market outcomes and employment opportunities. Working towards full employment has been an explicit policy objective. Two major achievements of the East Asian miracle are:-

Manufacturing export growth

Various features of East Asia's success in manufacturing are well known. In the 1960s, South Korea and Taiwan achieved industrial growth rates in excess of 10 per cent a year. The second wave of 'tiger' economies - Indonesia, Malaysia, and Thailand - has followed hard on their heels, making the transition from dependence on primary commodities to emerge as major manufacturing exporters. The share of manufactured exports in total exports has grown from less than 6 per cent in all three countries to 41 per cent for Indonesia, 61 per cent for Malaysia, and 77 per cent for Thailand. In the last 25 years, the share of the industrialized countries in global manufacturing value-added has fallen from 88 per cent to 80 per cent - almost this entire shift is accounted for by the rise of East Asian exports.

Export growth has been central to the rise in living standards experienced across East Asia .On a per capita basis, the value of exports has risen by 14 per cent a year in the 1990s, compared to a rise of 7 per cent for Latin America and a fall of 1.6 per cent for Africa. The difference is partly related to the composition of exports. Today, the share of manufactured goods in Africa's exports is less than 10 per cent, and dependence on primary commodities has hardly changed over the past three decades. During the 1980s, deteriorating prices for primary commodities erased around 5 per cent of sub-Saharan Africa's regional GDP. More broadly, dependence on exports characterized by low value-added has contributed to a situation in which the forty-four Least Developed Countries, accounting for 10 per cent of the world's population, have seen their share of international trade shrink by half - to 0.3 per cent of the total - over the past decade. Like sub-Saharan Africa, Latin America remains heavily dependent on primary commodities (which account for over one-third of export earnings), while its share of world manufacturing trade has stagnated at 4 per cent since 1970.

Success and failure in international trade have been major factor in explaining global trends in income and poverty. Because international trade has been growing faster than world output, it has acted as a dynamic engine of growth. East Asia has benefited because the region has successfully participated in global markets, so that the share of trade in GDP has been rising steadily. Moreover, most countries in the region have succeeded in raising the value content of their exports by moving into increasingly sophisticated, knowledge-intensive areas of production such as electronics, machine tools, and computing equipment. For most countries in East Asia, success in international trade and the development of more diverse and more sophisticated export structures has been closely related to success in attracting foreign investment. This began in the mid-1980s. But foreign capital inflows accelerated in the 1990s. Today, four countries in East Asia - China, Malaysia, Indonesia, and Thailand - account for around 40 per cent of capital flows to developing countries. At the other end of the spectrum, sub-Saharan Africa accounts for less than 1 per cent.

Growth and Employment

Between 1986 and 1993, employment in East Asia increased by more than 3 per cent a year - a figure which was well in excess of the rate of growth in the labor force. Rapid increases in manufacturing output were accompanied by rising real wages, averaging between 3 per cent and 6 per cent a year.

The 'poverty-reduction elasticity of growth' was a major feature distinguishing East Asia from other developing regions. East Asian growth has a high elasticity for poverty reduction, in part because the employment elasticity of growth is higher. The latter is an indicator of the relationship between value-added by manufacturing output growth and the creation of employment. For countries such as Mexico, Brazil, India, Pakistan, and the Philippines, each percentage point of manufacturing growth produces less than one-half of the employment creation experienced in countries such as Indonesia and Malaysia.

The reason is that the former countries have achieved industrial development by relying on capital-intensive growth. Perversely, employment opportunities have been restricted by policies penalizing labor-intensive industries, such as high import barriers for basic technologies and other imported materials needed to improve productivity. Small-scale industries with an export capacity have been forced to purchase from high-cost local suppliers, while over-valued exchange rates have further compromised their competitiveness. Employment and investment opportunities have been lost as a result. In East Asia the growth pattern is built on rising productivity, with the gains being passed on in the form of rising real wages, in Latin America, the pattern is one of low-productivity, low-skill employment, with stagnating real wages.

9

CHAPTER NINE

The Background to East Asia's Success

THERE has not been a single route to East Asia's success in developing a manufacturing base capable of sustaining rising real incomes and employment level. Still less, has the outcome been the result solely of macro-economic management? Social policies and the extension of educational opportunities have been crucial to economic growth. Four of the most important policy trends which can be identified are the following:

East Asia's Industrial Growth

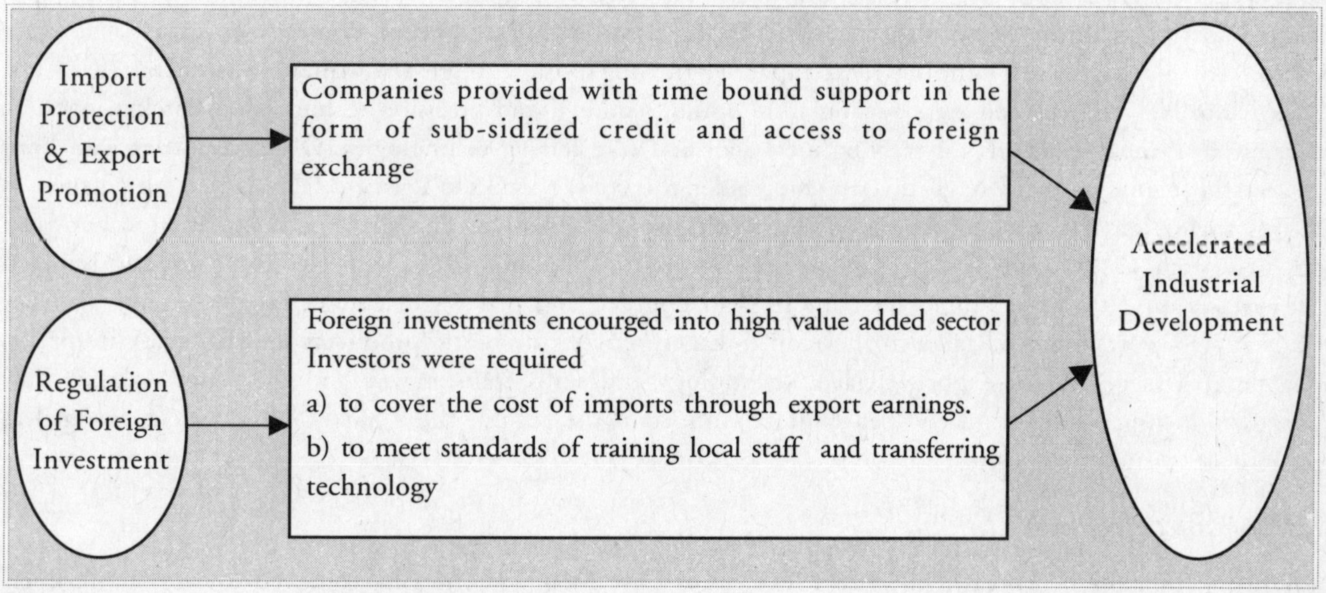

Selective Protection and Import Substitution

Most countries in the region developed their manufacturing base behind high, but carefully structured protective barriers. During the 1950s and the first half of the 1960s South Korea developed a labor-intensive manufacturing industry protected by direct import controls. However, trade policy was more liberal for the essential imports needed to increase productivity. After the mid-1960s, government policy combined import protection with export promotion, providing companies with time-bound support in the shape of subsidized credit and preferential access to foreign exchange. In both South Korea and Taiwan, exports surged after the early 1960s, yet the trade regime was a model of controls designed to promote domestic investment. Less successful as a model of import substitution was Indonesia, where average tariffs in excess of 500 per cent prevailed in the 1970s, shielding domestic companies that were uncompetitive and dominated by vested interests. Not only did the manufacturing sector fail, at this stage, to make a contribution to exports, there is evidence that it slowed overall growth rates and limited employment creation. By the early 1980s, subsidies to loss-making state enterprises accounted for 30 per cent of GDP. In Malaysia, the Proton car has become a symbol of the problems and potential in import substitution. This is the product of a joint venture between the state, domestic investors, and foreign trans-national companies. After the venture almost collapsed in the 1980s, by the early 1990s, domestic suppliers were providing 80 per cent of components and exports began to increase, providing a stimulus to investment and employment.

The Regulation of Foreign Investment

East Asia has been a focal point for direct foreign investment since the mid-1980s. Initially, the impetus came from the 1987 Plaza Accord, which devalued the dollar against the yen and gave Japanese companies an incentive to relocate to production sites with weaker currencies. Even before this, however, Malaysia had combined import protection with generous incentives to foreign investors under its New Economic Policy. These investors concentrated their activities in the electronics sector, where basic assembly operations led the export boom. One of the few requirements was that foreign investors cover the costs of their imports through their export earnings - a measure designed to protect the balance of payments and limit competition for local capital resources. In Indonesia, the surge in direct foreign investment after the early 1980s was accompanied by imports of new technologies which led to a dramatic increase in exports of textiles, clothing, and footwear. From different policy perspectives, both South Korea and Singapore tightly regulated foreign investment. South Korea prohibited foreign majority-ownership, reserving strategic industrial sectors for domestic investors who were provided with incentives for building local capacity. In Singapore, foreign investment was encouraged, but channeled into high value-added areas such as fibre optics, banking, and precision technologies. Foreign investors were required to meet standards for training local staff and transferring technologies. Other countries have been less successful in integrating foreign investment. Under the New Economic Policy, Malaysia provided generous incentives to foreign investors. At one level the results were impressive. The share of manufactured goods in exports rose from 12 per cent to almost 80 per cent between 1970 and 1990, with electronic goods assembled by Malaysia leading the way. Employment more than doubled and real wages rose at rates in excess of 3 per cent a year. However, Malaysia failed to develop linkages between domestic producers and foreign enterprises, which operated in an import-export enclave. Technology and skills transfer was limited. With Malaysia now facing growing competition from cheaper-labor economies in the region, these policy failures have emerged as a threat to continued prosperity.

Selective Liberalization

Starting from South Korea in the late 1960s, most East Asian countries have moved towards more liberalized systems, often under external pressure. Malaysia's Industrial Master Plan (1986-1995) relaxed the import controls of the New Economic Policy, although import barriers remained high and foreign ownership in domestic industries was limited. In Indonesia, liberalization started in 1983, but progress in this direction has been limited. Disputes between the Indonesian government and the industrialized countries over the protection afforded to the national car industry, underlies the growing tension between industrial policy in the region and international trade rules. In most areas of manufacturing, there has been little liberalization in the 1990s. It is a similar story in Thailand, although tariffs in most areas remain low by international standards.

As the above account suggests, there are dangers in attempting to draw universal lessons from East Asia. There are also dangers in over-stating the region's success. Thailand has been highly ineffective in developing the infrastructure for transport, training, and research and development, which was crucial to the success of South Korea and Taiwan. Failure to upgrade labor skills has reflected the deeper failure of state policy in education discussed earlier. In Malaysia, the economy remains heavily dependent on foreign capital and technology for its manufactured exports. Linkages between the export-oriented trans-national company subsidiaries and the rest of the economy remain exceptionally weak. Both countries have suffered from a failure to develop more integrated industrial structures. The same is true of Indonesia, where the bulk of foreign investment is concentrated either in unskilled assembly operations, which create jobs but offer little by way of skills transfer; or in extractive industries such as mining, which create few jobs and wreak environmental havoc, in return for increased foreign exchange earnings.

Reliance on cheap labor as the main source of export growth has also brought with it intra-regional tensions. Wages are higher in countries such as Malaysia, Thailand, and Indonesia than in Vietnam or China, raising the specter of a massive relocation of capital - and employment - as these countries develop. Some electronics companies have already relocated from Malaysia to Vietnam. So far, however, there is limited evidence of footloose foreign investors shifting the location of production sites. Even in labor-intensive sectors, the costs of establishing operations are relatively high, as are the costs of relocation. This is a deterrent to relocation, but if exaggeration is unwarranted, so too is complacency. Clothes retailers in Europe, for instance, are shifting their sourcing towards countries with lower wage levels.

High Savings

Another of East Asia's strengths has been its capacity for savings. Average savings rates in the region are around 30 per cent, compared to less than 20 per cent in Latin America. Low-income China saves an extraordinary 40 per cent of national income - middle-income Mexico saves 15 per cent. These high levels of savings have enabled investment to be financed domestically, reducing exposure to the problems of debt and dependence on aid which have characterized Latin America and Africa.

Specific Examples of Asian Countries

Taiwan and South Korea

South Korea and Taiwan appear to be twentieth century success stories. Both suffered through most of the first half of this century as colonies of Japan. They entered the 1950s in extreme poverty, with average life expectancy from 55 to 60 years, similar to India today. Yet both enjoyed growth rates of from 6% to 8% in their per capita GDPs from the early 1960s to the present. Unlike Brazil, much of this growth did trickle down to raise average wages and drastically increase living standards. Taiwan probably has the most equitable income distribution of any non-socialist lesser developed country, an income distribution pattern similar to that of Japan or Holland. South Korea's income distribution pattern is similar to that of France. Both have come close to the developed countries in terms of health and education.

Their development strategies were partially based on the production of labor-intensive goods for export. The obvious success of Taiwan, South Korea and the city-states of Singapore and Hong Kong has led to Export-Oriented Industrialization being heralded -- by the World Bank economists among others -- as the development model that others should follow. Before any country sets out to emulate the Asian dragons, however, they had best heed three warnings. First, both South Korea and Taiwan benefited from numerous particular elements that other countries cannot simply reproduce. Second, both are now at critical points as they are no longer low-wage countries and are rapidly losing manufacturing jobs to countries that are. Third, it is unlikely that the consumer markets of the rich countries can support very many dragon-style export miracles at the same time.

THREE BUILDING BLOCKS OF KOREA'S INDUSTRIAL DEVELOPMENT

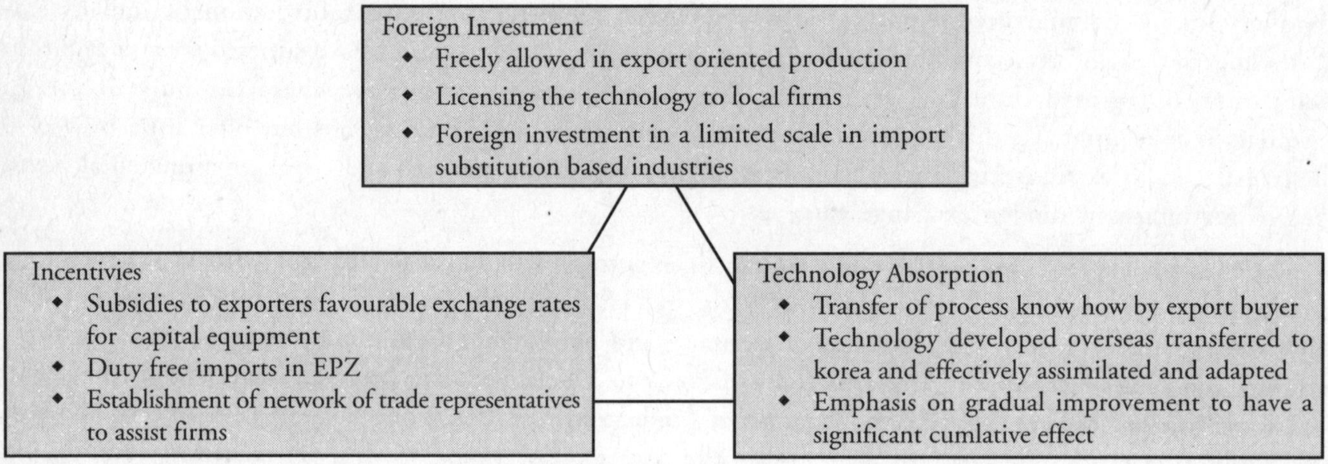

Early Post-Colonial Policies

The most important consequence of the limited power of the traditional elites was that both South Korea and Taiwan were able to implement extensive land reforms. The typical agricultural pattern in both countries was small farms owned by absentee landlords (each of whom usually owned numerous farms). The families that sharecropped the farms paid from 50 to 75 per cent of their crop as rent, leaving them only enough to subsist. Taiwan started the reforms in 1949 by limiting rent to 37.5 per cent of the crop. In 1953 farmland was turned over to the families that worked it in the "Land to the Tiller" program. The government also provided fertilizer, credit and technical assistance. South Korea, under United States pressure, completed a similar land reform in 1953 which effectively eliminated the power of the old landholding class.

The land reforms were partially responsible for the fact that income is more equitably distributed in South Korea and Taiwan than in most Latin American countries. They were also responsible for the rapid growth in agricultural productivity that both countries enjoyed. Additionally, the new land ownership pattern made the transition from farmer to industrial worker less traumatic than it usually was in Latin America:

Later, as industrial employment absorbed larger proportions of the labor force, a family employment strategy that included both farm and no-nfarm work remained important to the maintenance of working-class living standards. Return to the countryside is rarely a viable option for a Latin American worker confronting an employer's attack on his or her real wages, but a worker whose family owns a productive agricultural smallholding is in a different position. Indeed, the rural population of Taiwan chose to remain part-time farmers as late as the early seventies, even though most of them were deriving most of their income from non-farm work, indicating the importance of this unusually strong rural base in working-class economic strategies.

The land reforms were supplemented by other unique advantages. South Korea and Taiwan both exploited their geographic positions on the frontier of the cold war and received vast amounts of foreign aid. From 1946 through 1978 the United States extended $ 6 billion in economic aid plus $ 7 billion in military aid to South

Korea and $ 5.6 billion to Taiwan. During the same period, all of Africa received less than $ 7 billion in United States aid; all of Latin America received only $ 14.8 billion.

Both countries also benefited from large infusions of well-educated people. When Chiang Kai-Shek took the remnants of his government to Taiwan in 1949 he was accompanied by about a million mainland Chinese. Most of them were well-educated. At that time, the population of Taiwan was only six million. South Korea similarly benefited from the migration of many educated Koreans from North Korea.

South Korea and Taiwan also followed import substitution policies. They started the 'easy' stage of IS in the 1950s by manufacturing textiles, shoes and other basic consumer goods for their own limited markets under tariff protection. By the 1960s they applied IS policies to more complex goods such as steel and ship-building. There was little foreign control of key industries. In the 1950s United States firms had little interest in impoverished Asian countries when there were so many investment opportunities in Europe and Latin America. By the 1960s and 70s South Korea and Taiwan had strong enough governments to establish rules for foreign investment. And massive amounts of United States aid meant they were not as chronically short of hard currency as the Latin American countries.

South Korea allowed foreign firms to invest freely in production that was strictly for export, but import substitution investment was restricted to sectors that were clearly beyond the technological ability of South Korean firms; even in these sectors every attempt was made to license the technology to local firms before foreign investment was approved.

In addition to protective tariffs, the governments used other development policies pioneered by the Latin American countries. Hard currency was allocated by government agencies with the idea of directing investment into critical sectors. Government firms were established, particularly by Taiwan, in industries that were thought to be too large for the private sector and too important to allow foreign investment. The steel, ship-building and petroleum sectors were all dominated by state-owned firms in Taiwan.

Where South Korea and Taiwan set themselves aside from the Latin America countries, was in the way they promoted exports of manufactured goods. Subsidies to exporters, favourable exchange rates for capital equipment and establishment of a network of trade representatives abroad were among the policies used. They also set up special industrial parks called Export Processing Zones. Foreign firms could locate in these EPZs, import their parts and materials duty-free, assemble the goods using South Korean or Taiwanese labour, then export the product. The EPZs often included harbors and air-freight facilities. Buildings, personnel offices, guards, banking facilities and many other amenities were included. The idea was to make setting up an assembly operation as simple and as cheap as possible for foreign firms. It is not unheard of for a firm to be in full operation within six months of the time it makes a relocation decision.

By the 1980s, this approach gave the appearance of working. EOI delivered what IS could not: plentiful jobs in the manufacturing sector. By concentrating on production for export, South Korea and Taiwan overcame the limitations of their own markets. With low unemployment rates came – finally – rising wages. Both governments did what they could to stop or at least slow the wage increases. But by 1987 even South Korea's brutal anti-labor tactics could not stop a wave of strikes in the key heavy industries. Once workers saw their real wages substantially increase, they became better consumers. In 1988 alone, real wages in South Korea rose by 12%; in the first quarter of 1989 domestic sales of electronic goods leapt by 70%. South Korea and Taiwan had clearly ripened.

Both countries had to prepare for the risks of rising wages. Rising wages are at the same time a symptom of successful economic development and a cause, under the right conditions, of continuing economic growth.

But for a country that has industrialized rapidly by attracting low wage industries there is always the risk the ripening will be a brief phase followed by a rapid "rotting." First, foreign firms in search of low wages will locate elsewhere. Then foreign firms already in the EOI country will unbolt their machines and move on. Finally even the local capitalists producing goods under contract to Nike, K-Mart, etc. will be forced to shift production to low wage countries or lose their contracts. Taiwan lost 300,000 manufacturing jobs between 1986 and 1989 alone. Many Taiwanese businesses simply relocated to China in search of the type of docile low-wage worker that they remember from the Taiwan of the 1960s. Moving production is remarkably easy. The same South Korean entrepreneurs who were making shoes for Nike in South Korea fifteen years ago have shifted the most labor-intensive parts of their operations to Thailand (where their assembly-line workers are paid $ 130 to $ 160 per month), China ($ 90 - $ 110) and Indonesia ($ 40 - $ 60). Should wages go up in these countries; the Nike subcontractor will shift more production to Vietnam, Mexico and India.

The Asian dragons' strategy for coping with higher wages is to shift into exports which require more skill. They seem to have prepared for this by emphasizing education. By 1990, 45% of Taiwanese and 37% of South Koreans were getting some higher education. Taiwan's advertisements now stress an "abundant pool of scientists and engineers," available for about one-quarter of United States salaries.

The markets for manufactured goods in the rich countries are not infinite. Taiwan and South Korea had the advantages of being first and being relatively small. By the late 1980s South Korea was exporting over $ 15 billion worth of electronic goods per year. If much larger India were to be that successful, where could $ 300 billion in TVs, VCRs and stereo equipment be sold? In order for the entire lesser developed world to export as many manufactured goods per person as South Korea does, the rich countries would have to import manufactured goods equal in value to about one third of our GDPs. Protectionist tariffs and quotas will be in place long before any such thing could happen.

Nor will there likely be more opportunities to do what South Korea and Taiwan actually did: export the products of one industrial generation while developing the next generation under tariff protection. Mexico has gained more open access to United States and Canadian markets under the terms of the NAFTA, but is no longer able to protect the as yet underdeveloped capital equipment and computer sectors.

Sources of Technology and Know how?

The sources of technology were frequently buyers of output and suppliers of equipment and materials.

Foreign buyers contributed importantly through on-going programs to control and improve quality or through periodic visits to inspect production facilities. The buyers helped many exporters to achieve greater efficiency and lower costs by suggesting process improvements and improved management practices. Thus transfer of process know-how from export buyers had been a major contributor to process innovations that sequentially led to gradual improvements, the cumulative effect of which is quite significant. Koreans excelled in mastery of process technology. Technology, originally developed overseas was subsequently transferred or brought to Korea and then effectively assimilated and sometimes adapted by Korean industry. (Westphal et al, World Bank Staff Working Papers, Number 469). For example, Korean manufacturers have not only been able to manufacture capital goods that met global standards, but also adapted the product design to suit Korean conditions. Thus semi-automatic looms were adapted to fall between ordinary semiautomatic and fully automatic in terms of labour intensity of the weaving technology used.

For transfer of know-how, the domestic sources had an important role in the traditional sectors like textile than in non-traditional sectors but the difference was not significant.

There were high instances of innovation by both domestic and foreign firms. The foreign firms introduced the techniques for production of nylon yarn, polyester staple fiber, cement, production of small televisions, audio speakers, home stereo systems. The domestic firms introduced a wide range of techniques related to producing textiles, acrylic fiber, rayon yarn and staple fiber, techniques for producing pumps, transformers, diesel engines, industrial machines, automatic looms, electric motors, radial tires, iron and steel products etc.

The importance of local technological learning was significant in case of Korea. High industrial growth rates permitted rapid technological learning. The first plants were built on a small scale, much smaller than the size of the market or to exploit scale economies. But subsequently Korean technicians and engineers assumed greater role in project design and implementation and built plants, closer to or equal to world scale.

The Case of Japan: Role played by MITI

Japan's Ministry of International Trade and Industry (MITI) was formed in 1949 from the union of the Trade Agency and the Ministry of Commerce and Industry in an effort to curb post-war inflation and provide government leadership and assistance for the restoration of industrial productivity and employment. MITI has held primary responsibility for formulating and implementing international trade policy, although it has done so by seeking a consensus among interested parties, including the Ministry of Foreign Affairs and the Ministry of Finance. MITI has also coordinated trade policy, on issues affecting their interests, with the Economic Planning Agency, the Bank of Japan, and the ministries of agriculture, construction, forestry and fisheries, health and welfare, posts and telecommunications, and transportation. As trade issues broadened in scope, these other ministries became more important in international negotiations, so that in the late 1980s MITI had less control in formulating international trade policy than it had had in the 1950s and 1960s. The prime minister, the National Diet (Japan's legislature), and the Fair Trade Commission have also circumscribed MITI's operations.

MITI'S ORGANISATION STRUCTURE

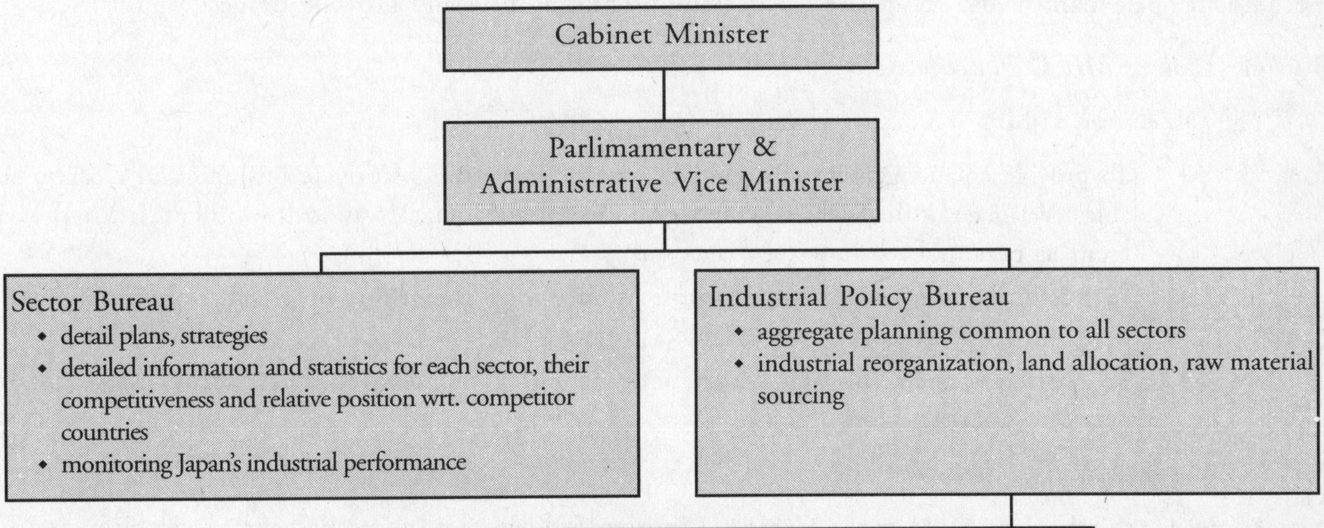

MITI has been responsible not only in the areas of exports and imports, but also for all domestic industries and businesses not specifically covered by other ministries in the areas of investment in plant and equipment, pollution control, energy and power, some aspects of foreign economic assistance, and consumer complaints. This span has allowed MITI to integrate conflicting policies, such as those on pollution control and export competitiveness, to minimize damage to export industries.

MITI has served as an architect of industrial policy, an arbiter on industrial problems and disputes, and a regulator. A major objective of the ministry has been to strengthen the country's industrial base.

It has not only managed Japanese trade and industry along the lines of a centrally planned economy, but it has provided industries with administrative guidance and other direction, both formal and informal, on modernization, technology, investments in new plants and equipment, and domestic and foreign competition. The close relationship between MITI and Japanese industry has led to foreign trade policy that often complements the ministry's efforts to strengthen domestic manufacturing interests. MITI facilitated the early development of nearly all major industries by providing protection from import competition, technological intelligence, help in licensing foreign technology, access to foreign exchange, and assistance in mergers.

These policies to promote domestic industry and to protect it from international competition were strongest in the 1950s and 1960s. As industry became stronger and as MITI lost some of its policy tools, such as control over allocation of foreign exchange, MITI's policies also changed. The success of Japanese exports and the tension it has caused in other countries led MITI to provide guidance on limiting exports of particular products to various countries. Starting in 1981, MITI presided over the establishment of voluntary restraints on automobile exports to the United States to allay criticism from American manufacturers and their unions.

Similarly, MITI was forced to liberalize import policies, despite its traditional protectionist focus. During the 1980s, the ministry helped to craft a number of market-opening and import promoting measures, including the creation of an import promotion office within the ministry. The close relationship between MITI and industry allowed the ministry to play such a role in fostering more open markets, but conflict remained between the need to open markets and the desire to continue promoting new and growing domestic industries.

A Closer Look at MITI's Functions

The Objectives of MITI's initiatives were:

- To provide encouragement and guidance to the initiatives of private business and create a suitable level playing field which would give critical advantage to industries, identified by government as potential for long term success.
- In R&D, the aim was to co-ordinate and encourage the efforts of private corporations rather than initiate large government funded projects.
- To create channels through which new technological ideas could be rapidly disseminated between Japanese firms.

MITI's Unique Well Knit Structure

At the highest level in MITI is the Cabinet Minister, followed by the parliamentary vice-ministers and an administrative vice-minister. This is followed by the bureau level, followed by the section chiefs and assistants.

The various industrial bureaus focus on detailed plans and strategies for various sectors. Detailed information and statistics pertaining to each sector's output, competitiveness and potential relative to international competitors are prepared, which forms the basis of Japan's industrial planning. These bureaus monitor Japan's performance sector by sector along dimensions such as demand growth, capacity utilization, factor input cost, productivity performance, market share, technology and import competition. By maintaining such high quality of statistical database, MITI actually performs the environmental scanning required for corporate strategy making. This information gathering and environmental scanning is made freely available to all companies, big or small.

Work at the level of these sector bureaus is complemented by the Industrial Policy Bureau, which carries out the aggregate planning common to all industrial sectors. This form of industrial planning includes industrial reorganization, industrial land allocation and raw materials sourcing and distribution. Here details are worked out for implementing particular policies and programs. The Industrial Structure division deals with the projected structure of industrial demand and output with policies aimed towards a particular targeted year, with provision for a rolling plan, which is continuously revised and updated. At a more specific level, the Business Behaviour Division works on areas like tax measures and labour issues to study the impact of changes in the industrial structure. Overall there are consulting committees, which provide input to the work of various divisions. The output is an integrated plan which acts as a common thread, pulling the diverse aspects together.

This matrix form of organization of MITI results in cross-fertilization of ideas and strategies and smoothen liaison and co-ordination with other government ministries. Another notable function performed by MITI is continuous consultation with the private sector companies and economic organizations.

Thus we find the organization of MITI as a matrix structure and detailed planning done by each of the divisions with the objective of diagnosing industrial environment on a continuous basis provides one of the cornerstones of Japanese industrial success.

The Case of China

Of all the countries with which India can be compared, China deserves special mention. This is because of the similarity between the two nations in demography and natural resources and that is where the similarities end. China has set an example through its implementation of economic reforms. Indeed, China, in its two decades of reforms has transformed its weaknesses into strengths and emerged as a colossus striding through the world of cut-throat competition. On the contrary, in India, hurdles posed by vested interests have often decelerated the reforms process. If India continues growing steadily at the current growth rate of 6 per cent it will take another 14 years to catch up with the current living standards of China.

China is the only country, the reforms initiative of which can be directly compared to those in India. This is because both the countries are not only neighbors but also share common demographics, climatic conditions and market potential. The economy has witnessed an unprecedented growth since the reform process initiated two decades ago. In the 23-year period of reforms, it has marched a long way to open herself. For nations like India, which are on the way to reforms, Chinese reforms provide crucial lessons. Insights gained from the Chinese experience with minor modifications can enable other developing nations to complete their reform process smoothly and efficiently.

Initial Conditions

From the day of its independence in 1950, China has emphasized on a socialist pattern of society. The Chinese pre-reform economic model was a descendant of the Soviet model. During the initial years of the socialist regime China had designed its economy as highly centralized and oriented towards heavy industry. This is in contrast to the emphasis of other developed nations like US and European countries where usually the thrust is on light industry and agriculture. It is only in 1957, that China realized the pitfalls of such a totalitarian system. Extending the decision-making powers of the provinces, the government laid emphasis on rapid internal development. Following a series of revolutionary measures like The Great Leap Forward (1958-60),

The Natural Disaster (1959-61) and The Cultural Revolution (1966-76), the course of economic development had taken a new dimension.

Until 1976, three key pillars can be identified with the People's Republic of China economic policy: collectivization of land, centralized control over accumulation and reinvestment of capital and state ownership of major industries and banks as well as small-scale enterprises.

Land reforms were the first step towards the development of a rural economy. Since socialist pattern of government prevailed in China, collectives were established to enable control over the means of production and prices of the products. Collectives also acted as pools of labor to develop rural infrastructure constructing the irrigation networks, roads and other transportation networks. Also, realizing the need for industrial development, 'people's communes' were formed to improve industrial infrastructure.

In contrast, India during the same period had emphasized on the economic development through careful planning process under the stewardship of the planning commission. The piece-meal approach to development showed good results in the initial stages. The mixed-economy structure never yielded expected results. Though both the nations had taken control over their economies after a prolonged independence fight, China has been able to make rapid strides into economic development through a series of measures. The history of China has one important aspect: they were able to give a revolutionary-bent to whatever they did. The revolutions like The Great Leap Forward, The Natural Disaster and The Cultural Revolution all helped the Chinese population to unite and fight for a common cause. On the other hand, except for the independence fight, there has been no common factor that united the people in India. Once independence was attained, the focus shifted to regional power - a fact that remains unresolved even today. The choice of a mixed economy has also impacted the economic development in the long run as emphasis was on development of public and private participation. This resulted in partial development of agriculture and industrial sectors.

The most important lesson that the history of China offers to any nation is realizing the need of internal support for development. According to Susmita Dasgupta, Deputy Chief Economist, Joint Plant Committee, "Economic reforms in China is an outcome of its economic growth for the Chinese economy grew much before the laws of the land opened up to accommodate new forces. It is important to understand the sources of economic growth rather than to draw false relationships between policies and their responses." Truly, China has opened its economy only after attaining self-sufficiency in agriculture and rural and infrastructure development. The communist ideology played a vital role in this regard.

Reforms Path

Despite certain commonalities existing between the two nations, China has been able to march ahead of India. The only differentiating factor is the way these countries addressed their reform process. Though the reforms in China had started much earlier than India, the transition from a closed economy to an open-market economy is notable. China had realized much earlier than India about the need for a pro-market approach rather than going ahead with a controlled regime. The conditions under which these two big countries (size) opted for reforms are a stark contrast. While Indian reforms were driven by crisis, the Chinese reforms emanated from the foresight of the Chinese leaders. Reforms in China were initiated without an intervening crisis. With Deng Xiaoping coming into power in 1978, the Chinese economy has opted for a pro-market economy. Deng realized that if China has to catch-up with the advanced nations he has to go ahead with reforms. Though there were no economic crises, imbalances were developing and income in-equalities were widening. The foresight of the Chinese leaders to recognize the long-term impact of these vital economic variables on the economy made their choice for reforms more concrete.

Liberalization of the Indian economy was initiated only after the 1991 balance of payment crisis and meeting the WTO specifications. To come out of the crisis, Indian government had to pledge its gold reserves with the IMF and obtain a loan. Only after the bitter experience, Indian leaders realized the importance of reform in Indian economy. Being crisis induced, the Indian reforms had focused on macro-economic stabilization. At the same time, the reforms were also aimed at achieving stability fast. On the other hand, China's reforms did not require any stabilization measures and no external pressure on targets to be achieved in a specific timetable. This was the unique characteristic of the Chinese reforms and to a great extent helped the Chinese authorities to approach the reform process through gradual experimentation, on the basis of priority.

However, the conditions under which both the countries opted for reforms are almost similar. Opines Susmita Dasgupta, "China's economic reforms took place much in a similar manner as ours; the government machinery failed to deliver goods to people. In fact, China's liberalization of agriculture precisely happened because the commune system failed to create adequate food stocks and famines caused havoc in the country." On the other hand, India had problems with both the agricultural and industrial sectors. While the agricultural sector suffered from low productivity, disguised un-employment and feudal structure of production, the efficiency of the Industrial sector was seriously battered due to the partial controlled regime.

Political Environment

There is a vast difference in the political environment prevailing in these two countries. The most significant is the fact that China is a communist state and India is a democratic state. The difference itself gives rise to vast differences in the way these two countries respond to problems. Truly, the political commitment has been a major success driver. The characteristic of communist pattern of government enables the state to take the important decisions and implement it. Unlike other democratic countries, this power of decision-making gives the advantage of moving swiftly. Whether the decision is good or bad, the state has the powers to enforce the law and the citizens are forced to accept the policies of the government.

In India, the democratic set-up proved to be more problematic rather than constructive. For every decision to be taken, the government has to balance the interests of the minority and majority of the population. Added to this is the complex inter-state political environment that makes the implementation of any developmental program a Herculean task. For any policy to be approved and implemented, it has to go through various levels and phases before it is finally implemented. However, by the time the policy is implemented, its effectiveness is lost. In contrast, the top four leaders of the communist party make policy decisions in China. Indeed, it has been said that Chinese leaders come off as much more pragmatic and decisive politicians compared to their counterparts in India.

Also the role of central government in China and India is very different. The Indian system gives the central government power to override the local authorities completely. In China, the central authority is derived from the local authorities. Most of the development programs in China are in charge of the local authorities, and in India, the central government owns most of the development programs. Thus, when China implements any program, the same program will be already running in the local units and village communes.

To find such a situation in India is very difficult. It becomes difficult in India, since the local and regional political parties are engaged in the war of vested interest rather than national interest. The political environment is such that whenever the central government attempts to take bold decision the pressures exerted by the other political parties will take it back. It is such that India takes a one good step forward but simultaneously takes three steps backward under pressure. It shows that implementation is more important than policy formulation. The differences among the parties in India are visible clearly by the functioning of the present 24-party coalition government. However, the single party rule in China has its own problems. However, the difference lies in the fact that the government is able to take aggressive decisions as compared to India. This resulted in mixed reactions as on one hand, it has been able to implement reforms, on the other its also enabled development of social unrest.

Public Sector Reforms

China's industrial growth is directly attributable to the performance of the state-owned-enterprises (SOEs) both in the pre-reform and after-reform phases of economic development. In the pre-reform era, the state-owned enterprises were able to garner the support of the government as they were seen as vital pillars of industrial development. In a state of controlled economy, these SOEs have been catering to the needs of the society by creating employment opportunities, products for consumption and infrastructure development. The overall contribution of SOEs to GDP was to the tune of 70 percent before 1978. In the absence of a private sector, the entire requirements of the state and population were taken care by the SOEs.

Once the economy was opened up, managing the SOEs became a problem and proved burdensome for the state exchequer. In the pre-reform era, the prices of the products were under the government control and the functioning was to enable the supply rather than profit making. The SOEs received huge grants from the government to meet their daily-requirements. In a controled economy, these sort of entities made sense. However, in an open economy, where competitive environment and private participation is increasing, these SOEs turned out to be white elephants. China has taken an altogether different approach to solving the problem of SOE's and reducing the burden on the state exchequer. Instead of opting for an outright privatization of these entities, the reforms aimed at attending to the problem in a phased manner by breaking the problems into smaller components. Among the various measures introduced, the key initiatives were introduction of a job responsibility system, shifting profit-remittance system into value added tax system, replacing fiscal grants with bank loans, allowing joint ventures with private and foreign partners, separating administrative functions from management functions, allowing the SOEs to issue stocks and allowing the employees to take a share of the disinvestment proceeds. These measures ensured that the reforms of SOEs are carried out in a smooth way. This also gave the Chinese authorities an opportunity to reduce their adjustment costs and maintain low unemployment rates. The removal of government connection with the SOEs has reduced the bureaucratic interventions drastically and facilitated the transition of SOEs from a regulated to a more de-controlled market.

In India, the growth of the public sector was quite dominating in the early '60s and '70s like China. Instead of working for profit making and self-sustainance, the PSU's worked the other way round and thus become the white elephants of the Indian economy. To solve this problem, the Indian government has opted for out-right privatization. Without solving the problem of political intervention in privatization process, the government embarked on privatization. The miserable performance of the disinvestment commission is attributable to this fact. However, in the last fiscal the Indian government planned to raise Rs. 10,000 crore from the disinvestment process, it failed in all aspects. The bureaucratic process along with political interventions is making the problem more complex. China's privatization has been far more ambitious than India's and has made much progress in implementing it. In this aspect, the Indian policy architects should ensure that they adopt a similar approach to China and attempt at doing things rather than talking to ensure success of PSU reforms.

Private Sector

In the pre-reform era, private sector in China was non-existent. Whatever industries existed they were either state-owned or small-scale industries. However, with the opening of the economy and liberalization of trade rules, private sector has become one of the hottest sectors. The contribution of SOEs to GDP has come down from 70 percent in the early 1970s to around 25 percent today. This is largely due to the development of private sector. In order to move from a controlled regime to a market driven economy, the most important thing was to develop a competitive environment. Macro-economic policies were modified to suit changes at the micro-level. This has been carried on three broad lines:

- Introduction of competitive and market oriented policies by shifting to indirect control from direct control, replacement of physical indicators by monetary indicators, using economic reform measures to adjust economic activities.
- Optimization of economic structure and balance the interests of individuals, enterprises and government.
- Maintaining economic and social stability: low inflation and unemployment.

The industrial firm policy has been modified to lay emphasis on specialization as opposed to completeness in the pre-reform era. Investments were encouraged in light industries and the banking sector was effectively utilized to channel the high savings propensity of Chinese people to private sector. The establishment of the Commercial and Investment Bank of China was to effectively channel the savings of rural and urban population to private sector.

The most important factor that propelled the private sector is liberalization of rural industry in the form of Township and Village Enterprises, a mix of collective and privately owned enterprises in rural areas. Also the establishment of a number of special economic zones, coastal open cities, economic and technological development zones had fueled the export growth. These special areas received various kinds of favorable tax and regulatory treatment. In addition, the amendments to the related labor laws have made the labor markets more flexible. The tax laws were adequately amended to give incentives to private entrepreneurs. Those enterprises associated with special economic zones were exempt from taxation and the tax rates were kept low to attract private participation.

In contrast, the Indian government right from the beginning has opted for a private sector growth along with public sector growth. Though the private sector growth was emphasized, the industrial policy has always been restricting the private sector growth. Key industries were reserved for the public sector and investment has been a constraint. Also, the government failed to create adequate demand for the products due to majority of population reigning under poverty. To a large extent, this has proved detrimental for the success of private sector. Added to this were the subsidies that the government has put in place thereby indirectly controlling the prices of many products and services. Poor infrastructure and large-scale inequalities in the developments of regions also played its part.

It is only after the liberalization of the Indian economy, the government has opted to remove trade restrictions and reservations that prevailed. But due to the sudden spurt in the mid '90s in service sector, the emphasis was placed on the service sector. Today, the service sector contributes more than 45 per cent to the GDP. Since the balance shifted from manufacturing to service sector in the recent past, reforms were not able to induce adequate growth to the economy though the level of activity was sustained. The Indian government should ensure that the balance between agriculture, manufacturing and service is struck and all the three key sectors contribute to the GDP growth. Though India has also in place similar TVEs, in terms of small scale and village industries, they can be compared only structurally while there is a lot of difference operationally.

To facilitate the speedy growth of private sector, the government is making moves to amend the tax system. The intention of the government to reform the sales tax system is a move in the right direction. Replacing the sales tax by uniform value added tax ensures the removal of various trade barriers existing among different states. Till the time these complicated laws prevail in India, the private sector cannot be provided with a competitive environment that can draw a lot of private participation.

Infrastructure Development

One of the areas where China has clearly out-performed India is infrastructure development. The infrastructure made available in China today resembles what most developing nations offer to the industry. Adequate railroads, roadways, connectivity to ports, and most importantly the proximity of the special economic zones to the resources and markets have made the reform process a success. Initially, the state has been concentrating on the development of the eastern provinces since they were more near to international trade hubs like Hong Kong. Encouraged by the results, the focus has now shifted to the western provinces close to the Indian border to exploit trading opportunities in India and other European countries.

In the pre-reform era, the emphasis was on rural development as part of the agriculture reforms. With the formation of Communes, the initial infrastructure as roads and railways was developed along with irrigation projects. This was benefiting the Mainland in both agriculture and industrial growth. With constant upgradation of the internal infrastructure after the economic reforms, China has been able to provide a good infrastructure support to its industrial policies. Apart from it, the needs of infrastructure development are dealt at the local level rather than at the central level. This has enabled the local authorities to identify the needs and plan accordingly.

In India, even after 47 years of independence, the infrastructure is pathetic. The road networks are insufficient and are of poor quality. Power supplies are not adequate and irregular in most cases. Water resources are poorly managed and communication network is very bad. On the export front, many exporters complain about the Indian ports as they are inadequate and hence they have to trans-ship their exports through Singapore or Sri Lanka thereby increasing the cost of the goods. Though the state governments and central government are talking about infrastructure development, they are unable to revive the situation due to their fiscal constraints.

However, in order to make the infrastructure the critical driver, India has to go for a regulatory change just like what China did. This involves two major steps. First and foremost, regulatory and oversight responsibilities for almost all infrastructure should be devolved to state level. This enables speedy development of infrastructure as states have a better idea of their requirements rather than central government. Second, it should be provided by privately financed projects. An example is the growth of cable TV industry which has grown much faster without government subsidy or because of the absence of regulation. If private players are allowed to enter and compete, the quality of the infrastructure will also improve along with the overall development. Unless infrastructure development is properly attended and addressed, realizing the dreams of doubling India's per capita income in the next decade will remain as a dream.

Foreign Direct Investment

Foreign Direct Investment in manufacturing facilities is one of the key requirements for creation of manufacturing capability as well as transfer and infusion of new technology. Processing trade led by FDI has become one of the key competitive advantages of China. Ever since the economy was opened up, China has been able to constantly attract foreign investment. This is mainly due to the enormous pains that China took to modify their land laws, property laws, investment regulations and provision of attractive returns on investment. The major portion of FDI was routed through Hong Kong. Though Hong Kong became a part of China recently, FDI continue to be routed through it. In spite of this criticism, China has been able to attract high FDI inflows as compared to any other emerging nation.

James A Dorn, Vice President for Academic affairs and China Specialist, Cato Institute, says, "Capital inflows depend on expected returns, which are affected by the property rights framework and regulatory policies. FDI in China has been mostly by overseas Chinese who have invested in export processing industries on the coast." These investments enjoy tax concessions and other incentives. China offers super-national treatment to foreign firms, who typically receive an exemption of income tax for the first two profitable years followed by three additional years of 50 percent tax reduction. Such benefits are not offered to domestic Chinese firms. Also the institution of free trade export zones enabled China to attract more FDI flows than any other nation. These zones give the foreign investor 'islands' within which they can operate under favourable conditions, without requiring the wholesale national reform, which often proves politically challenging and time-consuming.

On the other hand, India has not been able to gain access to even one-tenth of China's FDI flows. Though the liberalization of economy has been done, the attractiveness seems to be missing in the Indian context. One of the major hurdles in FDI is the conversion ratio. India is very poor in converting the proposals to actual investment and the numerous bureaucratic hurdles have driven away the potential investors. The recent saga of Enron's Dabhol power project highlights the structural deficiencies to a large extent. This is certainly a de-motivating factor for the potential investors. Despite the availability of a huge market, natural resources and pool of skilled labor, the failure in attracting FDI flows is directly attributable to the state of government trade and openness policies, transparency in business and legal systems, excessive bureaucratic controls and infrastructure facilities. These are the broad reasons pointed out by a group of consultants from Harvard Institute of International Development and Boston Consulting Group.

Latin America

Brazilian Style of Development - Charting the Progress

We take a detailed look at Brazil's development because of its similarity to Indian conditions.

Between the 1930's and the 1980s Brazil experienced economic growth rates, which were among the highest in the world. Industrial development has led to this economic growth.

Some of the characteristics of Brazil's industrialization are as follows (Ferraz et al, 1992):

- 'Infant' and inward directed industrial experience.
- Concentration of the benefits of industrialization to a restricted social group.
- Reliance on external funding for investment.
- Presence of national and multi-national firms along with state owned industrial organizations.
- Strong role played by the state as the promoter of industrial development

Initially Brazil's industrial growth had a strong inward orientation with import substitution being the prime development strategy. Such a policy was adopted because of the presence of a large domestic market and the possibility of utilizing potentially exploitable natural and human resources.

In the 1950s, a shift in the policy took place, a state planning apparatus was created and an explicit policy of industrialization was introduced, which laid emphasis on infrastructural projects. Also efforts were undertaken to build manufacturing capability and move away from the so called 'screw driver' technology and knock down assembly operations. But the fruits of Brazil's drive for industrialization primarily benefited the upper and middle classes.

In the 1970s the investment in the economy was financed extensively by foreign capital in the form of direct investment by multi-national companies and by foreign borrowing. Brazil was considered as an attractive destination of foreign capital and between 1974 and 1980 the country was heavily engaged in expanding its infrastructure and basic industries but the interest rates increased before maturity of the loans and the country faced a severe debt crisis.

Another unique feature of Brazil's development is what is called 'associated capitalism'. Here industrialization proceeds like that in advanced industrialized nations in terms of sectoral distribution and division of ownership between local and foreign interests, but the execution of the development strategies lie firmly in the hands of the state. In industries like automobiles the foreign firms accounted for roughly 30% of production but the computer industry was mostly domestically owned.

Brazil's primary goods export was not able to generate enough capital, necessary for sustained development. Because of the large inter-dependence among various industrial sub-sectors for large industrial projects, the state had to intervene and became the instrument for financial centralization, developing the infrastructure and controlling production of raw materials and basic inputs.

Institutional protection (tariff and non-tariff) against foreign competition, financial incentives for the establishment of productive units and regulation of labour market are the main stays of Brazil's industrialization. State intervention was aimed at developing an emerging industrial base. Selected sectors were established and Brazil's industrial products started to be traded internationally. Proportion of capital goods and durable goods increased from 10% in 1963 to 19% in 1985.

The rapid growth of Brazil's manufacturing exports has been one of the most important phenomena of the country's external sector. The dynamic performance of Brazil's manufacturing exports in the first part of the 1990s contrasts sharply with the near-stagnation of these exports in the 1980s. Much of this growth was attributed to the devaluation in the real exchange rate, the resurgence of economic activity in the rest of Latin America, where most of Brazil's manufacturing exports are directed, and the opening of markets under the MERCOSUR integration process. Equally important is the reduction of barriers on imported goods that has increased domestic competition and produced a restructuring of the manufacturing industries. This liberalization process has shifted the composition of exports away from resource-intensive manufactures. Although the trade and exchange rate policies implemented under the 1994 stabilization program have had a dampening effect on the demand for Brazil's manufactures, it is possible that the industrial restructuring instigated by these policies has had a favorable effect on the country's manufacturing exports.

Brazil's diversification policies have been remarkably successful in augmenting the participation of manufactures in the country's total exports, in reaching new export markets, and in developing a broad product base for exports. The Brazilian model in terms of the factorial and technological contents of manufacturing export industries is also worth mentioning. While the country's endowment of natural resources is shared by other countries in the region, the Brazilian model has characteristics that are unique, as reflected for example in the export of products with particularly dynamic growth. At the same time, intra-industry trade has risen sharply and intra-firm trade has almost doubled from about 5 percent in the early 1980s to 9 percent in the first part of the 1990s. These developments are favorable for the country's future export prospects, and are supported by a high degree of adaptability by firms to changing conditions brought on by the trade liberalization process.

According to survey results, these firms became more concerned with reducing production costs than in the past, and they started favoring specialization of production activities for both the domestic and foreign markets. These conditions point to the likely reversal of the Brazil's poor export performance during the '80s and the possibility of a sustainable dynamism in the future performance of manufacturing exports.

During the period of high inflation of the 1980s, in spite of the adverse conditions, investment by the private sector was strong during that period and it eventually led to a resurgence of manufacturing exports in the first part of the 1990s. Although industrial productivity has been positive during most of the last 25 years, the performance among industries has varied greatly. Recent improvements have been shown to be related to technological advances associated with the restructuring of industries, which has been spurred by the liberalization of trade, increased competition from abroad, and the government's privatization program.

Brazil's protective policy known as the `market reserve policy' has excluded foreign competition from the domestic market for micro-computers and other information-processing goods and services. Although other countries have also treated their domestic computer industry as an infant industry, Brazil's policy assumed unprecedented proportions that distinguished it from the strategies followed by other countries.

Assessment of Brazil's microcomputer industry provides important insights into the infant industry policy in its extreme form. The analysis shows that the eventual demise of the microcomputer industry was due to

(a) An excessively ambitious policy for the input industries that served to raise the costs of microcomputer producers,

(b) The inefficient strategy adopted by policy makers to promote competition in the product market, and

(c) A regulatory structure that gave excessive power to too few.

Problems

One of the lacuna of this industrial policy was that effort was concentrated in setting up production capabilities, attention was not paid towards efficient functioning of the industries.

Studies carried out in 1980's revealed very limited use of advanced technologies, to manage production within acceptable quality levels, and modern stock control systems. Repitive inspection was used but source of problems were not identified, production efficiency was low and industrial engineering practices were neglected. But auto parts industry showed better performance because of customers' involvement in component development (similar to India, but in India, it may have occurred a little late)

Studies in 1990s also revealed low quality levels, low inventory turnover and high lead times. Extent of diffusion of factory automation was low but again auto parts industry was the largest user of NC machines. Although the intention to use automated technologies was there, there was serious impediment to diffusion of technology because of the restriction on import of IT products. In a survey carried out in early 1990's positive future expectations were shown by industries like chemicals and electronics, because firms in these industries invested in knowledge intensive assets.

The Brazilian Quality and Productivity Programme

It was launched in 1990 to tackle deficiencies in quality and productivity performance, which was threatening future competitiveness and was exerting a cost of around 40% of industrial GDP.

The five main components of the programme were:

- Awareness and motivation building
- A Programme of development and diffusion of modern quality and productivity related management methods
- Human resource development
- Provision of technical services and institutional support.

Individual sub-sectoral programmes were initiated to give more specialized assistance. The main benefit from the programme was increased awareness but the structural changes were already in place because of increased competition. The most concrete contribution of the programme was in its associated legislation, granting tax exemptions on the purchase of machinery and equipment and a favourable tax treatment with accelerated depreciation. This resulted in increased investment in more efficient machinery and led to industrial recovery.

Mexico

The move from import-substitution to outward-oriented policies has been widely accepted as a pre-requisite for countries that want their industries to be cost-effective and able to adopt new technologies that will enable them to compete in foreign markets. The results of studies suggest that import licensing reforms did indeed lead to increased manufacturing exports, thereby supporting the hypothesis that import liberalization has been an effective catalyst for Mexico's manufacturing exports.

According to conventional theory, Mexico's trade liberalization should have led to an expansion of those sectors in which the country has a comparative advantage. However, results show that the patterns of resource allocation and factor intensity following trade liberalization have not been consistent with the factor abundance hypothesis of international trade. Credibility about the permanence of those trade barrier reductions may have contributed to the outcome. This credibility problem was one of the motivating factors behind the Mexican government's decision to join NAFTA insofar as the binding international trade agreement helped to ensure that the trade reform process would be perceived as permanent. As a result of NAFTA, the trade liberalization process is now helping businesses to real-locate resources in a manner that more closely reflects Mexico's comparative advantages.

The recent transformation of Mexico's industrial sector from one based on a natural resource-intensive process to one encompassing high value-added, technology-intensive processes reflects the globalization strategies of multi-nationals and domestic industrialization efforts. The expansion of in-bound processing zones in Mexico called maquiladora operations have had a remarkable growth performance in the past 20 years, thereby helping the country to achieve a dynamic comparative advantage and support its liberalization process -- particularly under NAFTA. The fact that maquiladora exports currently account for about one-fifth of Mexico's manufacturing export earnings underscores the striking success of this type of operation as a catalyst for the outward-orientation of the Mexican economy. However, their expansion has not been trouble-free and problems have arisen with the enclave nature of their operations that have inhibited forward or backward linkages with domestic industries and diminished the impact of technology transfers throughout the economy.

Maquiladora

In the 1960s the Mexican Government set up specific in-bound zones, similar to the US Foreign trade zones. Within these zones, foreign companies can operate facilities called maquiladoras. Materials and products can be moved in and out of these zones nearly duty free. A maquiladora zone builds plants to clients' specifications, maintain roads, sewers and power lines and also provide employee services like recruiting, training and preparing pay rolls.

The maquiladora facilities result in lower total costs of production for the foreign investors. This has resulted in numerous twin plants - one in US and the other in Mexico. The Mexican government also provides incentives in the form of duty free entry of machinery and equipment to be used in the production of export goods, no restriction on the type of goods, produced for export and authorization to lease land and facilities under a thirty year beneficial trust arrangement.

But poor infrastructure particularly telecommunication facilities, strong labour unions pose problems for the investors.

Some Broad Policy Lessons

There are no blueprints for successful industrial policy to be drawn from East Asia. As in other areas, however, there are some broad lessons of relevance to national and international policy formulation.

One lesson is the 'big bang' approaches to liberalization, in which trade restrictions are withdrawn across-the-board, financial systems deregulated, and foreign investment controls abandoned, are unlikely to succeed. In much of sub-Saharan Africa and Latin America, labor-intensive industries have collapsed under the weight of competition from imports. Much of the textile industry in West Africa falls into this category. Meanwhile, the deregulation of finance and investment can have the effect of concentrating growth in areas - such as financial services - where employment linkages are weak.

The pace and sequencing of reforms is important. So, too, is the composition of employment and production in the sectors subject to liberalization. The case of Mexico illustrates in extreme form how unbalanced liberalization and reckless financial deregulation can concentrate the benefits of liberalization on the rich and the costs on the poor. More generally, trade liberalization cannot be pursued successfully in the absence of adequate social policies.

Countries such as Thailand, Indonesia, and China began a process of liberalization after they had invested heavily in attempting to raise the skills levels of their work forces. In Latin America, trade liberalization has been pursued without sufficient attention being paid to policies needed to improve productivity through education and skills training. Trade liberalization has often cost jobs rather than created opportunities.

Investment policy can play a crucial role in raising skill levels. In South Korea, foreign investment was tightly controlled in the interests of developing indigenous capacity. In Singapore, it was regulated to ensure that skills and technology were transferred. The danger now is that new international rules designed by the industrialized countries in the interests of powerful trans-national companies will limit the capacity of governments to regulate foreign investment in the public interest. Under the Multi-lateral Investment Agreement on Investment (MAI) drawn up by the OECD as the framework for a multi-lateral agreement, governments will be prevented from establishing controls in areas such as profit repatriation, technology transfer, local-content requirements (under which investors are required to source their operations from local firms), skills training, and balance-of-payments provisions (i.e. controls requiring foreign investors to cover the costs of their own imports). These are the very controls which have enabled East Asian countries to successfully use foreign investment for the creation of long-term growth and employment. Surrendering sovereignty in this area will carry the threat of a wave of anarchic-capitalism, in which governments effectively lose the capacity to regulate economic life in the interests of their citizens.

Export opportunity is an often neglected component of the East Asian success story. The first generation of 'tiger' economies emerged during a period in which world trade was expanding and becoming increasingly open. Today, many of the poorest countries face a bewildering array of tariff barriers. Under the Uruguay Round these barriers were reduced. In areas such as textiles, leather, oilseeds, and other agricultural goods, import barriers remain a powerful deterrent to diversification. Removing that deterrent would do much to enable poor countries to trade their way into recovery.

International action is also needed to remove the mill-stone of foreign debt, which has been happily absent from the collective necks of the East Asian countries. The burden is most serious in sub-Saharan Africa. If

South Africa is excluded, the region's debt-to-export ratio is 327 per cent, compared to a World Bank sustainability ceiling of 200-250 per cent. High levels of debt overhang, which this figure reflects, act as a major deterrent to investment. Meanwhile, debt servicing - amounting to around $10bn annually - has the effect of reducing the foreign exchange available for essential imports, hampering the competitiveness of local industries. The Highly Indebted Poor Countries (HIPC) debt initiative could act as a spur to growth by reducing debt overhang and increasing import capacity. However, it provides insufficient debt relief over an unacceptably long time-frame. A further problem is the requirement that potentially eligible countries comply with IMF stabilization programmes, which have a poor track record in restoring growth and investment.

10

CHAPTER TEN

Framework for Proposed Manufacturing Policy

MICHAEL Porter in his book "The Competitive Advantage of Nations", reports on the results of a massive cross-national study of internationally competitive industries. In his study he has identified what it takes to succeed in the international markets and how can companies achieve competitive advantage. He also throws light on the possible roles that government can play in the development of technological and manufacturing capability. We therefore present a discussion of the insights provided by Michael Porter and then propose a structure and framework for the manufacturing policy of India to establish a competitive advantage globally. The issue of national competitiveness is examined below.

What is National Competitiveness?

National competitiveness has become one of the central pre-occupations of government and industry in every nation. Yet for all the discussion, debate and writing on the topic, there is still no persuasive theory to explain national competitiveness.

Some see national competitiveness as a macro-economic phenomenon, driven by variables such as exchange rates, interest rates and government deficits. But Japan, Italy, and South Korea have all enjoyed rapidly rising living standards despite budget deficits; Germany and Switzerland despite appreciating currencies; and Italy and Korea despite high interest rates.

Others argue that competitiveness is a function of cheap and abundant labor. But Germany, Switzerland and Sweden have all prospered even with high wages and labor shortages.

Another view connects competitiveness with bountiful natural resources. But how then can one explain the success of Germany, Switzerland, Italy and South Korea-Countries with limited natural resources?

More recently the argument has gained favour that competitiveness is driven by government policy; targeting, protection import substitution, and subsidies have propelled Japanese and South Korean auto, steel, shipbuilding, and semi-conductor industries into global pre-eminence. But a closer look reveals a spotty record. In Italy, government intervention has been ineffectual-but Italy has experienced a boom in world exports share next only to Japan. In Germany, direct government intervention is rare. And even in Japan and South Korea, government's role in such important industries as facsimile machines, copiers, robotics and advanced materials has been modest; some of the most frequently cited examples such as sewing machines, steel and shipbuilding are now quite dated.

Clearly none of these explanations is fully satisfactory; none is sufficient by itself to rationalize the competitive position of industries within a national border. Each contains some truth; but a broader, more complex set of forces seems to be at work.

The only meaningful concept of competitiveness at the national level is **productivity**. The principal goal of a nation is to produce a high and rising standard of living for its citizens. The ability to do so depends on productivity with which a nation's labor and capital are employed. Productivity is the value of the output produced by a unit of labor or capital. It depends on both the quality and features of products (which determines the price that they can command) and the efficiency with which they are produced.

PORTER'S DIAMOND OF NATIONAL ADVANTAGE

Why are certain companies based in certain nations capable of consistent innovation? Why do they ruthlessly pursue improvements, seeking an ever more sophisticated source of competitive source of competitive advantage? Why are they able to overcome the substantial barriers to change and innovation that so often accompany success?

The answer lies in four broad attributes of a nation; attributes that individually and as a system constitute the diamond of national advantage, the playing field that each nation establishes and operates for its industries. These attributes as represented in the figure below are:

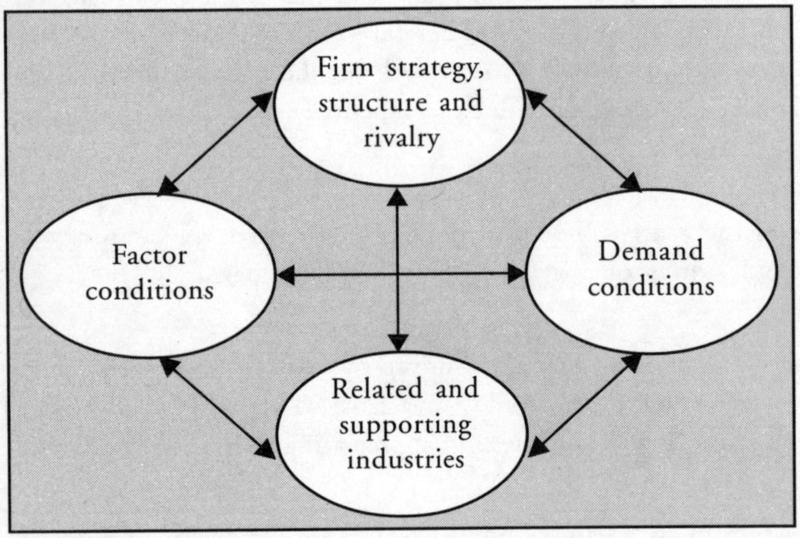

1. Factor Conditions: The nation's position in factors of production, such as skilled labor or infrastructure, necessary to compete in a given industry. In the sophisticated industries that form the back-bone of nay advanced economy, a nation does not inherit but instead creates the most important factors of production-such as skilled human resources or a scientific base. Moreover, the stock of factors that a nation enjoys at a particular time is less important than the rate and efficiency with which it creates, upgrades, and deploys them in particular industries. Nations succeed in industries where they are particularly good at factor creation. Competitive advantages results from the presence of world class institutions that first create specialized factors and then continually work to upgrade them. Denmark has two hospitals that concentrate in studying and treating diabetes- and a world-leading export position in insulin. Holland has a premier research institute in the cultivation, packaging and shipping of flowers, where it is the world's export leader.

2. Demand Conditions: The nature of home-market demand for the industry's product or service. Local buyers can help a nation's companies gain competitive advantage if their needs anticipate or even shape those

of other nations – if their needs provide ongoing "early warning indicators" of global market trends. Sometimes anticipatory needs emerge because a nation's political values foreshadow needs that will grow elsewhere. Sweden's long-standing concern for handicapped people has spawned an increasingly competitive industry focused on special needs. Denmark's environmentalism has led to success for companies in water-pollution control equipments and wind-mills.

3. Related and supported Industries: The presence or absence in the nation of supplier industries and other related industries that are internationally competitive. The third broad determinant of national advantage is the presence in the nation of related and supporting industries that are internationally competitive. Internationally competitive home-based suppliers create advantages in downstream industries in several ways. First, they deliver the most cost-effective inputs in an efficient, early, rapid and sometimes preferential way. Italian gold and silver jewelry companies lead the world in that industry in part because other Italian companies supply two-thirds of the world's jewelery making and precious-metal recycling machinery.

4. Firm Strategy, Structure and Rivalry: The conditions in the nation governing how companies are created, organized and managed, as well as the nature of domestic rivalry. The presence of strong rivals is a powerful stimulus to the creation and persistence of competitive advantage. This is true of small countries, like Switzerland, where the rivalry among its pharmaceutical companies, Hoffman-La Roche, Ciba-Giegy, and Sandoz, contributes to a leading worldwide position. It is true for the US in the computer and software industries. Another benefit of domestic rivalry is the pressure it creates for constant upgrading of the sources of competitive advantages. The presence of domestic competitors automatically cancels the types of advantage that comes from simply being in a particular nation-factor costs, access to or preference in the home market, or costs to foreign competitors who import into the market. Companies are forced to move beyond them, and as a result, gain more sustainable advantages.

Implications for Manufacturing Policy

Manufacturing policy of a country ultimately has to serve the objective of increasing the competitiveness of the country's products in the global market through higher productivity, innovation, lower costs and better quality. Such a policy cannot be created in isolation by the government and Industry actors' (manufacturers) also need to play their part of the role in order to achieve this goal. Thus our proposed manufacturing policy addresses the issue at two levels: - the role the government can play and what the industry players should perform. This is a view taken very strongly by Michael Porter also and the following sections present a brief outline of the different acts each of them should perform:

Role of the government

Government's proper role is as a catalyst and challenger; it is to encourage or even push companies to raise their aspirations and move to higher levels of competitive performance, even though this process may be inherently unpleasant and difficult. Government cannot create competitive industries; only companies can do that. Government plays a role that is inherently impartial, that succeeds only when working in tandem with favourable underlying conditions on other fronts. Government policies that succeed are those that create an environment in which companies can gain competitive advantage rather than those that involve government directly in the process, except in nations early in development process. It is an indirect, rather than a direct role.

Government should abstain from making mistakes like attempting to manage the industry structure, protecting the market too long, and yielding to political pressure to insulate inefficient retailers, farmers, distributors and industrial companies from competition. It might be tempting to take this route as it often takes more than a decade for an industry to create competitive advantage; the process entails the long upgrading of human skills, investing in products and processes, building clusters, and penetrating foreign markets. In the case of the Japanese auto industry, for instance, companies made their first faltering steps towards exporting in the 1950s-yet did not achieve strong international positions till the 1970s.

There are some simple, basic principles that governments should embrace to play the proper supportive role for national competitiveness: encourage change, promote domestic rivalry, and stimulate innovation. Some of the specific policy approaches to guide nations seeking to gain competitive advantage include the following.

- Focus on specialized factor creation: Government has critical responsibilities for fundamentals like the primary and secondary education systems, basic national infrastructure, and research in areas of broad national concern such as healthcare. Yet these kinds of generalized efforts at factor creation rarely produce the competitive advantage. Rather, the factors that translate into competitive advantage are advanced, specialized, and tied to specific industries or industry groups. Mechanisms such as specialized apprenticeship programs, research efforts in universities connected with industry. Trade association activities and most important, the private investments of companies ultimately create the factors that will yield competitive advantage.
- Enforce strict product safety and environment standards: Strict government regulations can promote competitive advantage by stimulating and upgrading domestic demand. Stringent standards for product performance, product safety, and environmental impact pressure companies to improve quality, upgrade technology and provide features that respond to consumer and social demands. Easing standards, however tempting, is counter-productive. When tough regulations anticipate standards that will spread internationally, they give a nation's companies a head start in developing products and services that will be valuable elsewhere. Sweden's strict standards for environmental protection have promoted competitive advantage in many industries. Atlas Copco, for example, produces quiet compressors that can be used in dense urban areas with minimal disruption to residents. Strict standards, however, must be combined with a rapid and streamlined regulatory process that does not absorb resources and cause delays.
- De-regulate Competition: Regulation of competition through such policies, as maintaining a state monopoly, controlling entry into an industry, of fixing prices has two strong negative consequences; it stifles rivalry and innovation as companies become pre-occupied with dealing with regulators and protecting what they already have ; and it makes the industry a less dynamic and desirable buyer or supplier. De-regulation and privatization on their own, however will not succeed without vigorous domestic rivalry-and that requires, as a corollary, a strong and consistent anti-trust policy.

Role of the Industry: Organizational Contribution

Ultimately only companies themselves can achieve and sustain competitive advantage. They must realize the central role of innovation and the uncomfortable truth that innovation grows out of pressure and challenge. It takes leadership to create a dynamic, challenging environment. And it takes leadership to recognize the all-to easy escape routes that appear to offer a path to competitive advantage, but are actually short-cuts to failure. For example, it is tempting to rely on cooperative research and development projects to lower the cost and risk of research,. But they can divert company attention and resources from proprietary research efforts and will all but eliminate the prospects for real innovation. Some kinds of company policies that will help are:

- Create pressures for innovation: A company should seek out pressure and challenge, not avoid them. Part of strategy is to take advantage of the home nation to create the impetus for innovation. To do that, companies must sell to the most sophisticated buyers and channels; seek out those buyers with the most difficult needs; establish norms that exceed the toughest regulatory hurdles or product standards; source from the most advanced suppliers; treat employees as permanent in order to stimulate upgrading of skills and productivity.

- Seek out the most capable competitors as motivators: To motivate organizational change, capable competitors and respected rivals can be a common enemy. The best managers always run a little scared; they respect and study competitors. To stay dynamic, companies must make meeting challenge a part of organization's norms. For example, lobbying against strict product standards signals the organization that company leadership has diminished aspirations. Companies that value stability, obedient customers, dependent suppliers and sleepy competitors are inviting inertia and ultimately, failure.
- Improve the National diamond: Companies have a vital stake in making their home environment a better platform for international success. Part of a company's responsibility is to play an active role in forming clusters and to work with its home nation buyers, suppliers and channels to help them upgrade and extend their own competitive advantages. To upgrade home demand, for example, Japanese musical instrument manufacturers led by Yamaha, Kawai and Suzuki have established music schools. Similarly, companies can stimulate and support local suppliers of important specialized inputs–including encouraging them to compete globally. The health and strength of the national cluster will only enhance the company's own rate of innovation and upgrading.

In nearly every successful competitive industry, leading companies also take explicit steps to create specialized factors like human resources, scientific knowledge, or infrastructure. In industries like wool cloth, ceramic tiles, and lighting equipment, Italian industry associations invest in market information, process technology, and common infrastructure. Companies can also speed innovation by putting their head-quarters and other key operations where there are concentrations of sophisticated buyers, important suppliers, or specialized factor creating mechanisms, such as universities or laboratories.

11

CHAPTER ELEVEN

Manufacturing Policy Draft

Recommendations for Government Policy

The proposed manufacturing draft in the following sections draws heavily both from the international experiences detailed earlier as well the framework listed in the previous section. The proposed reforms and actions required are segregated into different sections i.e.:

- Changes in the Regulatory system
- Labour reforms
- Infrastructural Requirements
- Industrial Cluster development
- Recommendations relating to the role of Ministry of Industry
- Recommendations for taxation laws
- Organizational Issues
- Institute - Industry Linkages
- Recommendations specifically for the SME sector

Regulatory System Reforms

Indian industry remains highly regulated by a powerful bureaucracy armed with excessive rules and broad discretion. As the pace of regulatory reform at the federal level accelerates, the focus of liberalization is gradually shifting to state governments that, under India's federal system, enjoy broad regulatory powers. The speed and quality of regulatory decisions governing important issues such as zoning, land-use and environment can vary dramatically from one state to another. Opposition from labour unions and certain political parties has slowed or halted important regulatory reforms governing areas such as labour, bankruptcy, and company law that would enhance the efficiency of private investment. Some broad recommendations that need to be implemented in this regard are discussed below:

Boosting the FDI

1. A core group needs to be set up at the center which would spearhead FDI investments in India's Manufacturing Sector. The core group would act as the single window clearance point

for all FDI investments into the country. It would comprise of representatives of the Industry, Government, and the Industry experts. This core group would serve as the apex body on all policy related issues and its working would not be dictated by the political masters. It would thus be able to take a long term perspective on all manufacturing related issues rather than being guided by short term political gains.

2. The core group at the center would do all the processing needed for the investments into the country. However there would be representatives of the state closely working with the core group at the center. The core group would leave it to the client to decide which state he wants to locate the industry in. At the state level this structure would be replicated. For all the clearances at the state level, the client would need to interact only with the state core group and not the departments within the state. The whole process would ensure that for the potential investors FDI proposal is processed in a very smooth manner.

Obtaining Maximum benefits from FDI

Foreign firms, especially TNCs, will shift their mobile assets across the globe to match them with the best immobile assets of host countries. The ability to provide the best immobile assets thus becomes a critical aspect of FDI policy. These include a world-class infrastructure, skilled labour, innovation capacity, and the agglomeration of efficient suppliers, support institutions and services.

FDI should be used as a vehicle for the development of new industries and for maximizing technology development. Due to lack of information, high transaction costs and high risks involved, the government can undertake the crucial role of targeted investment promotion to stimulate technology-based medium-scale foreign companies to cooperate with suitable domestic partners to promote technology-based supplier industries in such areas as capital goods, component manufacture and other high technology industries. Foreign investment was encouraged, but channeled into high value-added areas such as electronics, precision machinery etc. Foreign investors should also be required to meet standards for training local staff and transferring technologies.

The provinces must avoid an incentive race to raise financial incentives, reduce taxes or lower labour and environmental standards in their eagerness to attract FDI. The provincial governments should intensify cooperation among them to avoid being played one against the other by potential domestic and foreign investors. In addition, the central government's role will be crucial in coordinating their individual industrial policies to achieve maximum benefits from their relative comparative advantages and local resources. It should start by identifying regional potentials, specializations and priorities taking into account economies of scale, and plan the creation of industrial clusters and the provision of relevant information regarding business opportunities. It should also prepare common minimum national standards regarding labour conditions and the environment, and harmonize investment incentives, particularly concerning tax holidays and other fiscal incentives.

Encouraging the domestic manufacturing industries

An individual wishing to set up an industry has to get clearances from so many different departments ranging from electricity to water to pollution. We propose that a manufacturing cell be set up at each state level, which would handle all issues related to setting up of a manufacturing industry within the state. This cell would serve as the single point of contact for all approvals related to setting up of manufacturing industry within the state.

Special emphasis should be laid on industries which would help in promoting Downstream /Supporting industries.

Efforts must be taken to avoid the undue harassment by the inspectors of sales tax, pollution board, labour, water etc, who expect the small scale industries to buttress them at regular intervals. There should be prompt redressal against all bureaucratic oppression.

Labour Reforms

Archaic labour laws represent one of the most important inhibitors constraining industrial development and manufacturing growth. The specific actions related to labour which are recommended are:

Recommendations for Labour policy changes

The starting point of our discussion of labour policy change for India must be the fundamental principles upon which those policies are based.

1. **Employment laws should be realistic:** These laws must recognize the new ways in which work is being done so that they facilitate new work arrangements and enable India to remain competitive. They should also recognize the practical limitations of efforts to pursue social policy goals through the imposition of obligations on employers.
2. **Employment laws should be flexible:** They should allow businesses to adapt to the needs of their customers and the aspirations of their employees, and to provide the kind of innovative solutions to problems that make businesses and employees successful and capable of providing jobs.
3. **Employment laws should respect individual choices:** They should allow employees to make up their own minds about what kind of workplace arrangements best suit their needs.
4. **Employment laws should promote voluntary cooperation between employers and employees:** Employers and employees cannot succeed without each other's cooperation, and our employment laws should encourage (though they cannot require) the solution of problems through mutual agreement.
5. **Employment laws should be fair for all:** The laws themselves should not give special advantages to anyone and they should be administered in a manner that contributes to the confidence the community has in the essential fairness of the laws.
6. **Employment laws should provide a basic standard of protection for employees:** The laws should set a floor below which employment standards may not fall, to ensure a minimum level of employee protection.

Keeping this framework in mind, the following recommendations are provided:

Wages

- There should be one minimum wage fixed at the national level; state level minimum wages should be derived from this national minimum wage by using appropriate price indices. The minimum wage should be made payable to all unskilled workers irrespective of age and sex.
- The national minimum wage should be so fixed as to permit a worker to earn an income equivalent to the official poverty-line. Thus the minimum wage for agricultural laborers should be fixed with reference to the poverty line for rural areas and the minimum wage for industrial workers should be fixed with reference to the poverty line for urban areas.
- The national minimum wage should be fully indexed. The appropriate price indices are the

Consumer Price Index for Agricultural Laborers for rural areas and the Consumer Price Index for Industrial Workers for urban areas. Full indexation will obviate the need for frequent revisions.

- No separate minimum wage for the organized sector should be fixed by the Government. Determination of this can be left to the collective bargaining process.
- The interval between wage revisions should be long enough for productivity trends to be discernible; an interval of 5 years is suggested here. Wage revisions can then be based on the observed productivity trends over the last 5 years. This means that attention will need to be focused on product-wages in times of revision. Productivity in this context should be interpreted to mean productivity in a particular branch of industry.
- Wage revisions in industry should precede and guide pay revisions for service sector employees in general and government employees in particular. The reason is that it is virtually impossible to measure productivity of government employees. Even in non-governmental services, measurement of productivity is extremely difficult.
- Bonus should be viewed as a profit sharing scheme. A certain fixed percentage of profit, for example, can be set aside for bonus payment. If profit is zero or negative, there should be no obligation for an enterprise to pay a bonus. This system will introduce some amount of flexibility of labour costs without rendering labour incomes unduly unstable. However, standardization and transparency of enterprise-level accounting will be necessary for the system to be workable.

Employment Security and Labour Market Flexibility

- The reform program has to be so designed as to ensure that the exploitative labour relations of the past do not re-emerge. Essentially, what is needed is a shift of focus from issues relating to employment security to issues relating to income security. Employment security has so far been viewed as the only instrument available for guaranteeing income security. But the two can be de linked; a system of income security which does not involve employment security is quite conceivable.
- Changes in regulations designed to grant employers unrestricted freedom to retrench labour, provided that the rules relating to notice period and compensation payment are abided by, should be viewed as the medium-term objective of a reform program.
- Development of an income security system must precede changes in regulations and this is why changes in regulations have to be viewed as feasible only in the medium term; they should constitute the last step in the reform process.
- In the short run, the Industrial Employment Act should be suitably amended so as to facilitate transfers from one job to another within an enterprise or within a group of enterprises under common ownership or management. This would encourage employers to re-train and re-deploy.
- Efforts to develop an appropriate income security system are the critically important initial steps in the reform program. Such a system should incorporate three elements: payment of unemployment benefit to retrenched workers for a specified period; retraining of retrenched workers so as to enable them to acquire new skills (which may be needed either for self-employment or for re-employment as wage workers in organized industries); and active assistance to retrenched workers in job search.

- As regards payment of unemployment benefit to retrenched workers, two important issues need to be addressed. The first issue concerns financing; recurrent contributions from both employers and the government will be necessary. The second issue relates to specifying the period over which a retrenched worker will be entitled to unemployment benefit. Clearly, there has to be a linkage between the period over which a retrenched worker had been employed and the period over which he receives unemployment benefit.
- It is clear that new institutions would need to be built in order to develop an income security system. An integrated network of institutions, managed jointly by the government, trade unions and employers' organizations, would be ideal. The possibility of developing the network of institutions, currently being promoted under the National Renewal Fund, along the lines suggested above need to be seriously examined.
- When an income security system has been established and changes in regulations can be contemplated, the rules relating to notice period and compensation payment will need to be reviewed. For, these rules themselves might require alterations. Compensation payment of 15 days' wages for each year of service, for example, is widely thought to be inadequate.

Expanding Scope of Labour Redundancy and the National Renewal Fund (NRF)

The organized sector in India accumulated surplus labour over the years partly because the public sector viewed job creation as a social responsibility and partly because non-viable enterprises were kept going through subsidization in order to prevent loss of employment. The economic reforms have both undermined the logic of such policies and weakened the capacity of the government to provide subsidies. The hidden surplus labour, therefore, threatens to become openly redundant labour. Within this context, the NRF has been established to facilitate gradual and systematic reduction or elimination of the surplus labour while providing a safety-net for the affected workers.

The NRF programs include voluntary retirement schemes, programs for counseling, retraining and redeployment of the workers accepting voluntary retirement and area regeneration schemes for areas with high incidence of potential redundancy. So far, the programs have targeted only industrial enterprises owned by the Central Government. These account for less than 9 per cent of all redundant labour in the organized sector and 35 per cent of all redundant labour in organized industries.

There are three basic reasons for the poor performance of this program. First, the government has had no prior experience of implementing such programs. Second, it was implicitly and incorrectly assumed that the workers accepting voluntary retirement could be reabsorbed into organized industries. Thus while the different components of the NRF programs needed to be viewed as a package and implemented simultaneously, voluntary retirement schemes were actually implemented even before appropriate institutions for implementing the other programs could be developed. Finally, funds allocated for the NRF programs have been far too inadequate.

We therefore recommend that the government should actively involve the private sector in the functioning of the program and expand the scope to all the affected workers. This would also help in creating a virtual "safety net" for the labour force in the country.

Infrastructure Facilities

Presence of world class infrastructure is one of the basic pillars needed for manufacturing excellence. The earlier sections have portrayed the deplorable position of India and urgency of actions in this regard. The issue of infrastructure is a much debated one and improvements are required in all the aspects, we for this policy draft focus exclusively on the recommendations for the power sector primarily for the following reasons:

- The criticality of power for any kind of manufacturing activity
- The large scale mis-management on this front and distortions present.
- The extent to which government initiatives can impact the over all-situation.

A brief discussion of what is ailing the Industry is presented followed by the recommendations.

Power

High transmission and distribution loss, low utilization of installed capacity, high manpower costs and distorted tariff are the factors contributing to the sub optimal performance in the power sector.

Agricultural and domestic consumers consumed nearly one half of the energy supplied in 1998/99. Both consumer categories are heavily under-charged. On an average, supply to agriculture recovered only 12 per cent of the average cost of supply while supply to domestic consumers recovered only 54 per cent of the average cost of supply. Conversely, supply to industry recovered 122 per cent of the average cost of supply. If revenue realization is compared to the marginal cost of supply for these categories of consumers, the distortions become even more extreme.

Electricity prices for industry in India are higher than most developed economies, excluding Japan, and several developing countries. This distorted tariff pattern in India has been designed to cross subsidize domestic and agricultural supply by charging industry more than the average cost of supply. In most SEBs, the tariff for industry exceeds even the marginal cost of supply. This has resulted in self-generation being increasingly adopted by industry as a cheaper and more reliable source of power. But this does not solve the problems faced by the industry in general and further erodes the financial viability of the SEBs.

Recommendations for Power Sector

Tariff Rationalization

The method to be followed for determination of the Revenue Requirement of the utilities should be given utmost importance. The present existing Rate of Return method needs to be bolstered by elements of Performance Based Ratemaking.

The extent of under-realization of cost on the average requires an average increase of between 30 to 50 percent for the utility to break even at current efficiency levels. The impact of this increase has to be borne by customer classes, like agriculture and domestic, which have been traditionally subsidized. There is little scope for continuing to bleed the industrial sector and large consumers. The tariffs for industry and large consumers have increased over the years far beyond the cost of supply. With the overall demand supply imbalance, the quality of supply has steadily deteriorated. Voltage and frequency fluctuations are at an extreme level and cause significant though hitherto unquantified loss. Hence the move towards bringing agricultural tariffs to costs is welcome.

Tariff Design

Innovative tariff design which provides incentives for agricultural consumers, to install meters, through a preferential rate for metered supply as compared to un-metered supply is needed. This has been adopted by Gujarat. This was done by designing the flat HP based tariff in a manner that it translates into a higher per kWh charge as compared to the metered tariff. It can be used in a wide scale across the states.

Unbundling retail tariffs by allocating costs to the components of energy service is another exercise which needs to be explored. Customers need to be charged separately for the following services:

1. Bulk power purchase
2. Transmission system costs including transmission loss.
3. Distribution system cost including distribution loss.
4. Retail supply cost including metering, billing, collection and customer servicing.

Under each of the four major cost heads, costs can be further sub-classified as fixed or variable and charged for accordingly.

In comparison most retail tariffs today are single part and do not distinguish between the four separate components of retail supply. Charging these separately has the following advantages.

- It forces utilities to account for these costs separately which builds the data base needed in future for unbundling these services.
- Unbundling costs focuses the attention of stakeholders on those costs, which they can control. Generally customers have no control over the fixed costs of the utility. They can avoid or reduce only the variable costs. From the regulators perspective, fixed costs are inert costs which can be changed only once at the time that they are incurred and hence can be ignored beyond that stage. What is more vital from the view point of bringing about short run efficiencies are the variable costs which are changeable and hence controllable.

Performance Based Rate Determination

The PBR approach, while recognizing the revenue requirement of the utility, provides incentives for improving efficiency and reducing costs. The control aspects of regulation are thus sought to be replaced with a system of incentives and penalties through institution of industry wide norms. As a result, the PBR creates incentives, which are similar to those that would exist in a competitive market place. Such a system would reward efficient management while inefficient ones would sooner or later be thrown out. PBR also provides advance signals which efficient utilities can use to optimize operations. International experience in regulatory reform and implementation of PBR schemes suggest that significant efficiencies in investment are possible.

Encouragement for Alternative Sources of Power

Industries using alternative sources of power, particularly those in remote areas should be given incentives. Similarly firms capable of setting up captive power plants on their own should be encouraged but a ceiling can be put on power consumption. Incentives can be in the form of subsidy on the equipment used for power generation.

Natural gas is India's most important potential alternative to coal. There is need for widespread use of natural gas in power generation and in the industrial and residential sectors. Projected demand will require large volumes of gas pipeline and liquefied natural gas (LNG) imports.

Industrial Cluster Development

What are Clusters?

A cluster is a sectoral and geographical concentration of enterprises faced with common opportunities and threats which gives rise to external economies, favours the emergence of specialized infrastructure and services, and enables cooperation among public and private local institutions to promote local production, innovation and collective learning.

Some distinguishing features that industrial clusters should have are as follows-

- Geographical proximity of SMEs,
- Sectoral specialisation,
- Predominance of small and medium sized firms,
- Close inter-firm collaboration,
- Inter-firm competition based on innovation,
- A socio-cultural identity which facilitates trust,
- Active self-help organizations, and
- Supportive regional and municipal government.

Why Cluster Development?

Co-location of a large number of small scale manufacturing units boosts the effectiveness of policy programs targeted at manufacturing development because of the economies of scale and concentration advantages. The similarity of needs and support requirements, speeds up the dissemination of best practices because of the pervasiveness of demonstration effects, and allows for a distribution of the fixed costs of interventions among a large number of beneficiaries. This holds true for under-achieving clusters as well as for the best performing ones.

Sound economic reasons have been identified for the cause of development of clusters (Krugman 1991). Units operating in clusters derive a clear competitive advantage from:

- The proximity to sources of raw inputs,
- The availability of suitably customised business development services (BDS),
- The abundance of clients attracted by the cluster tradition in that industry, and
- The presence of a skilled labour force.

The Current Status in India

Within the SSI sector, an important role is played by the numerous clusters that have been in existence for decades and sometimes even for centuries. According to a UNIDO survey of Indian SSI clusters undertaken in 1996, there are 350 SSI clusters and approximately 2000 rural and artisan based clusters in India. It is estimated that these clusters contribute 60% of the manufactured exports from India.

Among the larger clusters, some worth mentioning are:

- Panipat accounting for 75% of the total blankets produced in the country;
- Tirupur, which is responsible for 80% of the country's cotton hosiery exports;
- Agra with 800 registered and 6,000 unregistered small scale units making approximately 150,000 pairs of shoes per day with a daily production value of 1.3 million dollars and exports worth US $ 60 million per year;
- Ludhiana, a city that is well known as the Manchester of India, which alone contributes 95% of the country's woolen knitwear, 85% of the country's sewing machines and 60% of the nation's bicycle and bicycle parts.

Due to better visibility in the form of machines, equipment and technologies, technology modernization has been the most sought after objective of cluster development in India. Technology Modernisation programme of SIDBI aims at modernizing the cluster through technology development, so that other firms can replicate. SBI Uptech programme, on the other hand had relied mostly on proven technologies that are easily demonstrable and replicable. Thus most of the technologies used are not radically new but upgradation of existing ones. SBI also ensures attention to specific areas like energy savings, quality control and pollution control and product mix.

UNIDO Cluster Development in India

Currently, UNIDO is implementing a project in India with the aim of developing capabilities at both the local and the national levels so to promote SSI networking and cluster development. This is done by:

- Assessing the competitiveness and organization of SSI clusters
- Assisting the clusters' actors (suppliers of raw materials, plant & machinery, buyers of SSI goods and services, testing laboratories and research agencies, industrial associations, training institutions, local government, financial institutions) in developing a common vision of what their cluster can achieve in national as well as international markets.
- Building up (through training, workshops and study tours), the capacity of cluster actors to implement such a vision.
- Providing advisory services at the policy level.

At the cluster level, the Focal Point forges working partnerships with one or more institutions (NGOs, producers' consortia/associations, providers of business development services). This is done to facilitate the implementation of various initiatives and to pass on the competence acquired as a result of UNIDO training and hands-on experience gained in the pilot cluster. The programme draws strength through collaborative support from national developmental institutions like Small Industries Development Bank of India (SIDBI) and recently office of the Development Commissioner, Small Scale Industry (DCSSI).

The implementation of the programme is monitored by a Steering Committee (chaired by Development Commissioner, Small Scale Industries, the Ministry of Small Scale Industries), whose task is also to strengthen the synergy between this programme and other Government initiatives aimed at the development of small-scale enterprises.

The challenge is how to make cluster development, an integral component of the overall manufacturing policy of the country and to extend the activities to more and more clusters.

Key Concerns & Issues

Despite the achievements mentioned above, the majority of the Indian SSI clusters share significant constraints like:

- Technological obsolescence,
- Relatively poor product quality,
- Information deficiencies,
- Poor market linkages and inadequate management systems.

Moreover, with the Indian economy on the path of liberalization, all SSI clusters (even the best performing ones) are increasingly feeling the competitive pressures coming from the international markets. Hence the existing mechanism of intervention and efforts towards industrial clusters needs a significant revamp.

Some other specific challenges for the industrial clusters in India are:

- Most of the cluster based enterprises are based in unauthorized locations for industries and operate from the entrepreneur's homes. Thus such units have problems in expansion and modernization due to space constraints and may also be denied registration as SSI units and thus access to credit and support facilities.

UNIDO's Cluster Development Activities

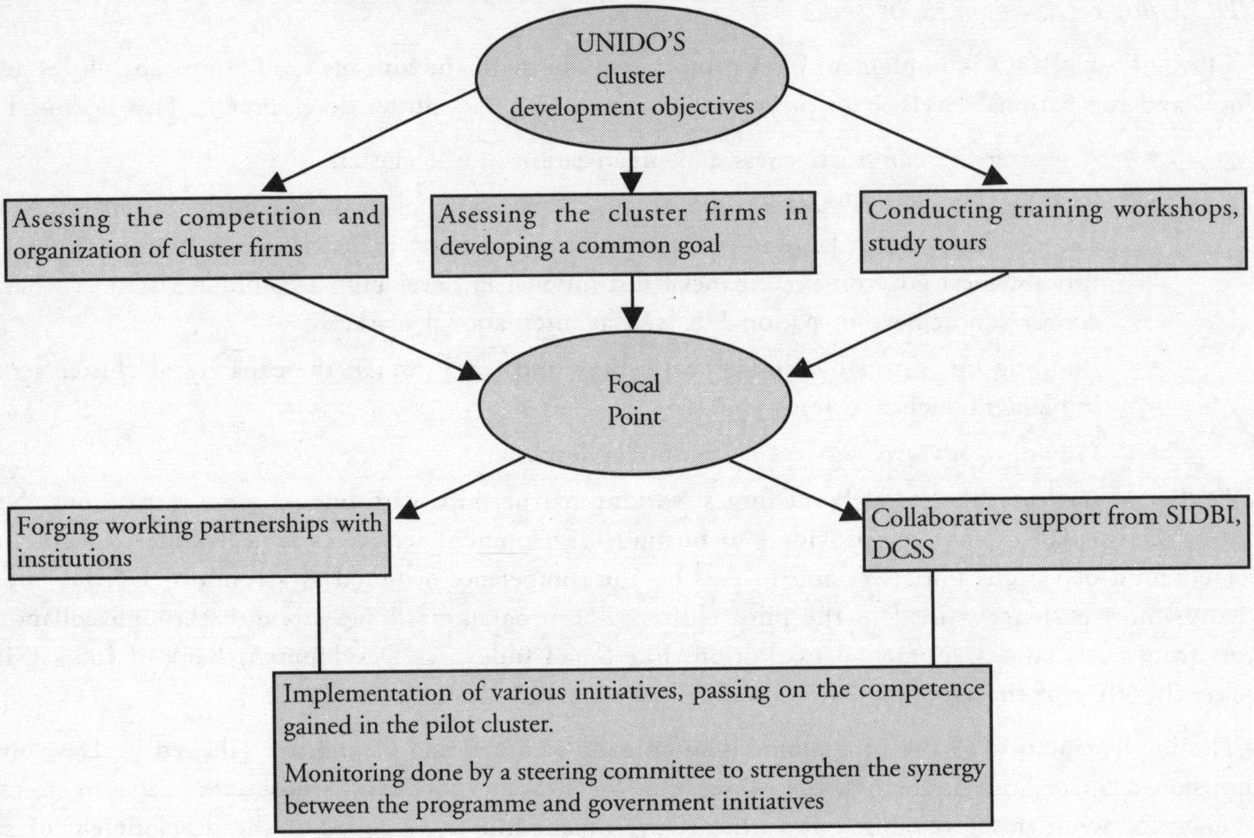

- Financial institutions do not have special policy or programme for providing support to units in a cluster. The assistance is provided on general policy guidelines, which fail to recognize the specific requirements of an industry or a cluster.
- FICCI, National Small Industries Commission, Maharastra Chamber of Commerce and Industry have started providing online information services to entrepreneurs. But very specific requirements like state government rules affecting their operations or knowledge about a specific type of machinery cannot still be provided through information networks. Also the information providers do not virtually exist in most of the clusters and local areas.
- New product development skills are very limited and most of the firms work on drawings or samples given to them by customers. The government institutions for developing new products and processes have not been able to fulfill the requirements.

Recommendations

Selection of Clusters

The criteria of selection of the geographical areas for cluster development and the industries to be promoted are a critical factor in the success of the program. Some of the issues that should be considered in this process are:

- Significance of the Cluster to the economy- the importance of clusters is judged on the basis of factors like employment, exports and output. The advantage of choosing clusters with high economic significance is that any change in the performance level of the units in the cluster will have a positive impact on the overall development of the regional or national economy.

- Replicability- the cluster chosen should have high replicability and dissemination potential of the benefits. Extent of linkages with other clusters is also very important, so that development in one will have positive impact on the connected clusters also.
- Viability - the viability of the cluster in terms of market potential is a pre-requisite for a cluster to be taken for development. The existing and emerging market potential should be assessed.
- Promotability- Clusters should exhibit dynamism and thus it is important to understand the aspiration levels of the firms in the cluster. Indicators like positive attitude of the entrepreneurs, existence of local infrastructure, adaptability of entrepreneurs to undertake modernization should be looked for before choosing a cluster.
- Potential for complementarity- When a programme is already initiated, there is need to look for opportunities for complementation. Thus a technology upgradation programme can be complemented by a marketing related programme.
- Subsidiarity- It relates to the scope of value addition. This is in contrast to the potential from complementarity where the lack of any ongoing development support for the cluster can be used as a criterion for selection. Such clusters will also have weak market linkages, understanding and involvement in the market framework will be limited.

Technology

There is a wide gap in the way small firms and development institutions have viewed technology. Entrepreneurs will go in for advanced technologies only if they perceive it as a means of acquiring more profit or gaining market access. They also prefer second hand machinery if it serves the purpose of catering to the markets, they wish to serve. But implementing institutions usually prefer new state of the art equipment. But Italian experience has proved it is the series of small innovations and not the state-of-the art technology, which is required for firms to get a competitive edge. SBI's Uptech Programme has been able to realize this lacuna and they have taken the right step.

Delivery Mechanism

The delivery mechanism of assistance by agencies and government has mostly targeted the individual units. There have been fewer initiatives to develop the capabilities of the private sector to start self-help programs. The focus should be to build strong institutions not only for the delivery of the policy based initiatives but also for sector specific services such as information provision, technology support and common facilities.

Linkages between service providers and cluster firms

Extension of support services should not be entirely supply-led or demand oriented. The local producers rarely articulate their own demands and they operate on a highly individualistic basis. It has been observed from UNIDO's experience that the inability of a cluster to seize ambitious business initiatives is more because of a lack of certain type of business development services like export consultancy, internet technology, joint product marketing, bulk purchasing etc. But these are also the areas where the potential suppliers are often willing to customize their services for the needs of a specific cluster and the potential customers are unaware of the needs of the services.

Hence the approach should be a simultaneous effort on the demand side (through the articulation of the demand by the SSE owners for BDS presently unavailable in the clusters) and on the supply side (through the design of BDS customised to the needs of the cluster SSEs and in line with the emerging vision for the cluster as a whole). On the same issue, it needs to be emphasised that capacity building cannot be confined to a once-and-for-all initiative such as the identification/provision of a BDS previously unavailable to the cluster SSEs. On the contrary, the aim is to enable the various cluster actors to tackle new challenges and to seize upon more ambitious visions on a sustainable basis. Such an objective involves, for example, strengthening the capacity of the SSE owners to convey their requests to the BDS providers, lowering the risk perceived by BDS providers in doing business with SSEs, enabling BDS providers and SSE owners to identify institutional partners able and willing to co-finance the development of new BDS, and identifying more effective procedures to assess BDS impact. This has been successfully demonstrated by UNIDO's initiative in the use of IT related services in the Ludhiana Cluster.

Subsidization

Considering the cost involved in providing specialized services, there may be demands for subsidization of services from both providers and receivers. But subsidizing may fail to sensitize the entrepreneurs of the economic viability and the benefits accrued from it by themselves. Also there will be possibilities of conflict if individual firm owners are provided subsidies.

Subsidies can be given to start-ups providing innovative BDS (e.g. access to new technology cum training, quality management). On the contrary, less support is to be given for BDS strengthening the commercial capacities of the SSEs or, more generally, to initiatives with an immediate impact upon the profitability of the latter (e.g. participation in fairs, product design). As far as possible, subsidies are to be given to the business associations and not to the individual firms to avoid conflict.

Product Choice

It is important that firms in the clusters should choose a niche which will be difficult for the bigger firms to cater to. While Indian companies have the competence to produce international quality products, they are not been able to compete due to lack of understanding of the non-economic needs of customers, lack of adherence to prescribed schedules etc. The development institutions in India have focused on the technology of production and there was little co-ordination with other specialized institutions providing services in the fields of design, export marketing, provision of finance etc. For the clusters to realize their potential it will be important for them to harness the competence of various resource organizations.

Sustainability

It may still be a little early to access the sustainability of UNIDO's initiatives. As in the case of new initiatives, cluster development can also get confined to specific regions as stand alone projects. Efforts should be taken from a phased sharing of responsibility to complete control by some government agencies in India.

The options for the future of Focal Point created by UNIDO

- Absorption by an existing national institution;
- Conversion into an autonomous foundation operated by a coalition of players from the private and public sector;

There could be an apprehension that clustering will limit the benefits of co-operation. It is certainly possible that the establishment of a common platform among a selected number of cluster SSE might become an obstacle to the inclusion of a larger number of partners into the cluster development approach. In fact, it is not uncommon for a cluster development project to begin its intervention in a cluster by involving a relatively small group of highly motivated and visionary SSEs (in order to produce visible results and eventually to earn the support of more SSEs). Such a strategy clearly runs the risk of splitting the constituency of SSE owners into 'leaders' and 'laggards' and to pave the way for the establishment of a latent conflict between these sub-groups, which might even worsen as the project gets increasingly disengaged from the cluster. Care needs to be taken so that such resistance does not surface.

Role of Ministry of Industry

Ministry of Industry should be responsible for businesses not specifically covered by other ministries in the areas of investment in plant and equipment, pollution control, energy and power, some aspects of foreign economic assistance, and consumer complaints. This span will enable the Ministry to integrate conflicting policies, such as those on pollution control and export competitiveness, to minimize damage to export industries.

The Objectives of Ministry of Industry's initiatives should be

- To provide encouragement and guidance to the initiatives of private business and create a suitable level playing field which would give critical advantage to industries, identified by government as potential for long term success.
- In R&D, the aim should be to coordinate and encourage the efforts of private corporations rather than initiate large government funded projects.
- To create channels through which new technological ideas could be rapidly disseminated between Indian firms.

Recommendations related to Taxation: Value Added Tax

What is VAT?

The value added tax is similar to the national sales tax, however, instead of implementing one tax of a certain percentage at the time of retail sale, there is a smaller tax, proposed at 5%, added each time the product is resold or when value has been added. For example a tax is added when a product is passed from a manufacturer to a wholesaler, and again from the wholesaler to the retailer. Most products would be priced higher than they are today because the cost of goods would include this built in tax.

Advantages

- A 5% tax is levied at each stage of production of goods.
- The total tax is built into the cost of goods.
- Exports are not taxed, while all imports are taxed.
- It allows multiple rates and exemptions on goods from other countries.
- No tax forms or tax returns will be needed for individuals.

Disadvantages

- It would be more complicated to administer and collect.
- It would also tax food and medicine as does the national sales tax.

In a VAT regime it would be the free market forces and not the tax rates which will determine the patterns of production, consumption and use of resources. The VAT system is intended to tax the consumption of goods by the consumers and not the businesses.

It is important to note that all the states levy an entry tax or octroi on select commodities and plan to continue them even in the VAT regime. This levy over and above the local sales tax creates tax cascading and result in higher consumer prices.

The VAT regime is supposed to come into force by April 2003. This is because in January 2002, the decision was taken to postpone the implementation of VAT. However, it is yet to be seen if this 12 month hiatus will be properly utilized to iron out the issues that have remain unanswered.

The Issues

The reasons cited for the postponement of implementation of VAT.

I. Lack of preparedness and the recessionary trends which have adversely impacted their revenues have been cited as the major reasons for skipping the deadline.

II. How revenue losses incurred by States due to switchover to VAT would be compensated is also yet to be concretized. States want all theses issues to be settled and legally concretized, before they can formalize their VAT Acts and go ahead with implementation.

III. Delay in effecting amendments to the Central Sales Tax Act to facilitate implementation of VAT. The CST has been proposed to be reduced to 2% recently.

Further a study of the draft VAT Acts published revealed that there was no uniformity in the different draft Acts on issues like taxation of deemed sales, definitions, input credit etc. It appeared, as though, that we will have 28 VAT Acts, each from respective states, replacing the existing Sales Tax Acts.

The implementation of a VAT based system can be a boom to manufacturing sector due to the advantages of pricing and taxing based on the value added and each stage in the value chain can take its share without motive or rather chance of profiteering.

Recommendations

Compensating Revenue Loss to States

The States would not agree to introduce VAT in the face of a fall in revenue unless the Central Government assures an acceptable proposal for compensation of possible revenue loss.

It has been reported that a floor rate of 10% and a maximum of 12.5% have been agreed to as the Revenue Neutral Rate by all the States (if this rate is Revenue Neutral, the States should not face any revenue loss! We think one should call this rate as a General Rate rather than RNR, as this rate may not be the Revenue Neutral Rate, if the claim of revenue loss by the States is true)

The revenue loss to States is likely to a major bottleneck. One of the avenues proposed is the levy of service tax by the States.

The ideal course would be to let the States choose the services to levy tax. The center should publish a list of services reserved for itself and a negative list of services not to be taxed. The States should be allowed to decide for themselves the services to be taxed.

Design a National Sales Tax VAT Act

There is no uniformity in the different draft Acts on issues like taxation of deemed sales, definitions, input credit etc. What will happen is that you will have 28 (or as many States that switchover to VAT) VAT Acts replacing the existing Sales Tax Acts. This will definitely be not a reform in the right direction. The need is for a central VAT Act on the lines of the Central Sales Tax Act or the Motor Vehicles Act. The broad principles and policies should be incorporated in the National VAT Act with powers to the State to define the schedules, exemptions, Special Additional Tax, etc and administrative provisions. Having a National VAT Act is definitely not an abrogation of the powers of the State as the powers of the State can be exercised by the State Rules.

Entry Tax by the States

Many people are of the view that minor indirect levy by the States like entry tax etc. should be dispensed with. If the levy of entry tax is insisted upon by the State to be continued on revenue consideration necessary provisions may be made in the National Sales Tax VAT Act.

Entry tax is essential because in its absence the purchases from outside the state would be tax free whereas those from within the state would be subjected to VAT. This would force the local businesses to absorb a part of the taxes in their profit or resort to dishonest means. Thus, the absence of VAT would create inefficiencies and waste in the economy and consumers would ultimately pay for this through higher prices.

A fully vatable entry tax levied at the rate equal to the VAT rate on sales within the state would not have any tax cascading. Under, this system the registered dealer who pays an entry tax on imports into the state should be entitled to a 100% tax credit of the entry tax paid on further sale of the commodity within or outside the state. This would also eliminate the concern of the land locked states that the entry tax may lead to tax cascading.

With a National VAT Act uniformity can be achieved on some of the following major issues.

a.) Uniform definition of "dealer", "sales", "turnover" etc.

b.) Uniform rates of taxes with powers to State to levy non vatable Special Additional Tax.

c.) Uniform dealer registration or identification numbers based on the income tax PAN as is now being done for Central Excise and Service Tax. This will help better audit trail.

d.) Uniform threshold limits at least with reference to inter-state trade.

e.) Common commodity classification for Sales Tax, Customs and Excise.

f.) Uniform book keeping requirements, self-assessment limits etc.

g.) Uniform format of invoices integrating the requirements under the Excise Act and Sales Tax Act.

h.) Uniform treatment of fines and penalties for offences.

i.) Uniform documentation for movement of goods

j.) Uniform treatment of concessions and deferral schemes of States that are similar with powers to States to decide on schemes that are peculiar to the particular State.

Permanent Consultative Body of State administrators for VAT

A National VAT Act is possible only if a permanent institution with representation for all the States is setup. This empowered Committee of State Finance Ministers can be the final policy making body. The tasks of designing the National VAT Act should be entrusted to a body of State Tax administrators. The State should be required to commit the same official for a minimum term of at least 3 years to ensure continuity and reduce politics-influenced transfers.

The committee should also have representation of Trade Bodies, Tax Lawyers and also Chartered Accountants for proper guidance on book keeping requirements.

Phasing out of Central Sales Tax and Other Levies

It is proposed that the CST Rate would be initially reduced and gradually removed fully and that the Center should compensate the States for the loss on account of any reduction in CST. The Center by levying service tax on more services can earmark funds to compensate the States for the gradual phasing out of CST.

The necessary amendment and time frame for the phased reduction to 0% should be incorporated in the CST Act so that this reform may not be affected in the future by any political change of guard at the center or by the realignment of priorities of the Center.

Give industry three months time to plan transition to VAT

It is imperative that all States finalize the VAT Law including finalization of rules, forms and procedures by 31st December 2002. A business with multi-location operations needs to finalize logistics, effect necessary modifications to software, train staff, decide on pricing policy etc. At least three months time is required by an industry to plan smoother transition to VAT.

The State Governments also need to put in place a VAT information system, train officers as well as trade and industry.

The focus has all along been only on the shape of VAT law and no serious thought has been given to the time required for the Governments and Industry to plan transition to VAT.

Recommendations Specifically for SMEs

- The State should take action for removal of procedural hassles and providing one-stop facilities that will allow the small producers to perform the necessary paperwork easily.
- It is necessary to provide the SME sector with a tool to identify the areas of strength and weakness in individual SMEs by comparing them with a 'peer group' of companies. These companies would have similar number of employees, similar turnover and preferably in the same industry sector. In case data is not available, process benchmarking can be done to identify the area of improvements. UK Department of Trade and Industry has already developed a Business Profile Benchmarking Instrument called UK Benchmarking Index.
- SMEs generally lack information on the technological capabilities available. Even when they are able to access the information on technology, their capacity to apply is constrained by the lack of financial resources and the skills that go with the new technology. It is to overcome the multi-faceted nature of the constraints faced by SMEs in applying and adopting technological improvements, assistance is to be provided in the form of package, which will inform them annually of the latest technological innovations in their area, the cost of advanced equipment and their benefits, possible sources and the financing needs to acquire that technology.
- Provide incentives for assemblers to identify potential suppliers for components, and to provide technical assistance to them to develop their capabilities, including the development of a vendor improvement and certification programme.
- Encourage local design and engineering capabilities to produce components based on client specifications or create original designs on the basis of agreed specifications

Providing Capital for the Industry

Inadequate, non-timely and lack of alternative avenues of finance are the major causes of under-utilization of manufacturing sector capacities. Unavailability of adequate working capital also seriously undermines the competitiveness of SMEs. Also because of higher transaction costs and absence of good credit assessment capabilities, interest rates for SSIs have increased substantially.

Recommendation

To facilitate SMEs to undertake diversification, technical upgradation and to abate the financial constraints arising out of delayed payments, there is need for alternate finance schemes like Factoring Services, Venture Capital, Credit Gurantee Schemes and options for Micro Finance. So there needs to be more synergy between commercial banks and SIDBI to promote such schemes at grassroot levels.

Banks should be encouraged to provide competitive environment and institute transparency in their procedures for project approvals. They should be urged to bring down the cost of raising funds for the industry to less than 10 percent.

Organizational Level Issues and Recommendations

It would be unwise to lay the entire blame of our lackluster performance in manufacturing on the government policies alone. Such an escapist attitude will miss some of the core problems that pervade the way our industries are run and organized and need immediate attention and correction. To enable the government policies to bear fruit we present a discussion of the current issues and the actions that need to be taken to correct them.

Key Issues

Inability to Draw on Technological and Support Services to Upgrade Manufacturing Capabilities.

In India most of the medium sized manufacturing enterprises do not have the capabilities to undertake development activities on their own and investment in R&D is abysmally low. Even in the cases where joint ventures are formed with international players, rarely has an Indian firm been able to assimilate learning. Thus these firms were not able to develop their own competitive advantage and outlearn the parent firm.

As we have seen in the case of Korea, though they depended on Japanese technology to a large extent, they were still able to adapt those technologies suited to their local conditions to carve out competitive advantage for themselves, particularly in industries like textile and ship-building. They benefited from the series of small improvements and learning accrued over the years.

Indian firms, barring a few have not been able to show such initiatives. But this shortfall could have been taken care of by a healthy interaction between industry and government support institutions. But in India, manufacturing firms have been rarely able to articulate their needs to the government technological institutions, while the institutions lacked in commercializing their activities. The initiatives of organizations like CII also failed to address industry specific needs.

Lack of Clarity Regarding Level of Automation Required

Some aspects of manufacturing are less amenable to automation because of the level of dexterity required. Automating those activities will not lead to any improvement but there are many activities which can be improved by process innovation and judicious use of technology. It may not require huge investments and many such areas of low scale automation can be identified from careful study of processes. Smaller firms in India have performed better in these aspects mostly because of the engineering skills of the entrepreneurs but such initiatives in larger firms have not taken roots. Such automation efforts have remained confined to specific shop floor initiatives but have not spread across the organization.

Clutter of Improvement Initiatives without Clear Vision

Pressure of competition post-liberalization has raised the awareness of management to undertake improvement initiatives. Lot of activities have been planned in a very short time without paying attention to the background work which is required to reap benefits from such initiatives. JIT, TQM, Six Sigma implementation plans were made but the rationale behind all these initiatives have not been communicated to the operating people. Thus though some of these initiatives fulfilled short term objectives, they could not be sustained in the long run once the initial enthusiasm was lost.

Proposed Initiatives and Actions

Institute - Industry Linkage

The current efforts at developing this interface suffer from major drawbacks some of which are mentioned below:

Product Barriers

- Institutional research is not related to industry's requirements
- Gestation period for basic research is very long
- Mismatch between industry's short term needs and institute's orientation for long term research

Institutional Barriers

- Industry's own in-house R&D capabilities do not take into account the cost advantages of institutional research
- R&D Institutions do not commercialize research results

RECOMMENDATIONS FOR INSTITUTE-INDUSTRY LINKAGE

Expand STEP's Initatives
- freedom to be given to academicians for involvement in this activity
- make venture capital available either from industry for financial institutions
- development of an industrial culture in the vicinity of the institute

Make the R&D system more demand oriented
- performance to be measured in terms of techno-economic impact of projects
- resource budgeting to be linked to economic pay-off at the end of project

Creation of Technology Resource Centre (TRC)
- to access current state in industries
- to identify areas of improvement
- to provide technical advice to adapts suitable technology
- to disseminate best management practices, launching pilot projects, monitor progress through Management Services Division (MSD)
- to act as a one stop shop to access a network of institutions to smoothen technology flow

We therefore propose a framework as discussed in the figure below for leveraging this valuable source of competitiveness.

- To start with, the goals of the R&D system should be made more demand oriented, which will help channelize R&D efforts towards generating output that is useful at the end of technology delivery process.

- The performance of the R&D sector, its institutions and its management need to be measured in terms of the techno-economic impact of its projects to the users and the industry as a whole.
- Similarly the resource budgeting for the R&D system needs to be related to the economic pay-offs at the end of technology delivery process. Technology needs of a developing country can be met by indigenous technologies, developed by the country's R&D system, by technology transfer and through importable modern technologies, that are embodied in equipment or processes, which are freely traded.

The selection criteria for indigenous technology can be as follows-

i) It should be able to generate competitive advantage of the local industry absorbing the technology against foreign competition.

ii) The industrial infrastructure or support industries should be present or need to be developed to make the technology viable.

Technology Resource Center (TRC)

In a developing country, much of the technology needs of the production sector will be on simple, modern technologies, already existing and able to be imported. So there should be some institution in the technology delivery process that will perform the function of propagating and help in assimilation of the technology, suitable to local conditions. This was very successfully performed by Korea.

There can be a Technology Resource Center (TRC) with the following set of responsibilities-

- Preparing the environment for technology transfer through various support services needed to apply the technology. It will involve assessment of the current state in the industry, identification of areas of improvement, which technology and in what form can lead to improvements and establishing information linkages for the access to that technology.
- Providing technical advice to install that appropriate technology.
- In case a set of principles or management techniques need to be changed, the assistance should come in the form of training in best practices, launching a pilot project in a particular firm, continuously monitor its progress, sustain its improvements and then spread it among other firms to adopt .There is no separate body as of date to look after the process improvement needs of organizations. One separate wing under CII, "The Management Services Division (MSD)" can perform this function across industries.
- TRC will act as a one-stop shop and will access a network of institutions with different functional specializations that will smoothen the technology flow to the industry.

Operationalizing TRC - A Step by Step Approach

CII can set up separate industry specific Technology Resource Centers. For convenience of implementation, it will have regional centers to have a better reach. MSD can also have separate focus groups to deal with specific initiatives like TQM, TPM, Kaizen etc.

The bigger sized firms should first take initiative to study their own processes and identify possible areas of automation. It may not be possible for TRC to analyse the processes of individual firms unless they come up with specific requirements. TRC can then provide the necessary guidance. TRC will have a bigger role to play for SMEs where firms in a particular industry will have similar needs. Efforts of TRC's can be much more focused where firms constitute a cluster.

It will be difficult for firms to identify and then prioritize which best practices they need to undertake. There will be possibilities to take up all such initiatives together without performing the background work.

Round tours of best operating plants in the country, both within and outside the industry, organized by TRC will first sensitize the entrepreneurs of the level of excellence that can be achieved. This will help in reducing the production orientation of firms.

Then training sessions have to be conducted both for entrepreneurs and employees of the firms, informing them of the minimum level of groundwork in areas of quality, productivity, inventory management which needs to be attained before launching a bigger initiative. Firms will be encouraged to bring them up to those levels with constant monitoring by TRC personnel.

Finally bigger initiatives like cell formation, implementing design of experiments, performing process capability studies etc need to be taken up with proper training inputs.

Science and Technology Entrepreneurs Park - A Step in the Right Direction

National Science and Technology Entrepreneurship Development Board has initiated a programme called Science and Technology Entrepreneurs Park. The idea is to create a necessary climate for innovation, for sharing of ideas, experiences and facilities and open up avenues for students, faculty, researchers and industrial managers to understand each other's requirements.

So far 15 STEPS have been developed, each with its own technology thrust areas depending on the strength of the host institution and the industry, they are serving. The STEP's are engaged in developing technologies as per the needs of the entrepreneurs and also in converting the technologies into viable business proposals.

This initiative has an immense potential to promote knowledge based industries but it should be remembered that the presence of an R&D institution with its scholars and researchers will not be able to develop high technology industries themselves.

The important factors to be noted for the long term success of STEP are as follows-

- Freedom should be given to academicians to enter into this activity and their involvement in this activity apart from their regular pursuits in the institute should be encouraged and should be considered as a performance measure.
- Venture capital should be made available. It can come from some sponsoring industry, who want to benefit from some upstream technology or should come from financial institutions like IDBI, SIDBI etc.

An industrial culture needs to be developed, preferably closer to the institute. So some of the proposed cluster development programmes should also consider this aspect of locating near a technology institute, participating in STEP.

Key Issue

Low Productivity of Labour

In India cost of machinery and equipment still exceeds the cost of labour and the emphasis has always remained on maximum utilization of equipment. Investment decisions also got dictated by this mentality and additional labour was preferred. This led to the inherent low productivity of labour. Thus the problem lies to a great extent in the way labour is utilized.

STEP- PROCESS AND ACTION

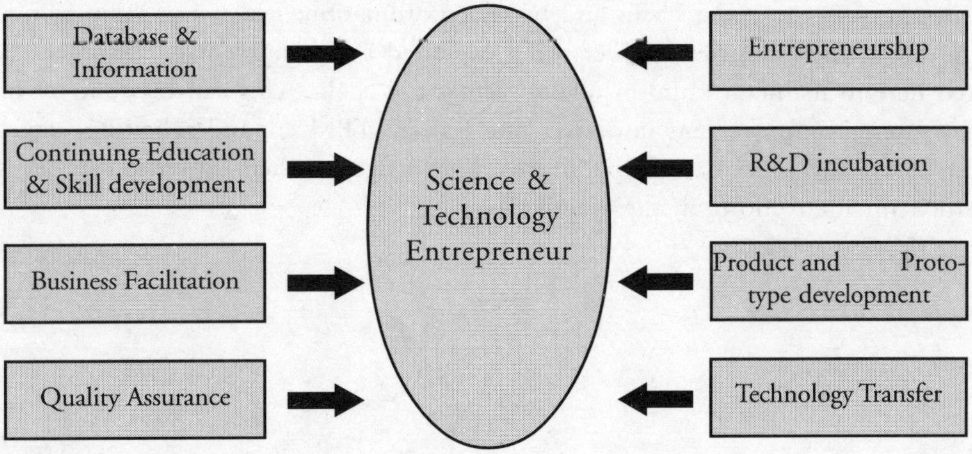

Flight of Talent from Manufacturing

Manufacturing as a career has failed to attract talented engineers because of the perceived lack of challenge. Manufacturing is seen more as maintaining the status-quo without scope of innovation and even less reward for innovation. The root of this problem again lies in production orientation of the management, which stifled improvement initiatives.

Proposed Initiatives and Actions

Changes in Engineering Curriculum

In the engineering curricula, students should be encouraged to take up live projects on operational problems, faced by firms. Students with a guiding professor can take up a specific problem, say of inventory management, performing design of experiments, developing simple hydraulic-pneumatic automation systems etc. These projects can be challenging for the students and will also benefit the firms. Thus students will get accustomed to the problems faced by Indian manufacturing firms, which will sensitize them to the challenges ahead.

Bigger firms have to take the initiative to induct bright engineers and involve them in improvement initiatives and challenging work.

As we have already recommended to form industry specific TRC's, which will require highest quality manpower both technical and managerial. TRC's should also recruit engineers and involve them in the technology initiatives with suitable remuneration.

Industry Initiatives for the SME Sector

SSIs can frequently benefit by utilizing each other's complementary skills and thus serve their customers better. In many cases, a particular firm may not have the necessary machinery or capability to perform a particular process. This activity is off-loaded to the one, which can perform it. If both the firms cater to the same segment in the supply chain, it can be beneficial when one of them gets a special order and relies on the other. This kind of trust and understanding should be fostered through local associations.

If firms operate in different segments, viz. one engaged in casting and the other machining the casting parts and in sub-assembly, then the casting firm becomes the supplier to the other and a separate tier of suppliers is formed. Original equipment manufacturers and final assemblers are increasingly classifying their suppliers in separate tiers and co-ordinating only with the systems assembler, who in turn deals with his supplier. This results in lesser inventory in the entire chain and better co-ordination. But it has some fallouts also, there is a frequent complaint about the lower-tier supplier being exploited by his immediate customer. The OEM's efforts should be required here to maintain uniform quality across the chain. This can be done by involving all levels of suppliers in the internal improvement initiatives like Kaizen, TPM etc and soliciting suggestions in product development. Suppliers many times have better process knowledge, so their expertise can be used and they can be suitably rewarded for their efforts in these activities.

BIBLIOGRAPHY

Books referred:

1. *Banerji Asit "An approach to the evaluation of Indian Industrial Strategy", Institute of Developing Economies- VRF memo graph series, 1990.*

2. *Gulati, Mukesh " Building Competitiveness Among Small Enterprises Using Cluster Approach" in Technology for Small Scale Industries - Current Status and Emerging Needs, Small Industries and Development Bank of India, 2001.*

3. *McMillan Charles J. "The Japanese Industrial System", de Gruyter, 1984*

4. *Menon, P.K.B " Co-ordination Among Institutions" in Technology for Small Scale Industries - Current Status and Emerging Needs, Small Industries and Development Bank of India, 2001*

5. *Rodrigues, Clovis W. "R &D and industry- some considerations from the Brazilian experience" in Research and Development - Linkages to Production in Developing Countries, ed. Mary Pat Williams Silveira, United Nations Science and Technology for development series,1985*

6. *Sandee Henry "Promising Small Scale Industries in Indonesia: A review of recent policies and programmes" in Small Scale enterprises in developing and transitional economies, Palgrave, Hampshire, 2002*

7. *Tuller Lawrence, W. "Doing Business in the Latin America and the Caribbean" Library of Congress Cataloging -in- Publication data, 1993*

8. *Westphal, Larry E., Rhee, Yung W., Pursell Garry "World Bank Staff Working Papers, Number 469", 1985*

Websites referred:

1. www.uchicago.edu
2. www.uscc.gov
3. www.cherry.gatech.edu/
4. www.jri.co.jp
5. www.indiafdi.com
6. www.mckinseyquaterly.com
7. www.bcg.com
8. www.indiainformer.com
9. www.indiatimes.com
10. www.indiainfoline.com
11. www.karvy.com
12. www.123india.com
13. www.samachar.com
14. www.businesstoday.com
15. www.dti.gov.uk
16. www.rspas.anu.edu.au/
17. www.epinet.org/
18. www.economics.anu.edu.au/
19. www.spookiepoo.com/
20. www.ilo.org
21. www.robarts.yorku.ca/
22. www.niep.org.za
23. www.accenture.com
24. www.iamot.org
25. www.domain-b.com/
26. www.fecolumnists.expressindia.com/
27. www.blonnet.com/
28. ww20.malaysiamanufacturers.com/
29. www.india-emb.org.eg
30. www.icfaipress.org/
31. www.cherry.gatech.edu
32. www.dti.gov.uk
33. www.fas.org
34. www.sidbi.com
35. www.smallindustryindia.com
36. www.state.gov
37. www.economictimes.com
38. www.business-standard.com
39. www.svat.com
40. www.WTO.org

Towards Sustainable Manufacturing Superiority

A Comprehensive Manufacturing Policy Framework for India

Indian Institute of Technology (IIT) - Madras

Authors : Dr. (Mrs.) M. Thenmozhi (Team Leader), Abhishek Jain, Mahesh Nair, Rohit, Sumon Dutta.

Contents

Towards Sustainable Manufacturing Superiority
A Comprehensive Manufacturing Policy Framework for India

(IIT - MADRAS)

1

CHAPTER ONE

Introduction

1.1 Why Manufacturing is Important?

While the Indian economy has been on a higher growth path through the 1990s relative to the previous four decades, the initial surge in growth has tapered off in recent years. The Indian manufacturing sector, a key component of the overall economy also reflects this larger macroeconomic trend, having been a major driver of higher economic growth in the mid -1990s and showing a sharp decline in performance thereafter.

Indian manufacturing sector performed badly as compared to the growth of manufacturing sector in China, where it constitutes nearly 40% of the GDP presently. Based on the experience of a number of western and newly industrializing economies, the evidence is overwhelmingly in favour of adopting an outward looking, international trade oriented economy as opposed to an inward looking domestic market focused one. With processed manufactures accounting for about 70% of world trade, global competitiveness in manufacturing is a national imperative. Of particular relevance to India is the fact that labour intensive exports constitute a 1.5 trillion dollar world market, in which relatively high wage rate economies such as Korea and Taiwan continue to hold significant share.

With 75% of India's working population of approximately 600 million people educated only to middle school or below, it is this labour intensive manufacturing sector that can potentially generate employment in adequate numbers to absorb the labour pool. The services sector, and in particular the knowledge based industries, while extremely important, are unlikely to provide opportunities to the poorly educated sections of our society.

Figure 1- GDP Contribution of Various Sectors

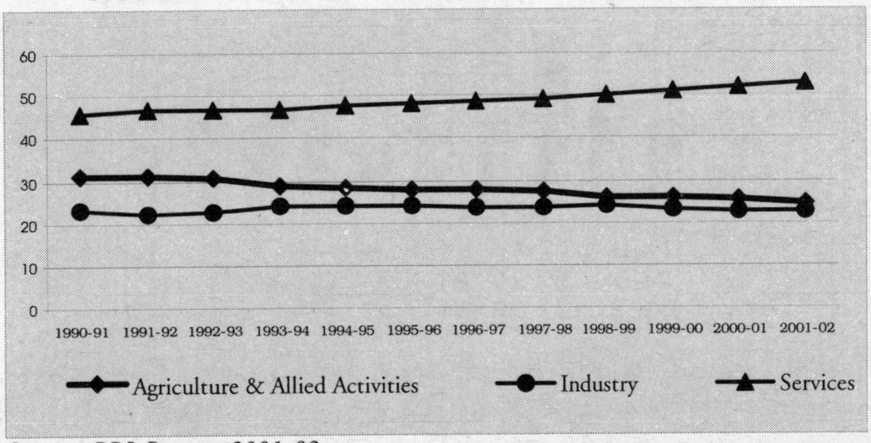

Source: RBI Report, 2001-02

The process of globalization only increases the imperative for a strong international trade oriented economy. In this context, the pattern of international trade points to the necessity of developing global competitiveness in the manufacturing sector.

The ever-growing economic sub-system, with countries competing with each other for greater and greater share of world market puts enormous pressure on the ecology. Making a policy for the development of manufacturing thus is an unenviable job.

1.1.1 Falling Share of Manufacturing

The earlier chart shows the GDP contribution of Services, Industry and Agriculture and allied activities in India. A gradual fall in Agricultural contribution to GDP is evident. Share of industries to the GDP has also shown a falling trend, in the last decade.

TABLE 1

Growth Percentage as a Contribution to the GDP

Quarterly Estimates of GDP (at 1993-94 prices)
Percentage change over from the corresponding quarter of the previous year

Sector	2001-02				2000-01			
	Q1	Q2	Q3	Q4	Q1	Q2	Q3	Q4
1. Agriculture and allied activities	1.1	6.3	7.6	7.6	1.8	3.9	-0.8	-4
2. Industry	2.5	2.7	3	3.4	7.6	6.4	7	4.1
Mining and quarrying	-0.3	0.7	3.1	3.5	4.8	3.6	4.3	0.9
Manufacturing	2.7	2.6	2.9	3.1	8.1	7.1	7.1	4.6
Electricity, gas and water supply	3.9	5.4	3.8	5.4	7.1	4.5	9.3	4
3. Service	5.1	6	6.6	7	6.3	6.9	4.4	2.9
4. GDP factor cost	3.5	5.3	6.2	6.4	5.4	6.2	3.4	1.5

Source: RBI report, 2001-02

FIGURE 2 - SECTORAL CONTRIBUTION OF GDP - VARIOUS INDUSTRIES

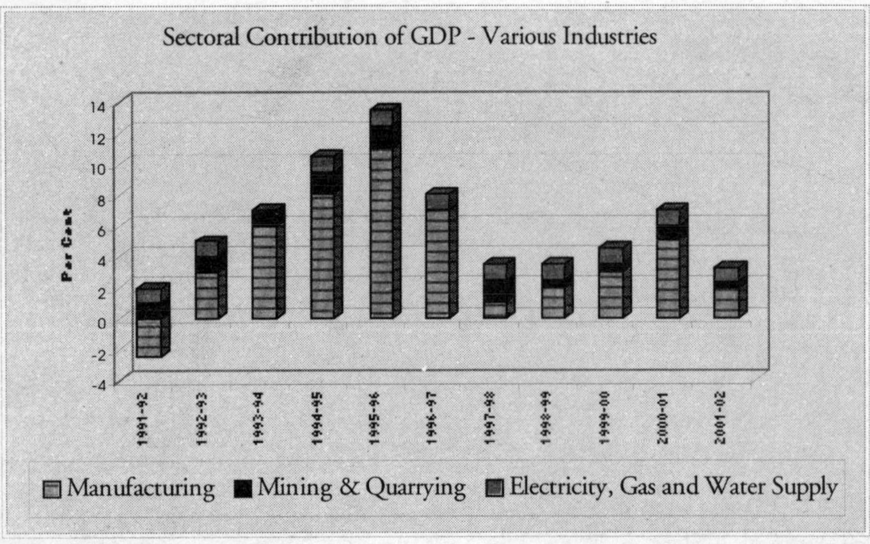

Source: RBI report, 2001-02

The figures clearly show that Industry as a whole has lost its share in all the four quarters of the last financial year. While services sector has stagnated in the last financial year, manufacturing has performed poorly in each quarter compared to financial year 2000-01.

The chart below shows that the manufacturing sector contribution to the GDP share of Industry in India fell after 1995-96, and has been picking up slightly in the last five years.

In 2001-02 there was again a sharp drop in manufacturing sectors share by nearly 1.5% per cent over the previous financial year. Thus, the Indian government needs to focus on revival and growth of the manufacturing sector.

1.1.2 Break-up of Manufacturing Sector Growth in 2000-01

The manufacturing sector can be divided into seventeen sub-sectors and it is observed that some of the sub-sectors have shown improved growth, some have had growth but which has decelerated over the previous year, and there are a number of sub-sectors that faced negative growth. Overall the contribution to GDP of manufacturing sector has come down. The following chart identifies the sub-sectors and their growth over the financial year 2001-02 in relation to the previous year and indicates the industries coming under accelerating growth, decelerating growth and negative growth.

TABLE 2

Differential Growth in Industries

Code	No.	Group of industry	Growth in 2000-01 (%)
	1	Beverages, tobacco and related products	12.2
	2	Transport equipment and parts	6.8
	3	Basic metal and alloy industries	4.3
	4	Paper and paper products and printing, publishing and allied industries	3.8
	5	Non-metallic mineral products	1.2
	6	Rubber, plastic, petroleum, and coal products	11.1
	7	Other manufacturing industries	8.9
	8	Leather and fur products	5.4
	9	Basic chemicals and chemical products (Except petro-products and coal)	4.8
	10	Wool, silk and man made fibre textiles	4.4
	11	Textile products (including wearing apparel)	2.4
	12	Machinery and equipment other than transport equipment	1.3
	13	Food products	-1.6
	14	Cotton products	-2.2
	15	Jute and other vegetable fibre textiles (except cotton)	-5.9
	16	Metal products and parts (except machinery and equipment)	-10
	17	Wood and wood products, furniture and fixtures	-10.9

Source: RBI Report, 2001-02

Code	Meaning
	Accelerating growth
	Decelerating growth
	Negative growth

1.2 Definition of Manufacturing

An accepted definition of manufacturing is as below:

> "Manufacturing comprises of mechanical, physical, or chemical transformation of materials, substances, or components into a new product".

Identified Industry-Groups of manufacturing sector in India are as follows:

- Food Products
- Cotton Textiles
- Wool, Silk, Textiles Products etc
- Jute Textiles
- Chemicals etc
- Non-metallic products
- Non-Electrical Machinery, Tools and Parts
- Basic Metals
- Electrical Machinery
- Transport Equipment.

To obtain a representative but detailed view of manufacturing, it is needed to conduct sector specific Analysis, which will largely follow the above framework.

1.3 Manufacturing and Environmental System

Externally imposed development disrupts human relationships and community life and causes significant hardships for the very people it claims to benefit. There is evidence that the world is experiencing accelerated social and environmental disintegration in nearly every country - as revealed by rise in terrorism, poverty, unemployment, inequality, violent crime, failing families and environmental degradation. Quest for short-term financial gain is causing disasters for the human race.

Figure 3 - Growth of 17 Major Groups of the Manufacturing Sector

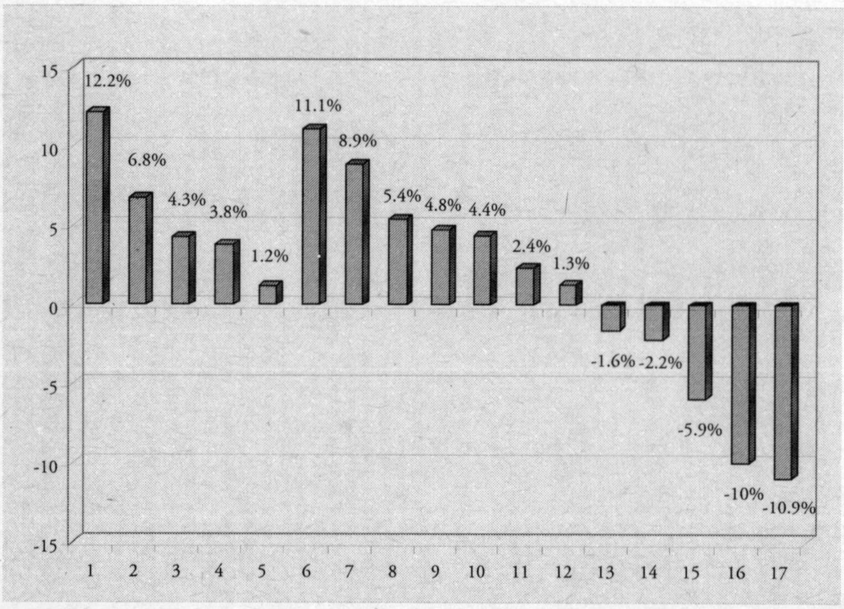

Kenneth Boulding's classic essay "The Economics of the Coming Spaceship Earth" (1968) justifies the need of an Ecological Revolution. The Cowboys of the American West lived in a world of sparsely populated expanses blessed with seemingly inexhaustible material resources. Everything was free for the taking, to be used and discarded at will. Compared to them Astronauts live on spaceships with a human crew and resources that are limited. Everything must be maintained in balance, recycled; nothing can be wasted. The measure of well-being is not how fast the crew is able to consume its resources but rather how effective the crew members are in maintaining their shared resources and the life-support system on which they depend. What is thrown away is forever inaccessible. What is accumulated without recycling fouls the whole. No one would think of engaging in non-essential consumption unless the basic needs of all were met and there was ample provision for the future. Boulding's analogy conveys a basic truth. Modern societies are practicing a "cowboy economics" in what is increasingly becoming a "spaceship world". Living like cowboys in a spaceship has tragic consequences:

- It overburdens the life-support systems, resulting in their breakdown and decrease in the level of human activity they can ultimately sustain.
- It creates intense competition between the more powerful and weaker members of the crew for a shrinking pool of life-support services.

Figure 4 - Transition to a Full World

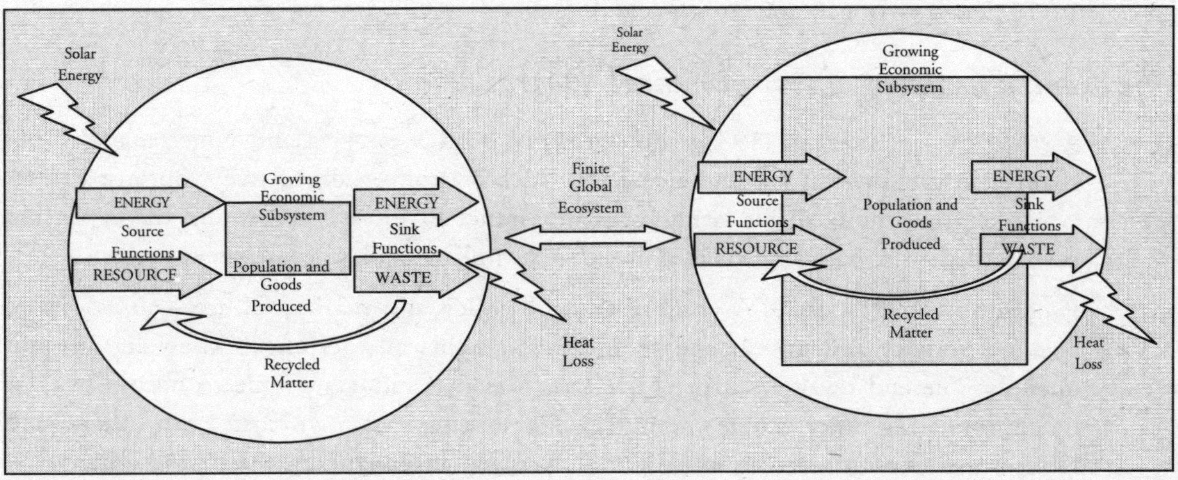

Source: Robert Goodland, Herman E. Daly, and Salah El Serafy, Population, Technology and Lifestyle: The Transition to Sustainability (Washington, D.C: Island Press, 1992), p. 5.

1.3.1 Growing Economic Sub-system

The figure shows that the "Growing Economic subsystem" has reached its limit in the second figure. The bottom line for our species is that because of population growth and the fivefold economic expansion since 1950, the environmental demands of our economic system have become dangerous. The first environmental limits that we have confronted, and probably exceeded, are not the limits of non-renewable resources such as oil and copper, as many once anticipated, but rather the limits of renewable resources and the environment's ability to absorb our waste referred to by ecologists as "sink functions." Evidence of our encounter with these limits is everywhere. Acid rain has damaged 77 million acres of forest in Europe alone. At the global level, each year deserts encroach on another 17 million acre of once productive land, the area covered by tropical forests is reduced by 30 million acres, there is net loss of 26 billion tons of soil from oxidation and erosion, and nearly 4 billion acres of prime agricultural land are abandoned due to salinization from irrigation projects.

Manufacturing operations are the chief cause of environmental pollution and resource consumption. A country's manufacturing policy should be made with the eco-system in mind. A developing country like India will like to develop fast but by undermining the ecological constraints long run sustainable development will be difficult.

1.4 Manufacturing and Sustainable Development

Sustainable development according to the Brundtland Commission Report is

> *"To ensure that development meets the needs of the present without compromising the ability of future generations to meet their needs."*

The Brundtland Report and later writers have identified poverty as one of the major constraints in achieving the goals of sustainable development. In most developing countries, like India, the poorest segments of the population occupy the least resilient and the most threatened environmental areas. But that does not necessarily mean that they will use ecologically harmful methods for their survival. Observers have concluded, "Poverty by itself does not cause environmental degradation," much depends on the policies of the government, local and national, options to the poor, and culture of the nation. So what causes environmental degradation are the exogenous shocks (flood, drought, collapse in commodity prices), which forces small-holders to follow unsustainable farming practices is a matter of concern for all.

1.4.1 The Root Causes of Environmental Degradation

- A study by Jagannathan (1989) shows that in both West Java and Nigeria, major observable changes in the physical system since 1970s such as, conversion of tree crop farms to food crop farms, clearing the bush for farming, disappearance of forest reserves and the expansion of human settlements, could be tracked back to stimuli from economic incentives.
- Panayatou (1990) states, "A combination of policy and market failures causes disassociation between scarcity and price, benefits and costs, rights and responsibilities, actions and consequences. The end result is an incentive structure that induces people to maximize their profits by appropriating other people's resources and shifting their own costs onto others, rather than by economizing on scarce resources and investing in enhancing their productivity.

In most cases, the root cause of these failures can be traced back to government policy and the nature of markets for natural resources. Thus the importance of a sound approach by the government is emphasized. The economic policies of the government will determine the sustainability or degradation of the environment.

1.5 Essence of our Manufacturing Policy

Indian manufacturing sector has huge potential given the natural endowments of the country. Use of state-of-art technology will bring efficiency in the production process. But the more important thing is to generate demand for "Made in India" products. Exploring untapped domestic and international markets will help in this endeavor. Indian firms tend to buy technologies rather than incubating them. Such, short term measures are counter-productive especially when global competitors enter the Indian market. India requires a "mental change" in the corporate world. This needs to be understood by the government and the community. A broad spectrum of developing countries in Asia, Africa and South America are testimonial to the fact that government policy failures are the root cause of unsustainable development. The need is to make our economic and industrial policies with sustainable development and environmental concerns at its heart. Our policy design is also an attempt on such lines.

Our policy takes a holistic view of the opportunities and threats that Indian industry faces in the 21st century. Focusing on manufacturing alone will not help the Indian industry. Thus, the policy seeks to establish links with infrastructure, education, environment and worker involvement.

Learning from the experiences of the other countries is essential for the Indian manufacturing sector. This helps make better and competitive policy. It also establishes links between countries for co-operation and mutual development. Our policy looks at examples from various countries.

Ecological responsibility of the nation is also very important. A developing country like India in its zest to grow fast industrially may follow a "dirty path" to progress, thus, creating environmental hazards for the community in general. Our policy's endeavor is to help define a "clean path" of growth for the manufacturing sector. The policy also stresses on value addition and self-reliance approach towards manufacturing.

2

CHAPTER TWO

International Comparison of Manufacturing

2.1 General Trend of the Ssector

Throughout the world, manufacturing currently faces a tough environment. The global slowdown has had a significant impact on manufacturers in all major economies - UK, US, Japan, and Germany (Chart below).

Figure 5 - Trend in Industrial Economies

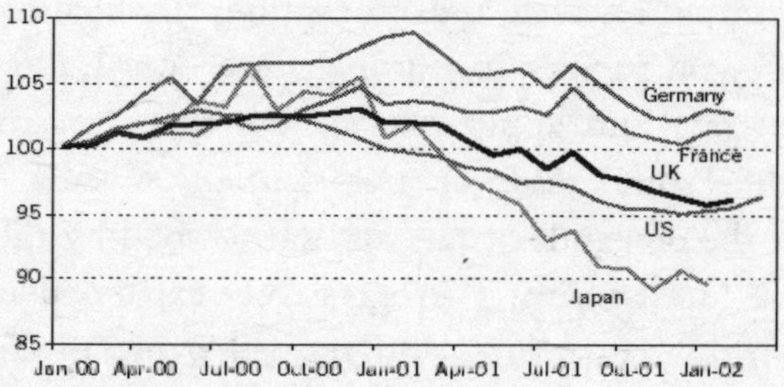

Source: National authorities, U.K

Figure 6 - Productivity Growth

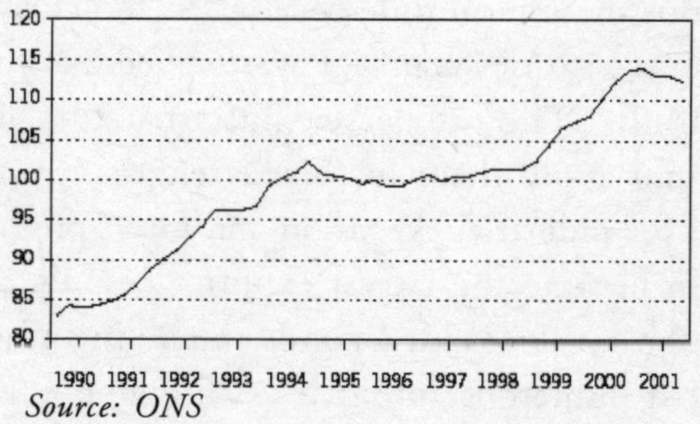

Source: ONS

- In the UK, manufacturing output declined 6.7 per cent over the 12 months from its peak in December 2000, with the loss of 150,000 jobs.
- US output fell 6 per cent
- Japan slumped 14 per cent and
- In Germany the reduction was 5 per cent.

The second chart shows that world over the productivity growth rate picked up sharply in the late 1990s, the middle part of the decade was marked by stagnation, which contributed to a sharp rise in unit labour costs relative to the competitors.

The following Chart shows

- UK manufacturing output as a proportion of GDP fell from 32 per cent in 1970 to 19 percent in 1999.

- In the US, output fell from 25 per cent to 16 per cent in the same period and 36 to 22 percent in Germany.

FIGURE 7 - PROPORTION OF MANUFACTURING GDP ACROSS THE GLOBE

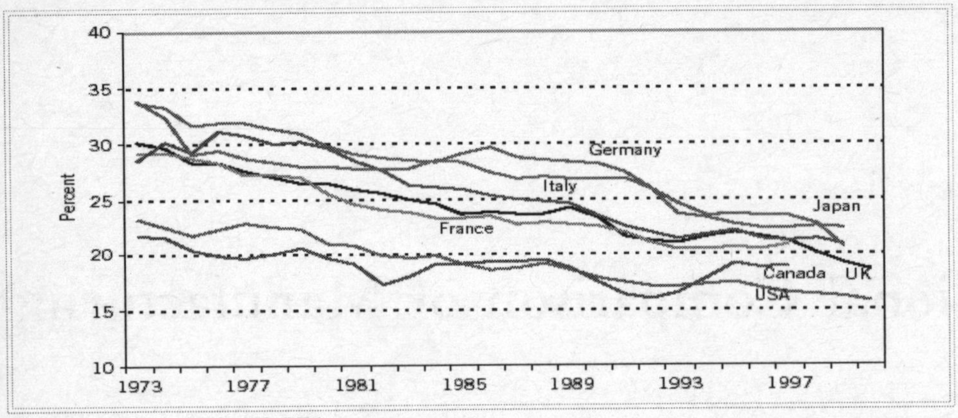

Source : OECD, ONS, US BEA

It is evident from the above data and charts that manufacturing is facing a tough time. Each country over the years has devised its own mechanism to counter recession and fall in trade. A study of the manufacturing policies of some other countries will help while making a manufacturing policy.

2.2 INNOVATION NETWORKS AND AGGRESSIVE LABOUR MODEL – SWEDEN

Sweden has done many things recommended by mainstream development economics:

- Its early land reforms stimulated agricultural productivity and provided capital and labour to emerging downstream industries.
- Its stable politics and reliable institutions minimized risks.
- Its efficient capital markets channeled resources to promising growth sectors.

But the critical difference lay in the environment of learning, research and innovation in which the process of resource-led development was embedded. Sweden was an equal partner in Europe's intellectual and scientific community. The intellectual infrastructure, both at an abstract and applied level, was there to absorb new ideas and apply them to their strengths, primarily based on natural resource area. Initially Sweden provided capital for industrial expansion and forest production to meet the demands of raw and sawn wood by the English. Sweden became the largest exporter of pulp and paper. The forest industry progressively expanded into more and higher value added goods, including pre-fabricated housing, furniture, and refined paper products, and acquired numerous subsidiaries throughout Europe to further process products.

Its phosphorous rich iron was processed using Bessemer process to support a new domestic industry. As in Britain and United States, Swedish mechanization was a slow process that implied ongoing accumulation of know-how and continuous interaction with the outside world.

2.2.1 Innovation Driven Manufacturing

Swedish manufacturing expansion was driven by innovations. Some of the examples of innovation are: -

- Steam Turbines.
- Centrifugal separators.
- Ball bearings.
- The adjustable spanner.
- The safety match.
- Air compressors.

- Automatic lighthouse technique.
- Various precision instruments

Moreover, Swedish trade unions were an enlightened lot. As early as the 1920s they strongly supported a productivity enhancing industrial policy. They placed a premium on continuous renewal of technology, plant organization, and machinery. The policy document, 1951 of the Confederation of Trade Unions stressed on competition. Pro-active policies to relocate displaced workers were also vouched for.

2.3 Manufacturing Policy – United Kingdom

Manufacturing creates a fifth of the national product of United Kingdom (UK). UK manufacturing also suffers from many weaknesses. The contry's policy makers are looking for restructuring the manufacturing policy of the country and have come up with a set of seven guidelines for the same.

Seven pillars of manufacturing success in UK are identified as:

1. Macroeconomic stability.
2. Investment.
3. Science and innovation.
4. Best practices.
5. Raising skills and education levels.
6. Modern infrastructure.
7. The right market framework.

2.3.1 SWOT Analysis

Strengths to be Exploited

- Strong science base and a culture that often produces new ideas, from domestic appliances to micro-chip design.
- The right economic environment consisting of sustained low inflation and lowest long-term interest rates.
- One of the world's largest markets.

Weaknesses

- UK on an average invests less than its competitors in Capital equipment.
- With some exception the R&D projects are not taken up.
- UK does not produce enough new and improved products as EU.

Opportunities

- Manufacturing is changing and UK needs to redefine itself as a provider of lifetime service around a manufactured product.
- The Internet is a major enabler and will initiate a paradigm shift.

Threats

- Presence of strong manufacturers in the neighboring countries of Europe.
- Effects of globalization and fierce competition from Asian manufacturers.

2.3.2 UK : Strategies for Future

- **Enabling Environment:** Government to improve the climate for manufacturing industry. A chief Manufacturing Advisor, appointed from the industry for a limited period to catalyze the necessary culture change across government will be one of the government initiatives.
- **Custom Driven Manufacturing:** UK's vision of Manufacturing is that it will be customer driven, high-value added, and customized product based industry. Mass customization and individual tailored products is thought to be the future of manufacturing.
- **Supply Chain Management:** A lean Just-in-time supply chain is of necessity in this model.
- **Strategic Alliances:** Companies should actively seek strategic alliances, often of necessity internationally, to broaden its services. They also need to sharply focus on technologies. Companies are looking to reassess their core competencies against globally competitive standards and benchmarks.
- **Environmental Standards:** The companies should lead in implementation of environmental standards. They should drive to reduce waste and demonstrate the sustainability of their operations.
- **Strengthen Relevant Research Areas:** Research providers and delivery systems must become more agile and approachable. The specific areas for future research focus are :-
  - Supply chain structure and web enabling the system.
  - Technologies for distributed manufacturing, micro-manufacturing and intensive processing, using less water and energy.
  - Technologies to enable the more effective and intelligent use of infrastructure – rail, roads and to enhance customer service.
- **Attract Young People:** Government and companies should aim to raise the awareness of manufacturing industry across society, improve the image of manufacturing industry and attract quality young people. Industry should aim to provide all school children and teachers with direct experience of the industry.
- **On the Job Learning:** Structure an open system that enables distance learning and then integrate this learning with on-the-job training. Educators and companies should group to provide a complete skill.
- **Self-Training:** Every individual needs to take advantage of the opportunities available inside the country to enhance and refresh his or her skill levels, both within employment and in his or her own time.

2.3.3 Strategic Planning by UK

The UK government has led an initiative that looks at what might happen in the future and what needs to be done now. The government believes in a strategic approach in which all plans and policies for various sectors create a synergy. Thus the government has set up the following panels to bring together the voices of business, government and people. The panels are: -

- Crime prevention
- Manufacturing 2020
- Built environment and transport
- Chemicals

- Defence, Aerospace Systems
- Energy and Natural Environment
- Financial Services
- Food chain and crops for industry
- Healthcare
- Ageing population
- Materials
- Retail and consumer services
- International communication and services

2.4 Manufacturing – Indonesia

Indonesia's competitiveness declined in the second half of the 1990s relative to other countries in Southeast and South Asia, three to four years before the south East Asian crisis. After spectacular growth of nearly 30% per annum in the early 1990s, export earnings slowed down to just 7% p.a., while those of the four major products (plywood, textile, garments and footwear) stagnated during 1993-97.

- Before the crisis, Indonesian manufacturers were already facing the prospect of a more difficult environment due to globalization of manufacturing production and liberalization (WTO, AFTA, expiration of Multi-Fibre Agreement and of guaranteed textile export quotas).
- The financial crisis, and the ensuing IMF-led restructuring programme, has greatly increased pressure on Indonesia to rapidly liberalize its domestic market, leading to the flooding of the domestic market of imported goods ranging from textiles and sandals, to motorcycles, hand tractors and consumer electronics.
- At a time when the domestic manufacturing sector has been greatly weakened by corporate indebtedness and inward foreign direct investment flows have virtually dried up, due in part to serious competition from equally reform-oriented countries, Indonesia now faces the real threat of de-industrialization.

A review of Indonesia's pattern of industrialization indicates that the factors contributing to Indonesia's lack of competitiveness both before and after the financial crisis are essentially the same, namely:

- The emergence of lower-cost competitors in Indonesia's traditional export products and markets leading to lower international prices, particularly of textile, garments and footwear.
- The inability of Indonesian producers to reduce production costs to match those of their competitors in China and other countries due to their heavy reliance on imported inputs.
- Indonesia's continued reliance on a small number of exports (plywood, textile, garments, footwear and electronics) and a small number of markets (USA, Japan, Singapore).
- Unlike China and India, Indonesia possesses a very small capital goods industry, and relied entirely on imported equipment and machinery.

In close collaboration and formal partnership with the private sector, the government intends to formulate and implement policies at three levels:

- **Industry-wide policies** will aim at reducing country costs and enhancing the transaction efficiency for business, including the necessary macro-economic, regulatory and business environment, the physical infrastructure, and the educated and skilled labour force under which manufacturing firms can invest and grow.

- **At the sub-sectoral level,** the government can support strategic initiatives and promotional activities in priorities industries, including trade and market diversification policies, industrial transformation programmes, industrial diversification and deepening, and strengthening industry associations.
- **At the firm level,** the government will aim to increase the manufacturing capabilities in a non-discriminatory manner in selected industries, including the adoption of total quality and productivity programmes, technological upgrading and investment in plant and industrial skills, using a network of private-sector led productivity centres, technical institutes and industrial extension services. Public-private partnerships in the formulation and implementation of industrial strategy can provide the necessary framework for the coordinated action of various government agencies involved in industrial development.

FIGURE 8 - R&D IN MANUFACTURING

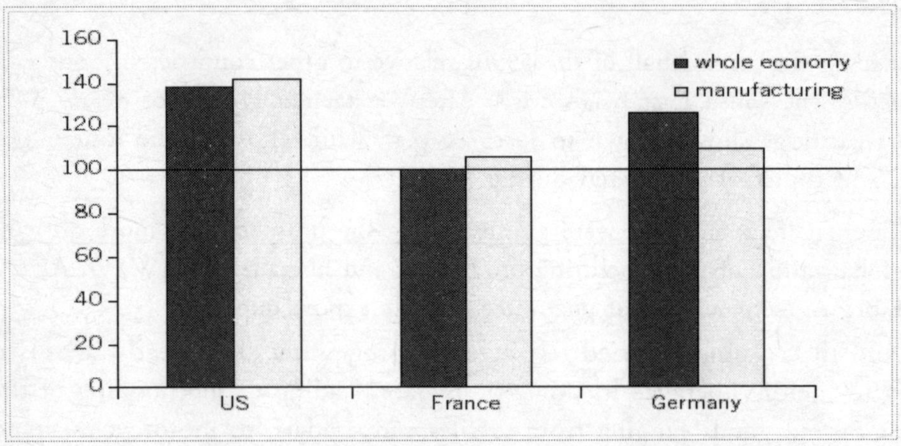

Source : Crafts and O'Mahorry Fiscal Studies September 2001 (Note: The R&D stock is the cummulative total of R&D spending over time, less the depreciation)

One of the most important determinants of the manufacturing operation is the kind of technology used. Research and Development is a tool to improve existing technology and to build newer designs and processes. A study of the world trend in R&D shows a lot of difference in approaches by developing and developed countries.

2.5 WORLD TRENDS IN R&D

The World Development Report of the Bank titled "Knowledge for Development" argues that "Governments in many countries have played a large role in the development and application of technology. In the past 50 years, among the handful of economies that have come a long way toward closing the knowledge gap with the global technological leaders, government was active in several, including Japan, Korea, and Taiwan, China.

- Korea followed a strongly interventionist and nationalist route, keeping FDI to a minimum and relying on other modes of technology transfer and a concerted domestic technological effort.

The Nobel laureate, Kenneth Arrow has also argued that

- R&D investment results in the creation of formalized knowledge. But the public good nature of knowledge makes it impossible for its generator to appropriate the full returns from this activity. This is known as knowledge spillovers.
- The existence of knowledge spillovers encourages private economic agents to under-invest (than is warranted by the social optimum) in R&D. Larger the spillovers, larger will be the desire to under-invest.

2.5.1 Trends in Developing Countries vs. Developed Countries

- While the developing countries are attempting to pare down the role of government in economic matters, often enough blindly in response to external pressures such as those from being a member of the WTO, the developed countries are putting in place a number of support measures that can help their private sector to commit more resources to R&D.

- The above view is supported by the efforts of the US and OECD governments in making public innovation policies.

Innovation is a key catalyst for growth. The aim is to increase manufacturing innovation performance, by

- Making the best use of the available science base.
- Utilizing technology from a range of sources, and by demonstrating the benefits which accrue to innovative companies.
- Both technological and non-technological aspects of innovation are keys to success. In particular, investment, skills and best practices, and close attention to customer needs, are essential if companies are to innovate successfully.

The Community Innovation Survey (CIS) shows that 3.2 per cent of UK manufacturing turnover was devoted to innovation activities in 1996, against a European average of 3.8 per cent. Companies with high R&D investment, like pharmaceuticals, biotech, oil and gas, and the food-processing sector, are the most successful all-round performers. The most rapidly growing manufacturing sectors tend to have the highest levels of innovation, as proxied by R&D.

2.5.2 Support for R&D in US

There is a consensus that Federal Research and Development spending stimulated much of America's technological leadership since 1945. Some prominent spin offs of governmental R&D efforts include:

- Automated air traffic control
- Microwave ovens
- Bio-technology industry
- Integrated circuitry
- Commercial satellite communications
- Jet aircraft
- Satellite-based global navigation and communication
- Super-computer and
- The internet.

The importance that US attaches to technology policy can be gauged from the following four indicators.

- There is complementarity between Federal investments in R&D and private sector investments.
- The rationale for this complementary relationship is based on the argument that as government makes more contract R&D money available to industry, firms increase their own R&D spending in hope of capturing more government funds.
- The fact that the US lobbied within the Uruguay round of trade negotiations to effect an upward revision of the degree of subsidizations in R&D expenditure of government is itself a clear indication of the importance that it attaches to public innovation policies.
- The main features and financial implications of the evolution of US technology policy with its emphasis on federal technology partnership programmes are summarized in the Table below. It is clear from the table that much of the support is in the form of intangibles, quite important in their own right.

TABLE 3

Key Features of Federal Partnership Programmes, USA

Fields	Co-operative Reserach and Development Agreement (CRADA)	Small Business Innovation Reserach (SBIR)	Advance Technology Programme (ATP)	Manufacturing Extension Programme (MEP)	Technology Reinvestment Project (TRP)
Project selection basis	The private sector partners are free to propose a research collaboration in area within the scope of a federal government lab's research mission	Government chooses mission topics on which it will consider proposals; applicants then propose specific projects.	All research areas are open to industry proposals involving high-risk enabling technology. Selection of proposals for funding is made by National Institute of Standards and Technology on the basis of extensive peer review of technical and economic issues.	MEP Centres provide help to all SMEs who seek it	Focuses on technology development and development issues of importance of the defence mission.
Government Financial Assistance	The labs may contribute human and physical resources to the collaborative project but may not directly contribute any funds. As a general matter, the lab contributes no more than 50 per cent of the total cost of collaboration	The funding agency pays the costs of the project, up to $ 1,00,000 for Phase I project and $ 7,50,000 for Phase II projects	Agency award co-operative agreements with required cost sharing	The entire programme consists of non-financial assistance	Co-operative agreement with required cost sharing
Non-financial Support	Technical collaboration is the mainstay of the agreement	None	Value added project oversight and technical support	do	Project oversight and technical support

Source : Broady (1996).

USA provides direct fiscal incentives for industrial R&D. Since 1954, industry has been able to deduct the full costs of R&D from income before taxes in the year in which they were incurred, while depreciating the costs of facilities and major equipment. Second, in 1981 with the passage of the Economic Recovery Tax Act of 1981, a series of special tax credits have been offered to firms that increase their R&D expenditure above previous levels.

2.5.3 Support for R&D – OECD

An OECD study titled "Public Support to Industry" has identified that the governments in these countries have been supporting technology creation and development through essentially three direct instruments, namely:

- Direct financing measures to support R&D activities of manufacturing enterprises.
- Civilian and defence-related R&D contracts.
- Support to intermediary R&D institutions

2.5.4 Critical Success Factors Identified for Innovation - UK

Government's Role

Government also has a key role in ensuring that.business is able to exploit results of the science base effectively. Knowledge transfer activities can overcome information failures between business and the science base. The Government also has an important role in fostering innovation, for example, in encouraging R&D collaboration and knowledge sharing thus enabling individual companies to capture knowledge spin-off from each other's research, and collectively to enjoy the benefits of economies of scale and scope in innovation. Similar benefits can arise from policies to reinforce the development of clusters. By locating near each other, innovative companies engaged in related activities can gain spillover benefits whereby the innovation activity of one company benefits others in the cluster.

FIGURE 9 - TRANSPORT INVESTMENT AS A PER CENT OF GDP

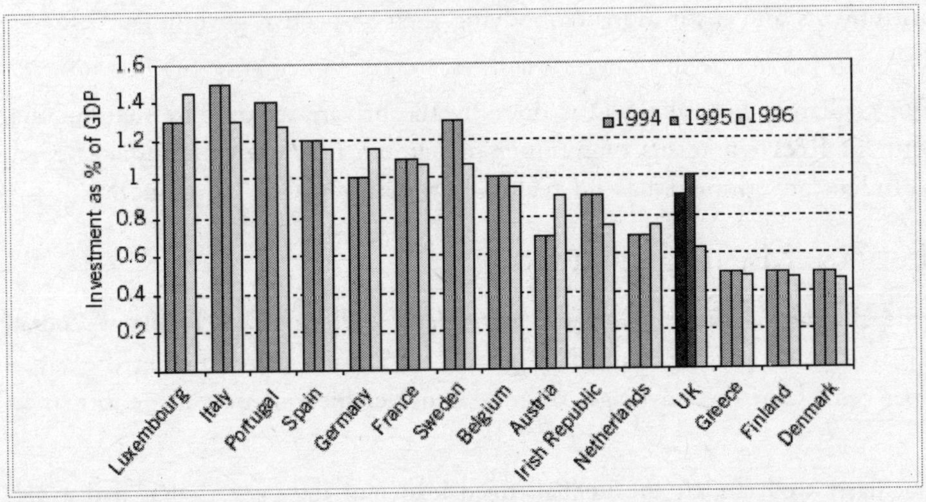

Source: Eurostat

Role of Universities

Universities have a vital role to play in the innovation process. The cornerstone of this is the role of universities in educating highly qualified manpower. This not only provides business with the manpower to undertake R&D and innovation (in its widest sense) but it is also the basis of the networks, which exist between universities and industry. International comparisons suggest that UK does relatively well at industry/science relations but there remains significant scope for improvement. The chart shows investment in R&D as a per cent of GDP worldwide.

Role of Labour

The quality of labour input is a key driver of productivity growth, both in manufacturing and in the wider economy. Higher-skilled workers are better able both to exploit the potential of physical investment, and to adopt new ideas and inventions.

Role of Public Infrastructure

Efficient public infrastructure underpins business competitiveness. Firms use public infrastructure, such as the road network and communication systems, to bring their products to the market. More efficient infrastructure lowers transport costs and facilitates more efficient distribution systems.

2.5.5 *What should the Indian Government do?*

This role is to be interpreted as initiating a set of policy instruments and institutions, which will aid the private sector to overcome the problem of apportionability of research output, and the consequent possibilities of under-investment.

- As far as basic research is concerned recent events and discussions have clearly shown that there is no other alternative but for governments to shoulder this in its entirety.
- As far as applied and developmental research is concerned, governments can put in place a variety of instruments through which its actual conduct in the private sector is strengthened.
- These instruments consist of-
  - Tax incentives to private sector to do more R&D.
  - Provision of adequate quantities of venture capital.
  - Financial aid in the organization and maintenance of technical standards
  - Administration of sharply focused partnership programmes in areas of high technology and the strengthening of the higher education sector.

The countries of East Asia, India and China, and some of the Latin American countries have relatively more advanced national systems of innovation. But with the exception of Korea and Taiwan, most of the countries do not have very clearly articulated instruments and institutions embodying their respective governments' forays into technology policies.

In India approximately one half of civilian industrial R&D is done by the private sector. The share of the private sector is in terms of the output of R&D in terms of number of patents taken within India is even higher at about 61 per cent. But the Indian government has no technology policy worth the name.

2.6 Competitiveness of Indian Manufacturing

In 1990s the rapidly globalizing developing countries (China, Brazil, Mexico, Philippines, Thailand, India, Argentina) measured in terms of increased trade participation have grown at 5.1 per cent (during the same time developed countries grew at 1.8 per cent). But these averages disguise considerable variation in performance among this group.

- China has done spectacularly well, and is the unchallenged leader of the pack. It has had a per capita GDP growth of 8.8 per cent during 1990-99.
- Thailand was another winner, inspite of the huge income compression due to the Asian crisis; it could still enjoy per capita GDP growth of 3.4 per cent during 1990s.

- Again, despite the crisis, Malaysia's average per capita GDP growth in the 1990s was 4.2 per cent.
- However, the per capita GDP growth of another relatively aggressive globalizer, Brazil, has only been around 1 per cent for 1990-99; and Philippines was 0.4 per cent.
- India, with per capita GDP growth of 3.7 per cent during 1990-99 is in the middle of the pack.

FIGURE 10 - PER CAPITA GDP GROWTH, 1990 -1999

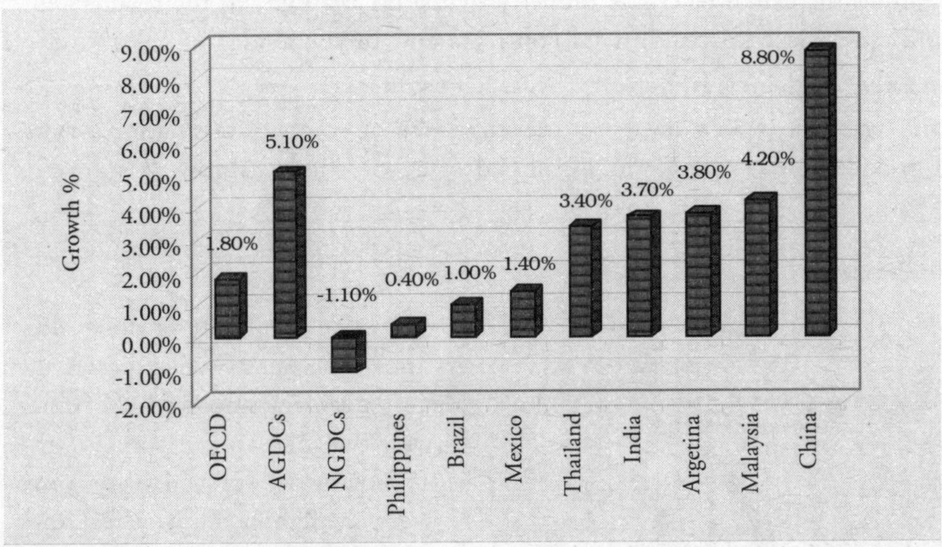

Source: *World Bank Report, 2000*
Note: *AGDCS are Aggressively Globalizing Developing Countries, NGDCS: Non-globalizing Developing Countries*

Foreign trade and investment is a necessary but not a sufficient condition to do well, greater investment and openness needs to be complemented with a host of other institutional factors and policies that can be classified under the broad "investment climate" at state levels.

2.6.1 India versus Others:- Cross-country analysis

A cross-country analysis can be done on certain driving factors for manufacturing in various countries. The following factors are compared below: -

- FDI as a share of GDP
- Governance
- Regulatory hassles
- Infrastructure and
- Labour cost

The above factors also contribute to the country's industrial output in comparison to other countries. The macro-economic factors such as stable macro-economic conditions, openness to trade particularly imported inputs, and a good rule of law is important. So, too are local market sizes and labour costs. On these last two dimensions, India seems to be well positioned. It has one of the largest domestic markets in the world and it has a large labour force available at relatively low cost. It also has well educated workers, particularly in areas of engineering and science. If these were the only determinants, India should be hugely successful at attracting investment.

In practice, as the graph shows, India does not receive the amount of FDI its market size would predict. India absorbs only 0.5 per cent of its GDP as foreign direct investment versus 4.1 per cent in neighbouring China.

2.6.1.1 Measure of Governance: Comparison with China

A work of Kaufman, Kraay and Zoido-Lobaton (1999), combines indicators from 13 sources under five categories that measure broad dimensions of governance. The categories include:

- Government effectiveness: which includes measures of bureaucratic delays, competence of officials, the quality of public service delivery and independence of the civil service from political pressures. These cover the elements needed for government to design and implement good policies.
- Regulatory burden: which includes the number of regulations within a market, the number of markets that are regulated, competition policy measures, and price controls. This provides a sense of how fair and market-friendly the business environment is.
- Rule of law: which captures the extent of crime, property rights, tax evasion and the legal system, its effectiveness. It indicates the enforceability of contracts and predictability of rules.
- Graft: which measures the frequency and size of irregular payments.
- Political instability and violence: which measures the incidence of coups, assassinations, riots, armed conflicts, and provides a measure of the likelihood of a violent overthrow of a governing party.

FIGURE 11 - THE GOVERNANCE DIAMOND, INDIA VS. CHINA

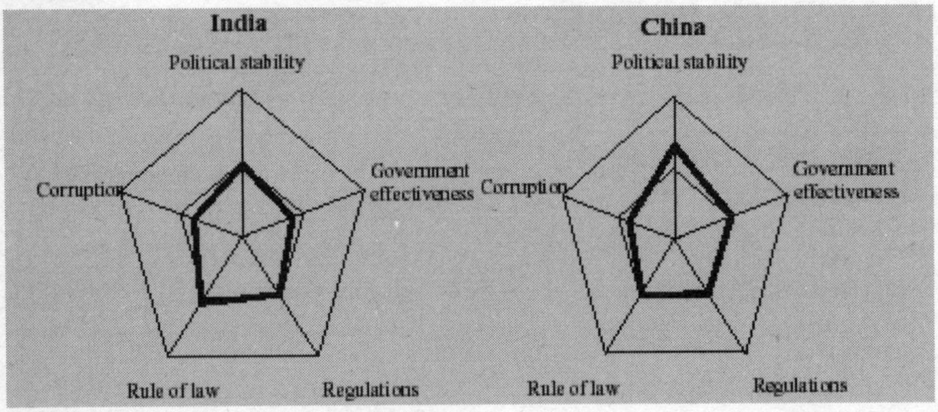

Source: Kraay and Zoido-Lobaton (1999)

The graph shows these measures. The outer web indicates the best performance, and the inner web the median measure for the 174 countries for which measures were available. Thus, the larger the country's web, the better is its governance measure.

India scores relatively well: near the median for political stability and regulatory framework, marginally better for rule of law, and somewhat less than the median for corruption and government effectiveness. The comparison with China does not reveal stark differences. China ranks better on political stability, somewhat better on government effectiveness and marginally worse on rule of law. Yet China received almost $ 40.3 billion as foreign direct investment in 1999, while India had to make do with $ 2.1 billion. Clearly, other factors are at work.

FIGURE 12 - REGULATORY HASSLES - COMPARISON

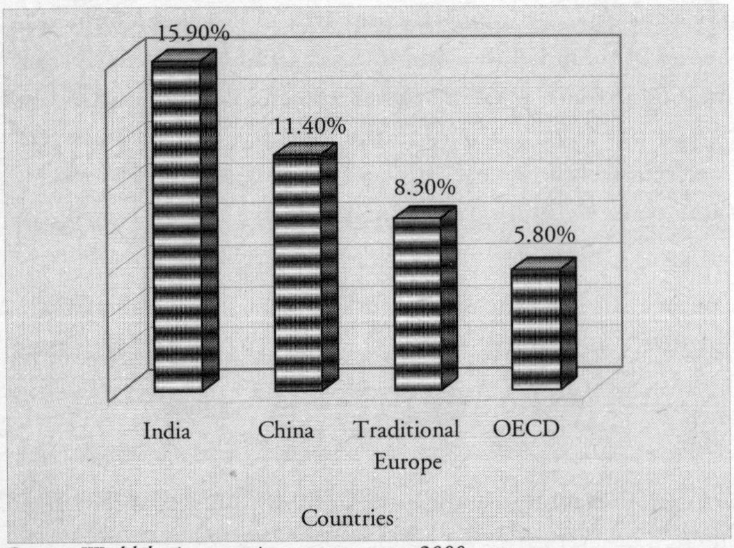

Source: World business environment survey, 2000

2.6.1.2 Regulatory Hassles

The World Bank World Business Environment Survey 2000 provides some additional clues. Two particularly useful measures are the amount of management time spent dealing with government officials and the frequency of making irregular payments. The amount of time spent by officials provides a direct measure of the regulatory burden faced by entrepreneurs. It also gives an indication of opportunities for irregular payments to be made. In the chart India registers the highest percentage figure. The figure show that in India 16% of management time is spent is dealing with government officials and regulatory authorities.

2.6.1.3 Infrastructure

Internet Hosts, Telephone Lines, Transportation Costs

Considering the number of Internet hosts per 10,000 people, the chart shows that on January 2000, India was at the bottom of the pack, having a fraction of the number in Thailand, Malaysia, or even the Philippines.

As far as the number of main telephone lines per 1000 people, India was at the bottom in 1998, with 22 line per 1000 people, compared to 37 for the Philippines and 70 for China.

Transportation is another area where India's infrastructure compares very unfavourably with that of its regional neighbours. Despite a longer route, shipping to the US eastern seaboard out of Bangkok is almost 18% cheaper compared to Mumbai or Chennai. Weighting for trade volumes, the overall transport cost advantage in shipping from Bangkok to the US is almost 23 per cent less. China enjoys a 13 per cent advantage in shipping garments from Shanghai to the US east coast, and a staggering overall advantage of 37 per cent. The major reasons for India suffering from such huge transport cost dis-advantages are delays and inefficiencies in Indian ports compared to their Asian counterparts.

FIGURE 13 - TELEPHONE MAINLINES PER 10000 PEOPLE, 2000

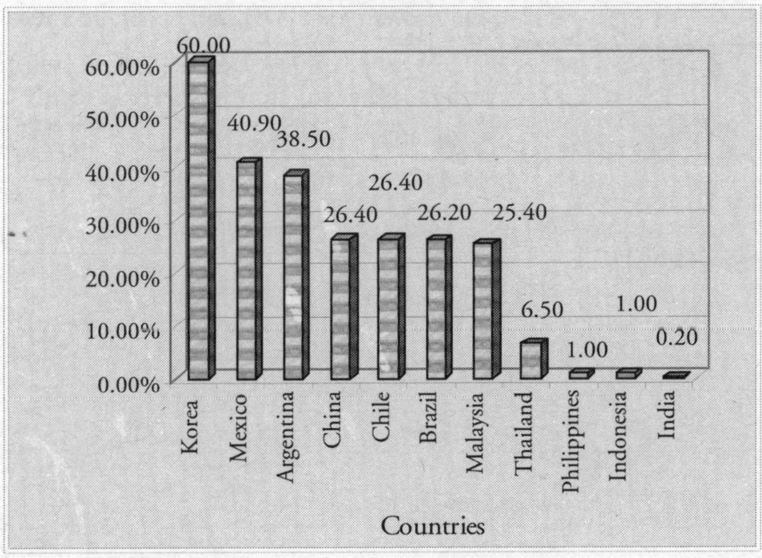

Source: The World Bank, World development report 2000/01

2.6.1.4 Labour Costs

The ratio of value added per worker in consumer electronics, garments and textiles compared across India, Philippines, Malaysia and Korea shows that India is not performing well.

The Firm Analysis and Competitiveness Survey (FACS) by CII, 2002 and World Bank showed that in value-added per unit of labour cost across the same three Industries, India is still worse off in consumer electronics. India is better than Thailand, Philippines and Malaysia in garments as well as textiles.

2.6.1.5 Share of Industry

India's low share of manufacturing in GDP is striking. India's share of manufacturing is lower than any of the so-called Asian tigers. India can be proud of its global achievements in software. But it will be difficult to attain a 7-8% annual GDP growth over the next decade only on the basis of services sector.

The FACS Firm analysis of competitiveness survey, CII and World Bank covering 1099 manufacturing companies in four major industries across 10 states in India showed that investment climate is very important for any developing country to boost its industry. Cross-country analysis shows that India enjoys a marginal advantage in value added per unit of labour cost.

However, the advantage is eroded by massive disadvantages in power costs, interest rate, delays at custom houses, infrastructural bottlenecks, and, to a lesser extent, regulatory hassles. Some of these constraints vary across states, with the better investment climate states being a bit more fortunate. The fact remains that the best investment climate states are still way behind the curve compared to India's South-east Asian and East Asian competitors.

The chart below shows India's share of Industry as very low. With a per capita GDP of $ 450 in 1999, India's share of manufacturing is lower than even Philippines.

FIGURE 14 - SHARE OF MANUFACTURING GDP - COMPARISONS

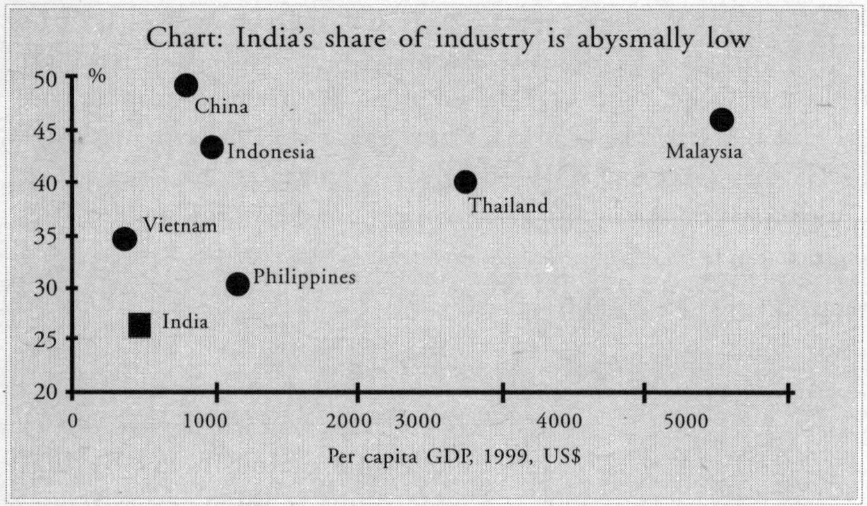

Source : FACS, CII and World Bank report

The reason why manufacturing in India is so low compared to the rest of Asia has to do with dys-functional policies, regulations and poor physical, and financial infrastructure. The above are also the things that constitute undesirable investment climate. Unless the right policies are made and followed India cannot expect a 7 plus per cent growth in GDP, and without such growth and a greater role of manufacturing, India will not succeed in eradicating mass poverty.

3

CHAPTER THREE

Global Competitiveness

3.1 Introduction

Why is it that some countries consistently outgrow others in terms of economic performance? Why is it that some countries, though gifted with natural resources, are not able to use that advantage? The answer lies in the overall competitiveness of the nation. It is competitiveness that makes a nation preferable to others in terms when it comes to trans-national investment decisions.

Globalization has brought unprecedented relevance to the perceived competence of a nation in developing itself. Since we as a nation aspire to be better than world-class in manufacturing, it will be useful to analyze where does India stand in terms of competitiveness.

3.2 Definition of Competitiveness

According to the Global Competitiveness Report published by the World Economic Forum, the definition of competitiveness is the set of institutions and economic policies supportive of high rates of economic growth in the medium term. Competitiveness represents a synergy of those factors that render a country with specific strategic advantages. It is important to note here that competitiveness is not merely the presence of comparative advantages, rather it tries to represent the willingness and the institutional support that makes it possible to utilize those advantages to the fullest benefit.

3.3 International Perspectives on Indian Competitiveness

The Global Competitiveness Report provides two broad types of measures - Growth Competitiveness and Current Competitiveness. The growth competitiveness index measures the factors that contribute to future growth of an economy, measured as the rate of change of GDP per person. These factors essentially explain why some countries can sustain a higher level of prosperity than others. The current competitiveness index aims to identify the factors that underpin high current productivity and hence current economic performance, measured by the level of GDP per person. These explain why some countries are improving their prosperity faster than the others.

3.3.1 Current Rankings and Future Indicators

A scrutiny of India's relative position with China in these two indices reveals some interesting insights.

In the 2001-02 report, the current competitiveness of India is ranked at 36 whereas China is at 47th place. What is surprising is the Growth competitiveness ranking which is a measure of future potential. China is ranked 39th, while India slipped 10 places from last year to 57th position this year. Thus, the relative growth potential of India has suffered a bleak outlook internationally. The following table provides a snapshot of areas where India's current and future potential and drawbacks lie.

TABLE 4

Growth Competitive Advantages, 2000

Indices	Ranking
Openness	
Export promotion	5
Finance	
Investment rate	14
Financial risk rating	14
Stock market	18
National savings rate	20

TABLE 5

Growth Competitive Disadvantages, 2000

Indices	Ranking
Innovation	
Intellectual property protection	48
Competitive advantage	48
Startups	
Starting a new business	39
Openness	
Capital account restrictions	57
Foreign access to capital markets	57
Import fees	59
Average tariff rate	59
Access to foreign capital markets	59
Finance	
Financial depth	45
Private share of banking system	49
Interest rate controls	53
Insider trading	43
Soundness of banks	45

In the growth rankings, India fares well in export promotion policies where a relatively high ranking (5) has been awarded, but openness as an oveall category has suffered because the other factors did not match up to the expected levels. Capital account convertibility, or rather the lack of it, has attracted quite a low rank. Average tariff rates, one of the highest in the world have also added to the list of disadvantages.

Depth of the financial system has come under scrutiny as well. Interest rate controls and ceilings and the corresponding high cost of capital are issues to be acted upon. Incentives for entrepreneurship is also pretty low, according to the report, since the facility for starting a business is not relatively helpful.

Table 6

Current Competitive Advantages, 2000

Indices	Ranking
Sophistication of Company Operations and Strategy	
Marketing	31
Licensing of technology	1
Production processes	40
Quality of the Business Environment	
Availability of suppliers	8
Legal framework independent of the state	21.5
Financial conditions and new businesses	26
Other Indicators - Government	
Median income tax rate	3
Value-added tax rate	12
Top marginal tax rate	12
Tax system encourages investment	22
Infrastructure	
Road indicator	2
Railroad indicator	15
Cost of air travel	18
Railroads	25
Technology	
Licensing of technology	1
Math and science education	13
Necessity of research by firms	18
Labour	
Union power	11
Institutions	
Litigation against government	16
Judiciary independence	24

Table 7

Current Competitive Disadvantages, 2000

Indices	Ranking
Sophistication of Company Operations and Strategy	
Exports to neighboring countries	49
Quality of the Business Environment	
Import fees	56
E-commerce	54
Government subsidies	54
Other Indicators	
Payroll tax rate	45
Tax evasion	45
Fiscal surplus/deficit	52
Government saving	55

Infrastructure	
Ports	51
Internet access	50
Telephone service	52
Roads	56
Telephone density	58

Current competitiveness advantages that India enjoys are in the areas of licensing of technology, availability of suppliers, and in tax rates. The good ratings here are due to the policies of economic dualism followed in the past, which catered to import of technology. A focus on letting small business grow has contributed to the presence of a large number of suppliers as well. However, such advantages are offset by the following factors where India ranks low. The competitive disadvantages include exports to neighboring countries which is limited since many of India's neighbors themselves are not exactly booming economies. China can become a high potential in this area, but specific targeting of the Chinese market has to be done. Government subsidies and tax evasion contribute to the inefficiency in governance along with budgetary deficits. Most importantly, as no ranking fails to note the dismal state of affair in the infrastructure sector.

3.4 The World Map of Competitiveness

The World Competitiveness Yearbook, published annually by IMD International, ranks nations based on certain Competitiveness Factors. These factors can be used in a detailed analysis and comparison of our nation's performance so far in relation to the inherent potential. The framework is as shown below:

Table 8

Factors Determining Competitiveness

Economic Performance	Government Efficiency
• Domestic Economy	• Public Finance
• International Trade	• Fiscal Policy
• International Investment	• Institutional Framework
• Employment	• Business Legislation
• Prices	• Education
Business Efficiency	Infrastructure
• Productivity	• Basic Infrastructure
• Labour Market	• Technological Infrastructure
• Finance	• Scientific Infrastructure
• Management Practices	• Health and Environment
• Impact of Globalization	• Value System

Source: IMD International

3.4.1 Comparison of India with Selected Nations

In order to get a holistic picture of India's position with respect to the rest of the world, the following countries have been selected.

TABLE 9

Global Benchmarks

Country	Comparative Analysis - Reason for Selecting
US	Global benchmark as a developed country
UK	Starting point of many of India's legal and contractual policies, Successful model of privatization
China	Proximity as a neighbour, Recent international attention
Brazil	Similar to India in economic and demographic structure, Developing country

A comparative analysis of the overall competitiveness rankings in each of these sectors with respect to India reveals the following insights.

3.4.2 Economic Performance

FIGURE 15 - COMPARISON OF ECONOMIC PERFORMANCE RATINGS

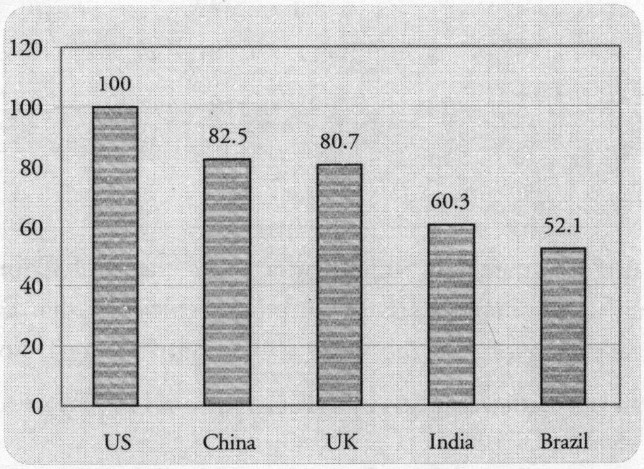

Ranking		Value
US	1	100
China	3	82.5
UK	6	80.7
India	23	60.3
Brazil	35	52.1

In the sphere of economic development, China has done quite impressively in the past two decades. In relation to China, India falls quite below in the rankings to 23rd position. The chief reason is that growth rate of GDP is consistently falling short of desirable levels, especially in the past five years. Secondly, the FDI in-flow has not been touching the full potential of doing business in India, while security and promise of steady returns are available for foreign investors in China.

FIGURE 16 - COMPARISON OF GOVERNMENT EFFICIENCY

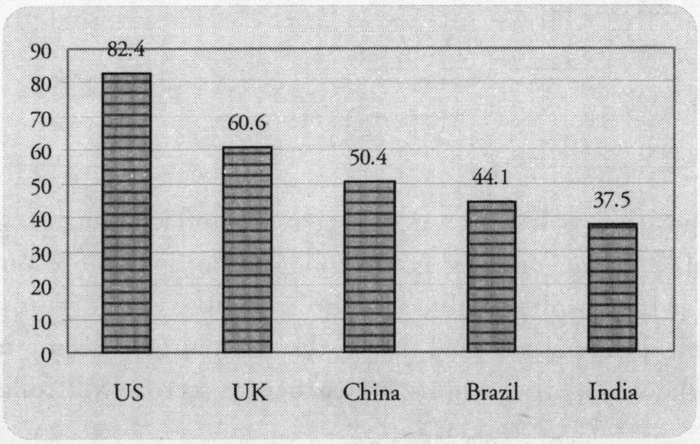

Ranking		Value
US	3	82.4
UK	24	60.6
China	35	50.4
Brazil	36	44.1
India	43	37.5

3.4.3 Government Efficiency

Efficiency of governance is a weakness area when compared with China and Brazil. Developed nations USA and UK are far ahead. What causes this huge disparity even though India is touted as the biggest democracy in the world? It can clearly be seen that the benefits of building a political system in which theoretically every citizen has an opinion are seriously undermined by the growth of bureaucracy and corruption.

3.4.4 Business Efficiency

FIGURE 17 - COMPARISON OF BUSINESS EFFICIENCY

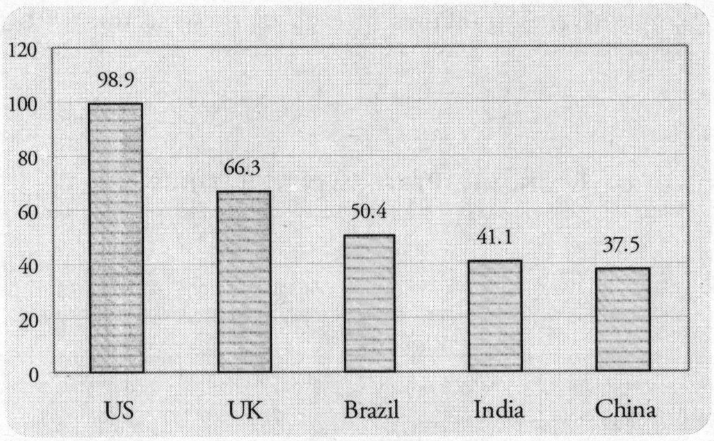

Ranking		Value
US	1	98.9
UK	17	66.3
Brazil	33	50.4
India	40	41.1
China	43	37.5

Efficiency of business or corporate efficiency is one parameter in which India ranks ahead of China. Even so, the ordinal gap is not substantial. In fact, the short advantage that we have incorporate efficiency that can be easily surpassed by China in the coming years since they are taking care of the other three contributing factors well.

FIGURE 18 - COMPARISON OF INFRASTRUCTURE RATINGS

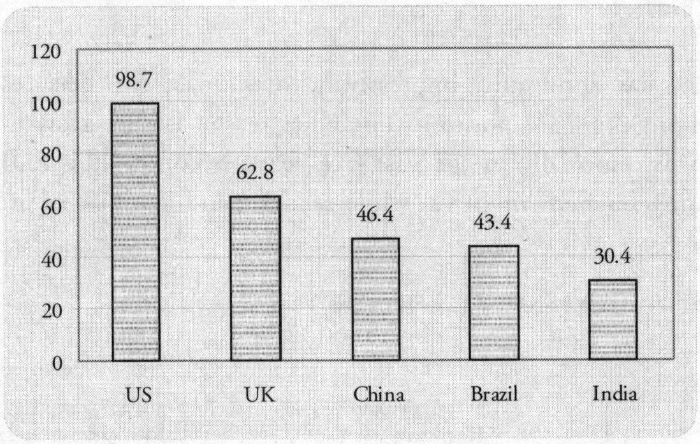

Ranking		Value
US	1	98.7
UK	21	62.8
China	35	46.4
Brazil	37	43.4
India	47	30.4

3.4.5 Infrastructure

The infrastructure ranking of India at 47th position is low, but does not really reflect the true gravity of the situation. On a scale of 100, the absolute value of the ranking is around 30. Comparable economies of China and Brazil are at least 50 per cent ahead, with absolute values in the mid 40s. Apart from all other factors, on the basis of rankings alone, it is possible to conclude that this is the most urgent sector to act on, for enhancing India's competitiveness. A detailed analysis of the various infrastructure sectors will follow in the subsequent sections.

4

CHAPTER FOUR

India's Developmental Strategy

INDIA has been following a protective policy framework till the liberalization initiative taken up in 1991. Most of our policies formulated had a socialistic perspective and now with the liberalization and globalisation reforms, the focus has to be shifted to a free society. The labor laws and regulations which were formulated with a different focus need to be amended to the changing thrust of the country. Though amendments have been made to the industrial policy, MRTP Act, FERA to FEMA, still a lot of reforms have to be made to make them suitable for a free economy. Liberalisation and Privatisation by itself will not lead to efficiency in operations of firms and growth of the economy until a proper competitive environment is created for firms to innovate, improve and compete with others to grow productively. The need now is for a comprehensive framework with competition policy, manufacturing policy, technology policy, infrastructure policy, industrial sector-wise policies and state-wise policies for leading the economy towards sustainable growth and development.

4.1 The Five Year Plans - Changing Emphasis

The Five Year plans provide important insights into the policy evolution after independence. An analysis of the India's developmental strategy during the various Five-Year Plan periods is discussed below.

4.1.1 First Five Year Plan: 1951-52 to 1955-56

The First Five Year Plan was formulated against the backdrop of the Second World War and partition of the country.

- It accorded high priority to agriculture, irrigation and power projects, in order to reduce dependence on food grain imports, solve the food crisis and ease the raw material problem particularly in jute and cotton.
- Almost 45% of the resources were allocated to agriculture, while industry got a paltry 4.9%. The strategy in the first five-year plan was to increase the development potential of agriculture, which will provide a platform for industrial sector to flourish.
- The plan also aimed to increase the rate of investment from 5% to 7% of national income.
- Though the first plan was conceived in haste, it was highly successful in as far as attainment

of objectives go. Agricultural production targets were surpassed, national income rose 18%, per capita income 11% and per capita consumption 9% during the plan period.

4.1.2 Second Five Year Plan: 1956-57 to 1960-61

In the second plan, which was formulated in an atmosphere of economic stability, agriculture was given a complementary role and emphasis was on the industrial sector especially heavy goods industry.

- Agricultural programmes were aimed at meeting the raw material requirements of industry besides meeting the food needs of the increasing population. Industrial sector was perceived as the leading sector that could enable the economy to grow at a rapid pace.
- The Industrial Policy of 1956 was formulated with a socialistic pattern of society as the goal of economic policy.
- The plan also aimed at a 25% increase in national income and a multi-fold increase in employment opportunities. Emphasis was laid on a more equitable distribution of income and wealth and an even distribution of economic power.
- The development programmes in other sectors envisaged close inter-dependence amongst all sectors of the economy.
- Compared to the first plan the second plan was only a moderate success. Unfavorable monsoons in 1957-58 and 1959-60 adversely affected agricultural production. Also the Suez crisis cut off international supplies leading to a jump in commodity prices.

4.1.3 Third Five Year Plan: 1961-62 to 1965-66

At the time of formulating the third plan, it was found that the limiting factor in economic growth was growth in agricultural production.

- Accordingly agriculture was given top priority again. Also allocation to power sector was enhanced considerably to 14.6% of the total outlay.
- Power generation was seen as an important factor in ensuring rapid growth of industrial sector, which could lead to self-sustaining economic growth.
- The relative weights of other sectors in the economy were broadly maintained at the earlier levels.
- Emphasis was on attaining self-sufficiency in agriculture and industry.
- The goal of import substitution was seen as inviolable and opportunities cost estimates were entirely ignored. In a zeal to prevent monopolies and to promote economic development in backward regions, unviable plants were propped up by subsidies.
- The plan aimed at increasing national income by 30% from Rs. 145 bn in 1960-61 to Rs. 190 bn by 1965-66 and per capita income by about 17% from Rs. 330 to Rs. 385. It also targeted a 30% increase in agricultural production and 70% in industry.
- These targets could not be achieved as the plan period was marred by two wars, the Chinese war in 1962 and the Pakistan war in 1965. Also, except for 1964-65, monsoon failed in all other years, with 1965-66 being the worst year.

4.1.4 Annual Plans: 1966-67 to 1968-69

The government was forced to declare a 'plan holiday' for 3 years, sensing the mood of pessimism that had been generated at the major failure of the third plan. In place of the fourth plan, 3 annual plans were introduced.

The economy, which was in the grip of a recession, had to grapple with another drought in 1966-67. Agricultural production rose marginally over the earlier year, but continued to remain much lower resulting in shortage of food grains. The lower agricultural production adversely affected industrial production in 1966-67. Recovery began in 1967-68, with a recovery in the agricultural sector. This was due to the twin effect of a good monsoon and the technological advancements like introduction of high yielding variety seeds and multi cropping techniques.

4.1.5 Fourth Five Year Plan: 1969-70 to 1973-74

At the time of formulating the fourth plan it was realized that GDP growth and high rate of capital accumulation alone might not help improve standard of living or help attain economic self-sufficiency.

- Emphasis, therefore, shifted towards providing necessary consumption benefits to the less privileged and weaker sections of the society through employment and education.
- The plan also aimed at accelerating the momentum of economic development and improving stability of food grain production.
- Share of outlay to agricultural sector was increased to 23.3% while that of industrial, transportation and communication sector was cut back. Family planning programme received a big impetus and was allotted Rs. 2780m against Rs. 250m in the previous plan.
- The fourth plan aimed to increase the net domestic product (at 1968-69 factor cost) from Rs. 290.7bn in 1969-70 to Rs. 383.1bn in 1973-74, reflecting a 5.7% CAGR.
- Foreign assistance was considerably reduced in the fourth plan to 12.9% of total resources, from 28.2% in the earlier plan period. Achievements in the fourth plan were way below targets.
- The last three years of the plan were quite disappointing; with problems arising from influx of refugees from Bangladesh and Indo-Pak war in 1971. Agricultural growth was only 2.8% p.a. against the targeted 5% p.a. coupled with successive failure of monsoons. The much-touted green revolution seemed to be lacking the spread effect. Industrial growth was only 3.9% against the targeted 8% p.a.

4.1.6 Fifth Five Year Plan: 1974-75 to 1978-79

During the fourth plan period the country faced severe inflationary pressures. The fifth plan therefore concentrated on reigning inflation and achieving stability in the economic situation.

- It also aimed at improving the quality of life of especially the downtrodden section of the economy. Several new economic and non-economic variables such as nutritional requirements, health, family planning etc were incorporated in the planning process.
- Investment mix was also chosen based on demand envisaged for final domestic consumption. Industry got the highest allocation of 24.3% of the plan outlay.
- In agricultural sector irrigation programmes attained prominence and outlay almost trebled to Rs. 39bn from Rs. 13.5bn in the previous plan. The plan targeted an annual growth rate of 5.5% p.a. in national income. Foreign assistance was reduced further to 7.5% of total resources, while reliance was placed on domestic capital and own resources.
- The fifth plan was terminated on the event of a change of government in its fourth year. Only the targets relating to food grains and cotton cloth were achieved in the fifth plan. The plan was introduced at a time when inflation was very high. The cost figures therefore had to be revised upwards and the targets had to be scaled down.

4.1.7 Sixth Five Year Plan: 1978-79 to 1982-83 and 1980-81 to 1984-85

The sixth plan, instead of the GNP approach to development, sought to achieve higher production targets with concomitant increase in employment opportunities for the poorest section of the society.

- Though the emphasis on irrigation and power and the shift in approach to development was laudable, the government lacked political will to forge ahead.
- The Congress Government on assuming power in 1980, formulated a new sixth plan, with a strategy to move simultaneously to strengthen the infrastructure for both agriculture and industry in order to achieve rapid economic growth.
- Within agriculture, emphasis was on rural development - the outlay was increased from 2% to 5.5%.
- Special programmes were designed to provide increased employment opportunities in the rural areas and unorganized sector and meet the minimum basic needs of the people. In the sixth plan industry achieved a growth of 6% p.a. compared to the 5.3% growth in the earlier plan period.

4.1.8 Seventh Five-Year Plan: 1985-86 to 1989-90

The first three years of the seventh five-year plan saw severe drought conditions, despite which food grain production grew 3.2%.

- Policies were aimed at rapid growth in food grain production, higher employment levels etc. Several special programmes like Jawahar Rozgar Yojana were introduced.
- Social sectors like welfare, education, health, family planning, employment etc. which were accorded low priority in the earlier plans, got a higher outlay - 21.6% in the seventh plan.
- Industrial growth also accelerated during this plan period to 8.7% p.a. Rural electrification increased significantly and covered 81% of the villages by the end of the seventh plan. Actual public sector expenditure in the seventh plan increased by 21.5% against the envisaged outlay.

4.1.9 Eighth Five Year Plan : 1992-93 to 1996-97

The eighth plan was launched immediately after a severe balance of payment crisis, which was accentuated by the Gulf War in 1990.

- Several structural adjustment policies were introduced in order to put the country on a higher growth path and remedy the precarious balance of payments situation.
- These included a substantial devaluation in the value of rupee, dismantling of licensing requirements, reducing trade barriers, reforms in the financial sector and tax systems.
- The plan targeted an annual growth of 5.6% in GDP and 7.5% p.a. in industry, while keeping inflation under control. Higher growth rates were envisaged in most of the sectors - manufacturing, agriculture, foreign trade etc. Achievements in the infrastructural sector, however, fell short of the targets.
- Deficit financing to meet the resource shortfall was curtailed to 4.6% of total resources as against 15.8% in the previous plan period. Government's own resources increased considerably to 42.2% of the total resources.

4.1.10 Ninth Five Year Plan: 1997-98 to 2001-02

It was found in the eighth plan that though the economy performed well, the benefits did not percolate down to the poor and disadvantaged sections of the economy.

- The key task in the ninth plan therefore was to improve the living conditions of the poor and provide them with adequate employment opportunities.
- The ninth plan makes serious efforts to raise the level of agricultural and rural incomes and target programmes at small, medium, marginal farmers and landless laborers.
- To increase real income the plan also aims to control growth rate of population. Allocation to power sector shot up to 25.4% of the total outlay.
- A growth rate of 7% in GDP has been targeted for the ninth plan, with current account deficit of 2.4% of GDP and savings rate of 26.2% of GDP.
- Reliance on own resources increased to 54.3%, while domestic capital receipts' contribution is slated to fall to 38.8% of total resources. Foreign assistance continues to remain low at 6.8% of total resources.

4.1.11 Trends in the Five-Year Plans

The analysis shows the targets set by each five-year plan. The plans were not able to meet their desired growth rates because of a number of factors. But in the ninth five-year plan some new line elements have been included in the governments plan outlay like rural development, science, technology and environment. A study of the percentage change in plan outlay of sectors like Agriculture, Irrigation, Power, Transport, Communication, Organized industries and mining shows the government's thrust areas are varied at different points of time.

FIGURE 19 - BREAK UP OF 5-YR PLAN OUTLAYS

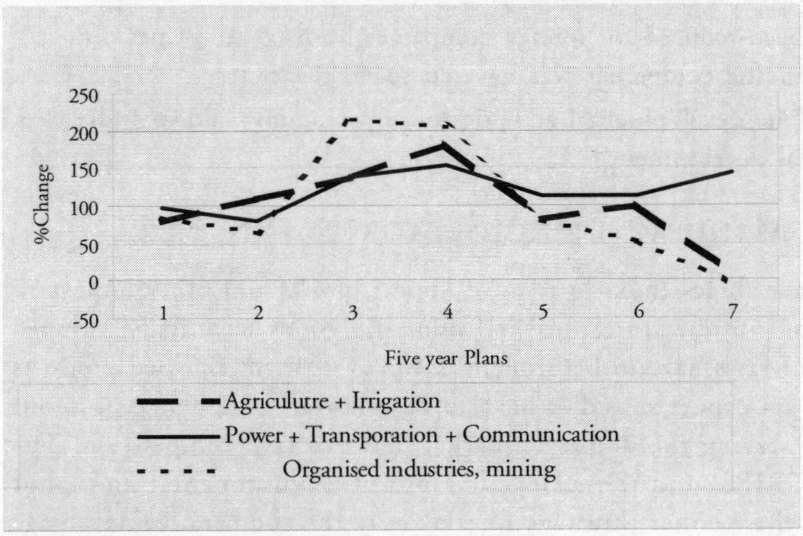

Source: Calculated from Indiainfoline

The importance given by the government to each of the following groups of industries: I- Agriculture and Irrigation, II- Power, Transport and Telecommunication and III- Organized industry and mining is shown by the graph. Though the value of outlay is not shown a variation in the percentage outlay planned for them shows that the government outlay for Agriculture and irrigation is consistently coming down. A number of factors may be responsible for it. But, the two together even in the ninth plan account for Rs. 944 billion. Agriculture outlay has been brought down in the ninth plan.

The highest planned outlay is for the Power, Telecommunication and Transportation sectors. Power alone has a planned outlay of Rs. 2219.7 billion. Actually, the power sector is a very cost-intensive one and requires huge investments. This clearly shows the stress government is giving on improving this sector. But money alone will not solve the problem of a deficit in power. The plan outlay for organized industry and mining is only Rs. 716.8 billion which is clearly the smallest among the three which is mainly due to the presence of a large number of private sector players.

4.2 Taxation - Incentives, Rebates and Allowances

The government has been providing concessions or Tax Holidays as they call to attract investments in thrust areas of concern and thereby lead to faster growth of economy.

The Government offers many incentives to investors in India with a view to stimulating industrial growth and development. The incentives offered are normally in line with the government's economic philosophy, and are revised regularly to accommodate new areas of emphasis. The following are some of the important incentives offered, which significantly reduce the effective tax rates for the beneficiary companies:

Five-year tax holiday for:

- Power projects.
- Firms engaged in exports.
- New industries in notified states and for new industrial units established, in electronics hardware/software parks.
- Export Oriented Units and units in Free Trade Zones.
- As of 1994-95 budget firms engaged in providing infrastructure facilities, can also avail of this benefit.
- Tax deductions of 100 per cent of export profits.
- Deduction of 30 per cent of net (total) income for 10 years for new industrial undertakings.
- Deduction of 50 per cent on foreign exchange earnings by construction companies, hotels and on royalty, commission etc. earned in foreign exchange.
- The corporate income tax has been reduced for foreign companies from 65 to 55 per cent (FY 1994-95). The tax rate for domestic companies was lowered to 40 per cent.

Incentives and concessions are critical for the developmental activities of any economy and in India also, it is being widely used for promoting industrial development.

4.3 WTO Implications for the Manufacturing Sector

The WTO regulation on multi-lateral trade forces India to remove import quotas and other important restrictions and to reduce import tariffs. Also, subsidies to export and domestic goods have to be stopped within the time-limits specified by WTO. Tariff rates have to be brought down to 40% for finished products and 25% for intermediate goods. In addition all export subsidies have to be removed by 2003. As a result all enterprises in India whether exporting or serving the domestic market, have to face competition. The non-trade issues of WTO such as TRIMS, TRIPS, and the stringent sanitary, environmental and labor standards also intensify competition. Some of these either threatens to take away the comparative advantage of Indian industries or to enhance the competitive advantage of new technologies. Hence, the manufacturing policy should provide for a comprehensive approach to develop industries inspite of competition from imports.

4.4 Competition Policy in India:- Issues

In India protection and controls are being replaced by a competitive and de-regulated open economic system. Such an initiative will attract more competition from outside the country. Moreover, private participation in previously reserved areas will increase.

4.4.1 Concept of Competition and Competitiveness

Competition: The 'competition' policy, rules are set for firms (or individuals) to compete within a sector within the geographical territory of the country or outside. The 'competition' policy is mainly concerned with easing of restrictive business practices (RBPs) that hinder free and fair competition.

Competitiveness: On the other hand, 'competitiveness' reflects keenness or urge to compete. It indicates the capabilities of a firm or a sector or a nation to compete successfully. In a similar vein, the 'competitiveness' policy refers to enhancing the strength of national firms (or industries) or a nation relative to foreign competitors or other nations.

4.4.2 Pre-Reform Restraints

In the pre-reform era (before 1992), various restraints to competition existed. They are as follows:

- Investment restraints due to licensing policy followed by the Government of India.
- Control over acquisition of economic power through Monopolies and Restrictive Trade Policies (MRTP).
- Public sector reservation for infrastructure and other industries creating monopolies in various areas.
- Product reservation for the small-scale sector.
- Government procurement policies favouring public and small sectors.
- Trade restrictions and high tariffs and
- Restrictions on foreign direct investment.

All these restraints have been relaxed now. One key issue in the current phase of transition is ensuring and managing competition, to derive the most out of liberalization.

4.4.3 Pre-conditions to Enhance Competitiveness

Ensuring of free competition is only one of the pre-conditions for enhancing competitiveness. The other, more vital, conditions are: -

- Promotion of investment in intangible assets.
- The development of co-operation between firms (especially between local and technologically superior foreign firms) and
- Proper functioning of institutions.

4.4.4 Suggestions for Competition Policy

- **Exemption from anti-trust:** Agreements covering technology development, standards etc among firms having moderate share (20-25 per cent) should be explicitly exempt from anti-trust liability. The market share can be higher in product groups where import competition is possible.
- **Necessity of Innovations:** Market definition should be tailored to the context of innovation and should focus primarily on the market for knowledge. The geographic market should be considered to be worldwide.
- **Joint efforts:** The law should explicitly include joint commercialization efforts to exploit innovation as part of R&D cooperation arrangements. So the efforts put in R&D can be exploited for commercial purposes.

- **Abuse of Cooperative alliances:** An administrative procedure should be created to evaluate cooperative arrangements among firms with higher market shares for their dynamic efficiency gains and anti-competitive nature. Processes could be such as to result in sufficient incentives for regulatory bodies to detect and prove cases of abusive cooperative arrangements.

The competition policy should be evolved in such a way that it ensures fair competition in the industry and enables firms to compete by improving efficiency.

5

CHAPTER FIVE

Manufacturing Sector
A Comparison among Indian States

EXAMINATION of industrial activity and investment flows across states reveals that some states lag behind with low levels of activity and investments. This brings out the need for the states to be competitive, with a more realistic approach towards industrialization. The concept of competitiveness suitably modified at the state level can be used to evolve a policy matrix towards this end. The role of state governments is crucial in states where investments do not flow in. These states have to break a sort of low-level equilibrium induced by perceptions of the investors caught up in tussle between 'historical performance of the state versus expectations from the state'.

5.1 Background

Competition is expected to usher efficiency and enhance growth. Competitiveness at the state level is important as states are forced to attract investments hitherto channeled by the centre. The proponents of economic reforms in India have conveniently bypassed the issue of widening inequality among states. Eyeballing the data on per capita net state domestic product (NSDP) clearly reveals that the state with the highest per capita income, Punjab, has an NSDP five times that of Bihar, which has the lowest per capita NSDP. Despite a decade of economic reforms there exists wide disparities in regional growth/development, which might get worse given the nature of policies pursued, especially with regard to industrialization, unless states equip themselves to improve their competitive positions.

5.1.1 Concept of Centripetal and Centrifugal Forces

- Centripetal forces are those tending to keep growth in certain confined packets.
- Centrifugal forces are those tending to pull away growth into new regions and new levels.

5.1.2 Manufacturing Activity and Investment Across States

In the Indian context the five states of Maharashtra, Gujarat, Karnataka, Andhra Pradesh and Tamil Nadu consistently register high level of growth. Central government has decreased the level of investment in states and there is more provision for states to manoeuvre in terms of industrial policies. This will further widen the disparities among states if the laggard states don't take adequate policy changes.

5.1.3 Disparities between States

The disparities existing between states are as follows: -

- 60% of the investments of private sector in manufacturing are cornered by the top six states. Maharashtra, Gujarat and Tamil Nadu head the list with more than 11% of the investments in each state. (CMIE, April 1999).
- From August 1991 to December 1999, the top six states accounted for nearly 42 percent of total FDI approvals. The share increases to 57% if Delhi is included.
- Top five states account for 60% of total LOIs and IEMs.
- Examination of the financial intermediation through banks and FIs shows that over the last decade 68% of the mobilized resources are distributed among the top six states.

The lop-sided development will harm the overall economy of the state because investments in the long run will yield higher profits but with increasing capital-output ratios or declining total factor of productivity.

5.1.4 Why is the Investment flowing to Specific States?

- The rate of return is not similar across states.
- Infrastructure development is taking place in a differential manner, with some states being more pro-active in their approach.
- Centripetal forces as explained by Krugman (1988) like forward and backward integration linkages in production and centrifugal forces (incentives given by other states, land grants, tax relaxation etc.) interact, with dominance of the former.

5.1.5 Investor Perception

Central policies of public investment have been acting against the equalization of infrastructural facilities. The importance of perceptions of investors in attracting investments is evident in case of Kerala. The state, in terms of investible surpluses, is endowed with reasonably good levels of funds as is evident from low credit-deposit ratios of banks. The demand for credit from the available stock of deposits is too low to spur production activities. This is in spite of good levels of infrastructure. Here, the major problem is the perception of the investors, who have been influenced by the character of labour in the state.

5.2 Elements of Competitiveness

Competitiveness is an alloy of numerous factors and measurement of factors is very important for improvement. The above format of "assets, processes and performance" provide a comprehensive framework for understanding the concept. Competitiveness in this framework is seen as the outcome of the interaction of three facets:

- A set of assets, which the economy would like to possess.
- A set of processes sequenced in such an order as to enable the efficient use of the assets.
- A series of expected outcomes culminating from the correct combination of assets and processes.

The differentiating criteria that determine competitiveness are listed below. Efforts to improve on these parameters within the individual state causes the overall business climate to change in favour of the investor.

TABLE 10

Parameters of Competitiveness

Criterion	Examples
Competitive assets	
Factor costs	Wages, material prices, etc
Human resources	Educational skills, quantity and quality of labour force
Industry infrastructure	Related and supported industry
Technological factors	Investment in R&D, etc
Demand conditions	General environment in the economy
Government	Supportive, incentive, facilitating
Intangible assets	Brands, relationships and intellectual properties
Leadership	Vision, values and ethics
Competitive processes	
Strategic management	Formulation of strategies
Implementation	Feasible projects, employee involvement, etc
HR development	Skill updating
Technological management	Collaboration, technology/transfer
Synergies	Networks and linkages
Competence building	Alliances and knowledge management
Competitive performance	
Higher productivity	Low employee to sales ratio
Human resources	Higher levels of employment
Quality	Standardisation of systems and Benchmarking practices
Lower costs	Cost reduction and cost control initiatives
Financial	Higher rate of return on investments
Technological	More patents
Intangibles	Perceptions of investors

Source: Adapted from Momaya and Ajitabh (2000)

5.3 Case of Firm vs. Economy: Role of State

In the case of a firm the assets necessary are limited and can be acquired, while in an economy, assets, which are given at a point of time, have to be harnessed with the appropriate policies. It is here that the role of state becomes crucial. The following are some policies the state government can make: -

- **Human resources and industry infrastructure,** state should provide educational facilities adequate to equip labour force with essential skills. This will increase labour productivity and reduce labour cost. Thus on the whole factor cost becomes competitive.
- **The competitive processes** have to be coordinated by states as market forces might fail to achieve proper coordination. This is where state governments plays a role as most of the policy decisions are left to them.
- **Environment:** - State governments need to create an "environment". The investor wants a secure and acceptable society. Attitude of the people matter.

Asked about the reasons for establishment of the "Hyundai" plant, near Chennai, the President, Mr. A.P Gandhi, said that, "We had the opportunity to establish our plant at various places in India, but one of the most important factor for choosing the south of India was that people here are perceived to be god-fearing and dedicated to work."

- **Continuous improvement:** Today's industrial products have smaller life cycles, are customized, and have to face the threat of imports. This demands continuous technological upgradation and innovations.
- **Collaborative efforts** are also important. New forms of organization of production centered on certain locations within the state might be of help in the process with formation of clusters.

The framework also brings out the development of linkages within the industrial sector, need for comprehensive labour policy, improvement of infrastructure facilities and committed political will.

Some states do not prove to be good hosts for traditional manufacturing, and rely on service-sector-led industrialization. This if, pursued vigorously, can establish linkages with manufacturing sector, e.g. information technology if strategically planned can lead to push production of hardware segment.

5.4 Benchmarking India against Asian countries

India is benchmarked against the experience of other Asian countries (including China, Thailand, the Philippines and Indonesia). The objective is to highlight the areas where the gap between India and the other countries is most important.

5.4.1 Special Economic Zones

The importance of Special Economic Zones is for both FDI and trade enhancement. The table summarizes the key findings of this overall benchmark on SEZ done by Ace Global Research, 2002.

TABLE 11

Special Economic Zones - Comparison

	India	China	Philippines
Number of Special Zones	7 (conversion of EPZs)	5 SEZs. 124 other special zones	148 proclaimed zones
Share of national exports	4.6%	45%	51%
EDI attracted	Not available	$ 275 bn	Not available
Export by foreign enterprises	Negative	41%	
Key Policy Differentiators			
Qualifying export % age	100%	70%	60%
Income tax treatment of zone units	Tax exemption cut off date 2009-2010 a low tax rate of 10%	Vary from zone to zone but provide specific period for tax exemption thereafter	Flat 5% tax on gross income. No other national taxes
Income tax treatment of DTA units near the SEZ	No special status (tax rate 15-24% instead of normal rate of 30%)	Special tax status for support units located in Zones cities	No Special status
Concessions on sales into Domestic Tafiff Area	Full customs duty on DTA sales, including on	Preferential customs duty on imports from SEZs into DTA priority sectors	Duty free supplies for all DTA units in value addition in zone units (including several basic industries)
Foreign ownership	Upto 100%	Upto 100%	Upto 100%

Source : Ace Global Research on Special Economic Zones. 2002

TABLE 12

Cost Competitiveness Factors - An International Comaprison
Benchmarking Cost Competitiveness Factors (data for 2000-01)

	China	India	Indonesia	Philippines	Thailand
Minimum Wages $/yr	–	408	241	1472	1159
Manufacturing Wages $/yr	729	1192	3054	2450	3868
Manufacturing Value $ added/person/yr	2885	3118	5139	10781	19946
Lending Interest Rate	5.8	12.3	18.5	10.9	7.8
Real Interest Rate	4.9	6.7	6.7	4.0	5.9
Gross national savings as % of GDPs	39.7	23.5	21.5	28.9	30.3
FDI as % of Gross capital formation	9.5	2.1	-16.6	15.2	12.2
FDI stock as % of GDP	30.9	3.6	–	14.9	17.5
Marginal tax rate individual	45	35	35	32	37
Marginal tax rate foreign corporate	30	40	30	32	30
Total tax revenues as % of GDP	6.8	9.6	16.5	13.9	14.1
Direct taxes as % of total tax	6.8	35.8	64.5	44.3	34.0
Domestic indirect taxes as % of total tax	6.5	5.2	6.2	5.0	7.7
Import duties as % of total tax	6.6	26.7	2.3	20.7	12.3
Tariff rates	_	408	241	1472	1159
Simple mean tariffs all products	16.0	33.0	8	8	17
Simple mean tariffs, primary products	17.0	31.0	6	12	22
Simple mean traiffs, capital goods	0.0	33.0	9	7	16
Share of lines with international peaks	43%	93%	11%	9%	46%

Source : World Development Indicators. 2001

5.5 COST COMPETITIVENESS AND ITS FACTORS

The other important aspect is the comparison of the overall factors of cost competitiveness. The most significant observation is the high level of tariffs and the tax revenues in India when compared to other Asian countries.

TABLE 13

Import Policies - Comparison

Description	India	China	Indonesia	Phillippines
Share of world trade	0.2%	3.2%	0.38%	1.2%
Share of national export % ($bn)	15% ($ 6.9 bn)	32% ($ 83 bn)	17% ($ 10 bn)	73% ($ 29 bn)
Share of national import	11% ($ 5.5 bn)	41% ($ 92 bn)	72% ($ 9 bn)	52% ($ 18 bn)
Tariff range in sector	39-182%	20-129%	0-200%	0-30%
Simple mean tariffs all products	33%	16%	8%	8%
Simple mean tariffs, primary products	31%	17%	6%	12%
Simple mean tariffs, capital goods	33%	0%	9%	7%

Source : WTO Annual Statistics 2001, and World Development Indicators 2001

Engineering Sector Performance

The above point is reinforced by the comparisons in the specific sector of mechanical engineering, given its multi-sectoral linkages with the entire gamut of manufacturing activities in India. The most striking comparison in capital goods policies of India and some benchmark countries is in the share of engineering goods exports of these countries as a result of aggressive import policies on capital investment, especially in China.

- Overall taxation is much higher in India, but the exemption periods are in line with other countries.
- India has stricter labour laws than benchmark countries with regard to employment of contract labour and retrenchment for permanent workmen. Major amendments in the labour laws are in an advanced stage of discussion. However, the minimum wage in the manufacturing sector, though higher than China, is much lower than other benchmark countries.
- Time lag, defined, as the period between request for investment license and effective starting date of operation (e.g. import of the equipment), is much longer in India (Source: FICCI study, 2002).
- Average tariff rates for international goods and equipment goods are more than twice the level of other Asian countries.
- Average dispersion of tariff rates within broad categories of product is not higher in India, but the number of peak lines is much higher than elsewhere.

The UNIDO Scoreboard provides a measure of competitiveness. It focuses on manufacturing and a small number of structural variables, so it can be used for cross-country comparison in our case. It is a simple tool that countries can use to assess their position with respect to industrial performance. Competitive Industrial Performance (CIP) index focuses on the national ability to produce manufactures competitively. It is composed of four basic indicators of industrial performance. They are: -

- Manufacturing Value Added (MVA) per capita.
- Manufactured exports per capita: It measures ability of national industry to keep pace with technological change at least in export products.
- Share of medium and high-tech activities in MVA: The higher the share, the more technologically complex is the industrial structure of a country, the more competitive is the country's industrial performance.
- Share of medium and high-tech products in manufactured exports: This is taken separately because in large import-substitution developing countries, the structure of MVA tends to be more complex than that of exports.

5.6 Competitive Industrial Performance (CIP) Index

Industrialized economies

A comparison of the CIP index during 1985 and 1998 shows that Switzerland led the industrialized countries while Ireland has shown the biggest improvement and Germany and US are near the top of the scale.

Developing economies

Among the developing regions East Asia led in both years and also had the greatest improvement in ranks over the period. Singapore led East Asia in both the years. Its nearest rival was Tiwan Province of China, ranking

14th. Most of the economies improved their positions; China is now ranked 24 places ahead at 37. India has been ranked 50th. Malaysia has also shown improvement from 30th to 22nd position now.

Singapore has attained the 1st position due to its high exports per capita. Philippines improved from 60th to 25th rank on the basis of the large share of high-tech products in its exports. China also is in an improved position largely due to technology intensity of its manufactured exports.

Least Developed Economies

The 12 least developed countries in the sample are near the bottom of the world rankings, along with Algeria and three African countries, Cameroon, Nigeria and Ghana. Bangladesh and Nepal have had stable rankings.

TABLE 14

CIP Index Rankings - LDC

Region or country group	Rank		
East Asia and the Pacific	1998	1985	Economy
	1	6	Singapore
	15	19	Taiwan
	18	22	Korea, Republic of
	22	30	Malaysia
	25	45	Philippines
	30	18	Hong Kong SAR
	32	43	Thailand
	37	61	China
	49	65	Indonesia
South Asia			
	50	50	India
	60	55	Pakistan
	69	71	Sri Lanka
	73	74	Bangladesh
	79	79	Nepal

Rank			Index Value	
1998	1985	Economy	1998	1985
1	6	Singapore	0.883	0.587
2	1	Switzerland	0.751	0.808
3	15	Ireland	0.739	0.379
4	2	Japan	0.696	0.725
5	3	Germany	0.632	0.635
6	5	USA	0.564	0.599
7	4	Sweden	0.562	0.633
8	7	Finland	0.538	0.494
9	8	Belgium	0.495	0.489
10	12	UK	0.473	0.426

Source: UNIDO Report 2002

FIGURE 20 - CHANGE IN MVA AND SHARE OF MEDIUM AND HIGH-TECH ACTIVITIES

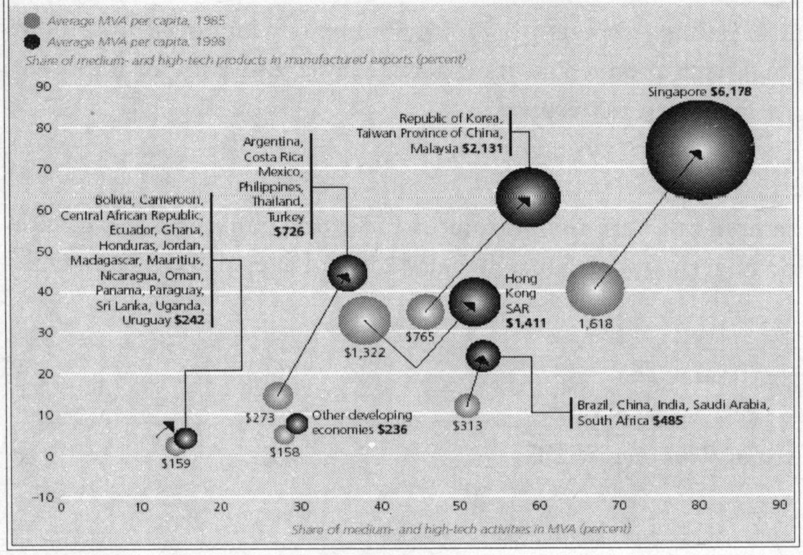

5.6.1 Changes in MVA and Share of Medium and High-tech Activities in MVA

The chart below shows the improvements and the falls of various countries in terms of their ranking. Here the Manufacturing Value Added per capita is shown for Developing countries. Overall there are seven clusters in this region of the world.

India and China lie in the same cluster and show marginal improvements while Singapore, Malaysia, and Republic of Korea have jumped and improved on both accounts. (Figure above)

5.6.2 Export Performance

The CIP index identifies the leading and the lagging exporters - as assessed by manufacturing exports as a share of MVA and medium and high MVA suggest an interplay of several factors in explaining export performance. The main factors are the size of the domestic economy, competitive capabilities, the nature of the trade regime and the intensity of participation in integrated global production system.

India lies among the bottom fifteen exporters when we consider the manufactured exports as a percentage of the total MVA. Size matters in developing world, but trade policies and capabilities probably play a larger role. Thus weak export performance of India reflects not just the large market size but also the legacy of strong inward looking policies, the small presence of transnational corporations and competitive weakness in manufacturing.

In case of China, despite its size, it does not appear among the bottom 15 for medium and high-tech exports as it does for total manufactured products. Its medium and high-tech sector is apparently far more competitive than the rest of its manufacturing. Foreign enterprises in special export zones contribute a lot to the medium and high-tech products category.

TABLE 15

Manufactured Exports as a Percentage of Total MVA

Top 15 exporters		Bottom 15 exporters	
Economy	Value	Economy	Value
Singapore	529.5	Russia	54.0
Belgium	338.5	Turkey	51.9
Malaysia	371.4	Morocco	51.0
Hong Kong SAR	245.2	China	47.2
Holland	225.0	Australia	46.3
Ireland	222.3	Uruguay	41.9

Contd...

Hungary	213.0	Japan	41.4
Philippines	197.2	India	40.5
Mauritius	187.0	US	38.4
Costa Rica	174.4	Colombia	32.3
Czech Republic	159.2	Argentina	26.5
Sweden	158.6	Brazil	25.7
Canada	154.3	Ecuador	22.0
Taiwan	144.2	Peru	15.5
Denmark	143.4	Egypt	11.2

5.6.3 Linking of Structure of Exports to Growth

Econometric analysis suggests that technological structure of a country's exports affect the growth of that country's manufactured exports. The analysis done by UNIDO covered 80 countries for both 1985 and 1998. It was found that the average share of medium and high-tech products for large exporters in 1985 was eight times that of small exporters.

5.6.4 Linking Industrial Performance and Environmental Sustainability

Not enough data is available to check the relation. But the reason is that it is not easy to assess why the pollution contribution from one country is less. It may be due to the better measures used in the country or due to the lesser extent of industrial activity in the country.

One interesting fact that is revealed in the analysis of Carbon Dioxide emission is that countries that have had large jumps in CIP ranks (China, Hungary, Malaysia and Thailand) have also had substantial increases in CO_2 emissions relative to GDP. This suggests that rapid industrial growth can raise the propensity to pollute, at least until some level of industrial maturity is attained. This inverted-U-shaped relationship between emissions and industrialization is of concern and a learning point for India.

5.7 Inferences for India

India led South Asia in the CIP ranking in 1998, though its rank had changed little since 1985. During this 13-year span India's industrial structure remained static, and the country was overtaken in the CIP ranking by such economies as China and Indonesia, mainly because of their greater orientation towards technologically complex export industries.

India had annual growth of MVA (5.1 percent) lower than the region's average (5.5 percent), and its MVA per capita in 1998 ($ 65.2) was similar to that of Cameroon ($ 64.6) and Nigeria ($ 62.2). Its manufactured exports reached only $ 26.4 per capita in 1998, less than in Senegal ($ 34.5), Cameroon ($ 34) or Kenya ($ 28.3).

There are various policy imperatives for enhancing India's competitiveness some of which may be as below.

- Technology Upgradation

Historically, India has followed the path of low technology, low productivity manufacturing. Although this blemishes India as a maker of low quality goods rather than premium products, such an approach has several advantages.

- *Focus on Core Competence*

A low productivity approach is based on the exploitation of sources of immediate comparative advantages (abundant labour and/or resources). This is the first step to be taken for a smooth transition from agriculture to services via manufacturing as characterized in the economic history of all countries which have subsequently established strong positions in manufactured exports.

- *Steady Improvement Lessens Undesirable Social Impacts*

In addition, there can be excellent synergies between the low and high productivity growth paths. Exports built up initially with technologically simple products, provide the foreign exchange needed to sustain investment in more sophisticated sectors. As the rate of learning and productivity increase in those sectors depends on the rate of investment in them, low technology exports must be regarded as an essential part of a strategy of learning and technology upgrading.

- *Employment*

Low productivity exports create more employment per unit of output. In periods when international trade grows slowly (unlike the past twenty years or so), this may be an important consideration in terms of income distribution. The qualification is that low productivity production often requires that wages be held down to maintain competitiveness.

With the recent changes in global trade, however, sticking on to such an approach will compromise the insistence on quality. Moreover, continued importation of technology cannot lead to strategic leadership in the long run. In order to foster innovation and entrepreneurship, from the government's part increased participation in R&D is necessary.

The annual growth of its manufactured exports (11.6 per cent), while healthy, was less than the region's average (12.4 per cent), which was in turn lower than the average for all developing economies (13.3 per cent) and for East Asia (14.5 per cent).

What really distinguishes India from other countries in South Asia and from Africa is the depth and complexity of its industrial structure.

- Medium and high-tech products accounted for almost 60 per cent of India's MVA in 1998, compared with less than 10 percent for its neighbors.
- In fact, India ranked 12th globally in the technological complexity of MVA, ahead of Taiwan, Brazil or China–a legacy of its prolonged import substitution strategy emphasizing heavy industry.

But this strategy has also left India's diverse manufacturing sector with a large technological base but with competitive gaps relative to world frontiers. India's manufactured exports have a strikingly different structure than its manufacturing production, with low-tech products (textiles and clothing) and resource-based activities (diamond cutting) accounting for 62 percent.

The gap between the two structures reflects the competitive gap in large sections of Indian industry. India's most dynamic export activity 'software' is not captured by the figures, which exclude services. This competitive gap in manufacturing is the result not only of past trade policies but also of weaknesses in industrial drivers.

- The country's skill base is weak and deteriorating relative to those of competitors. The absolute numbers are, of course, large–India accounted for 16.3 percent of tertiary enrolments in the developing world in 1998, and it ranked only behind the United States, the Russian Federation and China in tertiary enrolments in technical areas. But the number of tertiary students enrolled in technical subjects is small relative to India's population, and in contrast to the numbers in most other regions, it has declined since 1985. Still, the absolute number of technically qualified people is large, allowing India to foster nodes of skill-intensive activity. That explains the presence of software exports and of some competitive firms in complex activities. The real skill constraint will arise when India tries to upgrade technologies in a large range of activities to compete in liberalized markets.

- Deficiencies in infrastructure are widespread. More striking is that India is falling further behind its competitors in this driver of industrial performance.
- Moreover, enterprise-financed R&D was the same on a per capita basis in 1998 as in 1985 and declining as a share of GNP. Inward foreign direct investment has risen significantly in recent years, but the growth has not been enough. In sharp contrast to foreign direct investment in China, little in India goes into export-oriented manufacturing. That puts India at risk of losing out on links with the most dynamic part of industrial export activity, integrated production systems in technology-intensive activities.

5.8 Study of Industry Sectors

5.8.1 Analysis - Food and Beverages Industries

Introduction

India is the world's second largest producer of food next to China, and has the potential of being the biggest with the food and agricultural sector contributing around 26% to India's GDP. The sector includes processed food, fruits and vegetable processing, milk products, packaged food and drinks and liquor.

Strengths

- India is the second largest milk producing country with anticipated production of about 78 million tons during 1999-2000.
- India has a livestock population of 470 million, which includes 205 million cattle and 90 million buffaloes. The country produces about 450 million broilers and 30 billion eggs annually. With over 8000 km. of coastline, 3 million hectares of reservoirs and 1.4 million hectares of brackish water, India has vast potential for fishes from both inland and marine resources.
- The country is self sufficient in grain production and is the second largest rice producer in world, with a 20% share.

Weaknesses

- Quality standards in India are below international standards.
- Packaged food consumption is low in India.
- Wide variety of fish resources found in Indian inland waters, coastal areas and deep seas comprising India's Exclusive Economic Zone, still remain grossly under-utilized.
- Lack of initiative from individuals and small groups of entrepreneurs to market the food products internationally.
- Villagers and farm owners are not aware of the technologies used world wide in food production. Thus they loose the opportunity of getting a good bargain out of value-added produce.
- Lack of an international marketing network for this sector.
- Lack of cold storages and refrigeration facilities in small towns. Lack of industrial and commercial refrigerators.

Opportunity

- The total food production in India is likely to double in the next ten years and there is an opportunity for large investments in food and food processing technologies, skills and equipment, especially in areas of Canning, Dairy and Food Processing, Specialty Processing, Packaging, Frozen Food/Refrigeration and Thermo Processing. Fruits and Vegetables, Fisheries, Milk and Milk Products, Meat and Poultry, Packaged/Convenience Foods.
- The government of India has given a lot of support to the Food Processing Industry sector, with a number of fiscal relief and incentives, to encourage commercialization and value addition to agricultural produce.
- Production of milk and milk products has become very organized in India. Some plants are coming up for producing lactose, cassein and improved cheese varieties.
- For packaged products the demand has been rising at a good pace and there is enough latent market potential waiting to be exploited through developmental efforts.
- Government initiatives through tax reduction in potable alcohol opens opportunities for MNCs to bring their international brands to India.

Strategies

- A large portion of agricultural and other products produced in the country perish due to lack of a serviceable refrigerated storage and distribution network. As a result of this, meat, poultry, fresh produce and dairy can only be distributed within limited geography and need to be consumed in a short period. Also, sensitive medical requisitions like vaccines and pharmaceuticals are suffering due to the lack of refrigerated distribution. Refrigeration industry in India has to make industrial refrigerators to meet the requirements of export oriented food product based firms. Given the poor transport infrastructure in India, it is absolutely essential to have excellent refrigerating systems. The manufacturers need cold storage, refrigerated transport, and retail display and preservation units. The systems need to be rugged because power fluctuations often occur.
- With over 8000 km. of coastline, 3 million hectares of reservoirs and 1.4 million hectares of brackish water, India has vast potential for fishes from both inland and marine resources. Units mostly exist in the small-scale sector as proprietary/partnership firms or fishermen co-operatives. Entry of organized corporate sector in preservation, processing and export of coastal fish will help the fishermen co-operatives in these cost-intensive operations.
- Fruits and vegetable processing offers potential for high value addition and potential entrepreneurs can earn a lot from such products. By processing of fruits and vegetables Indian producers will get proper price for their products, entrepreneurs will have good business opportunities, masses will get employment and export of processed fruits and vegetables will increase. While processing fruits and vegetables, Indian entrepreneurs need to take all possible care so as to ensure quality and standard of their products.
- India has been in the export market for fruits and vegetables for almost 30 years. Popular items for export include mango chutneys, pickles, fruit juices, canned and dehydrated mushrooms, frozen and canned fruit and vegetables. Government has taken up a number of developmental measures and the future of the Industry looks bright. Multinationals entering the food industry will help India to develop an international marketing network as they have their brand

loyalties all over the world. This will enable Indian products in reaching all over the world in the form and packing required.

5.8.2 Analysis - The Textile Industry

Introduction

The Textile Industry is the largest industry of modern India. It contributes about 4 per cent of GDP, 20 per cent of total industrial output, provides employment to about 20 million people and, together with carpets and handicrafts, has a share of 38% in total value of exports.

Strengths

- Government support for this sector is ample. It provides lots of flexibility for the manufacturers.
- Natural endowments namely, climatic conditions and soil is good for the raw materials like cotton.
- The Textile Policy of 1985 provided tremendous impetus to the sector, raising the share of textiles to 13% of value added domestic manufacturing of the country; and to one third of the export earnings of the country.
- Cheap and abundant labour and handicraft skills are further strengths.

Weaknesses

- Low productivity.
- The quality and availability of raw material for manufacturing has been poor.
- Lack of mechanization across the sector.
- Very little visible use of IT tools within the sector.
- Lack of strong market linkages.
- Lack of collaborative research projects

Opportunity

- Handloom industry is unique to India in the sense that the industry has not only survived but also grown over the decades due to its inherent strengths like flexibility of production in small quantities, openness to innovation, low level of capital investment and immense possibility of designing fabrics.
- Use of information technology for the rapid up-gradation of production process.
- The phasing out of MFA (Multi-Fibre Arrangement) by 2005 will benefit India as the exports of textiles and clothing will increase.

Threats

- Presence of China in the world market requires focused marketing incentives, which is not possible because Indian firms lack collective initiatives.
- Cost factors like Indian salaries and wages (30 to 60 per cent higher than China and Pakistan) can render Indian products costlier and hence less competitive.
- Modernization involves automation, which is likely to displace labour.

Strategies

- There are three sectors in the textile industry - mill sector, power loom sector and handloom sector. The latter two are jointly considered under the "decentralized sector". The share of decentralized sector has increased from 24 per cent in 1950-51 to 95 percent in 1998-99. Despite a 58% global share of looms, consisting of 3.5 million handlooms and 1.8 million power looms, technology still remains backward. This sector is critical to the survival of the Indian textile industry and its export thrust. The powerloom sector has shown a greater growth. This has to be supported to capitalize on the good export potential post MFA. The support needed is removal of supply side constraints (like higher cost and poor quality of raw material).
- Large-scale mechanization can render a lot of laborers jobless. Automation of looms can help one worker manage 48 looms, while today he can manage only 6. The excess labour can be absorbed elsewhere in the sector only if the demand for Indian produce goes up. Exports intensive manufacturing is the focus, which this sector requires. Moreover, the yield per hectare of raw materials like cotton has to be increased by atleast 10-15%.
- Collaborative research between private and government firms is required. The research can be in terms of development and customization of IT tools for supply chain management in India. Collaborative research is required to augment availability of quality wool. FDI flow into the sector will help such initiatives. Government should encourage and promote it.
- Also there should be active co-operation and partnership of the State Governments, Financial Institutions, Entrepreneurs, Farmers and Non-Governmental Organizations in the fulfillment of these objectives.

5.8.3 Analysis - Automobile Industry

Introduction

Automobile industry contributes 7% of Manufacturing GDP and has 15% Export contribution in manufactured products (CMIE, 2001). In the financial year 2001-2002, the Automobile Industry recorded a growth of 13% over FY 2000-2001. However, it was a year of stagnant sales, except for some distinct growth in individual segments, particularly, two wheelers. Government has taken up steps towards alignment of import tariffs on components with ASEAN levels. Incentive for R&D expenditure (especially in the context of environmental norms) is also being provided. Further impetus to infrastructure development and encouragement of projects in rural and semi-urban areas are the new developments.

Initiatives

- Prices of diesel and petrol brought down by around 50 paise and Re 1 per litre respectively, from March 1, 2002.
- Excise duty on CNG increased from 8% to 16%.
- Excise duty on Diesel engines up to 10 HP increased from 8% to 16%.
- Peak import duty on components lowered from 35% to 30%.
- Peak import duty on steel related items lowered.
- 5 % National Defence Surcharge imposed.
- Plan outlay for roads increased by 39%.
- Additional depreciation at the rate of 15% allowed on new plant and machinery acquired on

or after 1st April, 2002 for setting up a new industrial unit, or for expanding the installed capacity of existing units by at least 25%.

- Allocation of Rs. 2,500 Crore for the Pradhan Mantri Gram Sadak Yojana (PMGSY).

Strengths

- Acceleration in the infrastructural activity including the Golden Quadrilateral, fleet rationalization and improved availability of easy finance have contributed to a robust 41% growth in the HCV sales.
- In overall terms, exports grew by a healthy 60% in April-July 2002 compared to the same period last year (SIAM). In cars a large amount of investment is also taking place, which has raised annual production capacity to 750,000 cars (of all sizes) in 2001. Auto components manufacturers have been slowly gaining global recognition and maintaining a certain level of exports despite the recent downturn.

Weaknesses

- Technology employed for assembling is not of world standards.
- The companies like TELCO, Ashok Leyland with huge manufacturing base in India have concentrated on commercial vehicles alone. Diversification may be a choice but it is difficult.
- A comparison of vehicle penetration in different countries reveals that India lags significantly behind countries with similar levels of GDP per capita, such as Pakistan or Nigeria.

Opportunities

- In 2001 there were 44 million vehicles and by the end of year 2002 experts estimate that number will grow to 64 million. This growth is expected to be in countries such as: China, India, The Pacific Rim, South Africa, and South America.
- Other future endeavors include low emission cars, which are expected to provide expansions in sales.
- The World Wide Web, which allows consumers to thoroughly research and purchase automobiles from home, has become an essential ingredient to any successful automakers arsenal, and will continue to play a large role in the 21st Century.
- Worldwide, the customer preference is moving towards larger, safer cars.
- Indian markets continue to be largely under-penetrated, and there exists good potential for long-term growth.
- The sector was de-reserved from SSI list in 1998-99 budget so that the implements can be produced with better design and technology that can help in farm mechanization. This will help in enhancement of agricultural productivity in the country.
- Despite the relatively low cost of labour in India, the automotive industry has not yet contributed significantly to exports and accounts for only 2 per cent of all Indian exports, compared to 14 per cent in Brazil and 6.7 per cent in Korea.

Threats

- Implementation of WTO rules will attract Chinese companies.
- The GOI policy towards depreciation norms, excise duty sales tax etc will have a bearing on the demand for MUVs and CVs. The levy of 12% uniform sales tax already has and may in the near future play a role in the performance of the industry.

- The car multinationals (which have much deeper pockets than the Maruti joint venture) are willing to sustain heavy losses in order to build up a commanding position in the market by slashing prices and living with spare capacity, until (perhaps a generation hence) India becomes a major car market.

Strategies

- Stringent labour legislation is the main external barrier to productivity growth. It has prevented plants from reducing surplus labour in the past. Due to the effects of unionism and labour laws, companies that have been operating automotive manufacturing plants for many years have decided not to staff their new plants with surplus workers from their existing plants. To shield new plants from the old culture, these companies hire inexperienced workers from vocational schools and continue to induce surplus workers to leave with VRS. But this may result in loss of most specialized workers too. The government should liberalize labour laws by simplifying procedures for retrenchment. Currently, the process is complex and companies need state government approval to retrench workers. This approval is often denied for political reasons. The government should establish a system that allows companies to retrench employees by giving them a standard severance package. In the UK, for example, companies have to make a redundancy payment of between 1 and 1½ weeks' salary for every year of service. Such a system would enable companies to reduce the workforce without having to seek the agreement of unions and workers. Further, companies should also be allowed to select the employees they wish to retrench.
- Import restrictions remain a barrier to productivity growth by protecting Indian-produced cars from competing with models produced in global best practice plants. The government should gradually phase out tariffs by 2010. This will ensure that the industry is increasingly exposed to best practices, while giving new plants sufficient time to increase both scale and utilization and close the productivity gap. The time frame for the removal of tariffs should be determined on the basis of the plants' output volumes, as that is the key variable for viable operations in the presence of competition from imports.
- The complexity of tax and labour rules, customs procedures and other interactions with government authorities forces Indian automotive companies to hire people solely to deal with these unnecessarily cumbersome tasks. While this is admittedly only a secondary issue, red-tapism can also cause work stoppage. For example, customs clearance for urgently needed parts often takes up to 7 days because of red tape. Phasing out of tariffs will help expose the industry to international best practices.
- In a competitive market, post WTO, the government will be hard pressed to retain shares in car making firms (like Maruti) and other companies. Disinvestments at an early stage will help to get a good price for the company's share. And thus the unnecessary control will also be relinquished.
- Poor infrastructure lowers demand and raises the cost of road transport. It also raises damages during export delivery. Efforts by government for infrastructural development will also help automobile manufacturers.

5.8.4 Analysis - Auto Ancillaries

Introduction

The Auto Ancillaries industry consists of about 402 players with direct employment of 2.5 lakh people.

With an investment of over $ 2.3 billion, the industry caters to about 24 automobile units in India. The output of the industry is around $ 4 billion and it exports goods worth of $ 417 million.

Strengths

- The domestic auto ancillary industry contributes up to 87% of the automobile sector needs, while the rest is imported. The demand for ancillaries arises from OEM market (25%), replacement market (65%) and the export market (10%).
- The industry has been growing at a CAGR of 19.6% for the past five years.
- Globally, auto industry is more into out-sourcing, and the trend is reflected in India which has led to a proliferation of ancillary units in India.

Weaknesses

- The industry is scale sensitive and hence larger volumes are needed for cost effectiveness and improving quality.
- The industry is capital and labour intensive.
- It is highly fragmented. As a component itself constitutes a segment, there are more than 6,000 medium and small players in the fray.
- The capacities are nowhere near the global standards. With each MNC entry into Indian auto industry, they bring many new component vendors.

Opportunities

- The prospect of exports from India is bright on account of its cost advantage.

Threats

- The main competitors for the sector are Malaysia, Taiwan and other South Asian countries
- These countries have cost advantages on the basis labour cost, productivity, larger volumes and after sales service.
- The above countries being small players, technology dependency on foreign players exists and R&D is negligible.
- Dependence on OEM.
- In replacement market, spurious products from the unorganized sector and cheap imports from China pose great threats.

Strategies

- Despite reduction in duties by successive governments, relatively high tariff barriers continue to offer a degree of protection to the Indian industry. But the competitiveness of the Indian firms suffers on this account. The government should look towards removal of these tariffs.
- The lack of know-how in certain areas has to be quickly improved. In the near future with the entry of more MNCs the local manufacturing capabilities can help them get large and sustained offers. So, the focus should be on learning the best and the most marketable technologies. The lower cost of labour may help the firms become providers of products worldwide for MNCs. They need to take up initiatives to export their products into international markets. Since the plants located in India are very far from the developed markets of the USA, Europe, and Japan, component suppliers incur significant transportation and inventory carrying costs in exporting products to global markets. Their situation is worsened by the poor Indian infra-

structure, which leads to frequent power interruptions and long delays in supply. These companies need to adopt innovative techniques to cope with these uncertainties. One model, which can work for them, is that of participation of large auto firms. The larger firms can provide both technological and economic support in the initial stages. Moreover, the backing given by the Indian units of the firms can help raise the reputation of the component manufacturers in the international market.

5.8.5 Analysis - Electrical Equipments

Introduction

- In India, it has a history of several years. Manufacturing and infrastructure sector needs an able and proactive electrical equipments industry for development. The demand depends upon power programmes, industrial requirements, urban and rural demands. Indian Electrical Equipment Manufacturers Association (IEEMA) monitors the 15 segments of the industry.

Strengths

- Today the industry has the capability of manufacturing most of the equipment and in fully meeting the needs of the country.
- Strategies of joint-venture with reputed foreign companies have given good results.
- Globally, the trend is towards using high-end technology and Indian electrical equipment industry has the ability to manufacture nearly all ranges of equipment.
- Indian transformers in 800KV class are world class. India is exporting transformers to over 50 countries.

Weaknesses

- Last three years it has faced problems of lower production levels, delayed payments and diminishing profit margins.
- It is either stagnant or reporting negative growth with companies operating at less than 50% capacity.
- Taxation policies of the government affect this sector. Import duties have been reduced but state and local taxation, high interest rates have to be resolved too.

Opportunities

- A three-fold increase in exports is targeted by BHEL, one of the major players, by 2006-07.
- India can be positioned as a regular source of components and products for global Original Equipment Manufacturers (OEMs), Engineering, Procurement and Construction (EPC) contractors, and other multinational companies not only for intermediate and semi-finished products but also for complementary equipment for their projects.
- Overseas demand can be met by the currently under-utilized capacity of Indian companies.

Threats

- A major problem is cheap imports from China and Nepal. The Indo-Nepal treaty allows imports from Nepal at nil duty. With economies of global scale production China does have an edge particularly in low value high volume goods.

Strategic Action Points

- Among the developing countries, next to China, India is the only country with good technological and human resource capability, so we should exploit it. Project exports orders are very large and profitable. Local laws come into play if sites have to be set-up overseas. Moreover, the projects are long-drawn. But the advantage is that equipments are in operation for long periods in which companies will have proprietary control. So all the spare and after-market servicing is controlled. Long-term credit at competitives rate is required for such projects.
- Identifying the "star" products can consolidate product exports. Then investing is necessary to make the products competitive at international level. External customer perception surveys can be done to study the market attractiveness of the products. Qualification and registration of products with major production companies worldwide will also make the products internationally attractive. The problem in product exports is that individual orders can be low.
- Another initiative can be to become the global sourcing agent to collaborators. So the collaborators will make some major things and will depend on the agent to provide the other parts. Even in major equipments sub-assemblies can be sourced.
- Raw materials are imported and hence foreign exchange fluctuations can cause problems. So the need is to make an internal hedge by using exports and enhanced physical exports to make dollars that are equal or more than what is spent.
- World trend in one of the most important segments, Switchgears, demands compact, small, low weight equipment with high quality high reliability, less maintenance and easy accessibility. Indian manufacturers have to meet all these expectations of the user industry.

5.8.6 Analysis - Copper

Introduction

India has huge reserves of Copper-590 million tonnes (containing about 8 million tonnes of metal), located in Rajasthan and Bihar. The production is 0.24 million tonnes of refined copper per year. The average share of India's production and consumption of Copper is 0.03 and 1.2% of the world. The demand of refined copper in India was estimated, as projected in ninth five year plan, to be 3.35 lakh MT in 1996-97 and 4.87 lakh MT in 2001-02 (assuming 8 per cent growth).

While developing countries account for a little over 50 per cent of world ore production, their share of annual consumption of refined copper is only 32 per cent. But, the average share of India's production and consumption to the world production and consumption are about 0.03 per cent and 1.2 per cent respectively. In India, demand for the auburn metal is expected to rise by three per cent every year. So it is the right time for the government to go for disinvestment.

Strengths

- Reserves of copper ore in India are huge and can be exploited. 64 years of reserves as per Geographical Survey of India at present rate of consumption.
- The potential demand of copper is estimated to increase at 3% every year in India.

Weaknesses

- Targets are fixed much below the installed capacity of production and initial grade of ore. Even then there are shortfalls.

- For steady production growth in mines, timely development of ore body along strike and depth is vital. However, there is no planned direction on mine development.
- The utilization of the mining equipment varies between 6-50%.
- Preventive maintenance has never been focused upon.
- Power shortage is also a problem.
- There is more focus on ore production than development.

Opportunities

- In India, demand for the metal is expected to rise by three per cent every year.
- Supply in the Asian region falls at least 1.5 million tonne per annum (tpa) short of demand.
- The world annual consumption of refined copper is about 10 million MT with the annual growth rate of 1.5 per cent.
- Major players are restructuring and getting ready for competition.
- Government is positive towards disinvestment, which can help the private players to make a focused bid to correct the weaknesses.

Threats

- The major copper producing countries are USA, Japan, Chile, Zambia, Poland, Belgium and Zaire are very competitive and will seek opportunities in India.
- The MNCs have better technology.

Strategic Action Points

- Modernization of production processes is required. This is necessary to reduce the gap between the production and demand. It will also reduce production cost and make the industry competitive vis-s-vis global players.
- The demand for copper will continue to be met by the discovery of new deposits, technological improvements, efficient design, and by taking advantage of the renewable nature of copper through reuse and recycling.
- Restructuring by mergers and acquisition is a new effort towards consolidation. This is the first step towards building a "Metal Powerhouse" in the country. Such an initiative should help develop a strong Indian player with the capability to tap foreign markets too.
- Copper is in expansion mode in India and so a policy of backward integration by going into mining of copper can help attain long-term sustainability.
- Industry should explore options such as setting up units in countries where power costs are far lower than in India, and backward integration into mines abroad. Further the sales can be done in fast growing markets like China.
- Aluminum and Copper both are asset-intensive businesses, commodities and LME-led. And the mindset, of sweating the asset, is the same. An effort to consolidate their operations will help develop a strong sector.

5.8.7 Analysis - Gems and Jewellery

Introduction

It contributes nearly 22% of export in manufactured products. The government has allowed private (and later public) bonded warehouses for import, stock, export and sale of rough diamonds in the EPZs. Export of gems and jewellery surged to an all-time high in the first half of the current fiscal even as it broke all previous records, according to the Gem and Jewellery Export Promotion Council (GJEPC). Total gems and jewellery exports jumped by 25.18% during April-September 2002/03 to $ 4.10bn from $ 3.28bn in the same period last year. The rise in the total gems and jewellery exports is attributed to the remarkable increase in the Indian exports of cut and polished diamonds to the US, the world's largest market of gems and jewellery. India exports to U.S, Hong Kong, Belgium, Israel, Japan and Thailand. India is the largest consumer of gold, processing over 800 tonnes annually.

Strengths

- India is the leader in the field of cut and polished diamonds-55% of world market by value.
- Diamonds also have 80% share in total gems jewellery exports.
- There has been widespread introduction of modern technology. India has developed a variety of the most innovative shapes at affordable prices.
- India poses the prime position in supply of cut and polished colour stones to the world. Most of the emeralds sourced world-wide are procured from India.

Weaknesses

- The industry sources diamonds and other gems from outside.
- The operations in high priced jewellery market of Indian firms are negligible.

Opportunities

- Coloured gem stones have also shown growth with exports being $ 205 million in 2000.
- The use of diamonds as gift items and a life time product is being highlighted and thus prospects of demand boost is an opportunity. World sales of Gems and Jewellery is increasing but jewellery is most in demand.

Threats

- Its nearest competitors are Israel and Belgium, who have steadily recorded growth.

Strategic Action Points

- India has been sourcing diamonds from Belgium and De Beers (London). In order to further consolidate the leadership position India should seek opportunities to establish direct trade links with mining countries (Canada, Congo, Russia) and companies.
- Forward integration is possible from manufacturing diamonds to making jewellery. With the strength of conventional knowledge of various communities in making jewels the product with an "Indian brand name" will be appreciated in the world markets. The marketing strength of large firms can be used for this purpose.
- Use of technology in manufacturing is of great importance. Indian firms will have to learn the skill from current clients like De Beers. Moving up the value chain will help Indian firms to earn and develop.

5.8.8 Analysis - Lubricants

Introduction

The size of the market was one million tonne in 2001and is growing at the rate of 10% per year. The global average is 2%. This is the smallest and the most diversified segment of the petroleum industry. The domestic demand is being completely met by indigenous production. The Indian lubricant market has been opened completely to private and foreign manufacturers. The total market value is Rs. 5,500 crores.

Strengths

- The Indian lubricant market is growing at a rate greater than the global growth rate of 3%.

Weaknesses

- The industry does not have varied products to offer.

Opportunities

- India is the sixth largest consumer of lubricants in the world. Automotive oils have a 60% share. The remaining 40% is used as industrial lubricants.
- The domestic consumption is around Rs. 6,000 crores and is met from domestic sources. The demand is to grow to 1.5 million tonne by 2006-07 (Source: Basic statistics of Indian Petroleum and Natural Gas).

Threats

- It's nearest competitor is Israel.
- Environmental concerns have been raised all over the world.
- Emissions standards set is also a challenge to be met by lubricants.

Strategic Action Points

- New generation passenger cars, increasing use of sophisticated machinery working under stringent parameters, exacting environmental conditions has led to demand of lubricants with enhanced features. The switch to less polluting fuels and lubricants is imperative.
- With the introduction of stringent laws relating to environment the biodegradable lubricants have become important.

5.8.9 Analysis - Machine Tools

Introduction

The expected output is Rs. 1,50,000 crores from this sector. Being an integral sector, growth of the machine tool industry has an immense bearing on the entire economy, especially India's manufacturing industry. It is even more crucial for development of the country's strategic segments such as Defence, railways, space, and atomic energy. The Machine tool industry in India comprises about 450 manufacturers with 150 units in the organised sector. Ten major companies of this industry contribute almost 70 per cent of production in India.

- The International Centre for Advancement of Manufacturing Technology has initiated a number of programs for machine tool industry. Awareness programme and Technical sessions on CNC retrofitting etc. Experts from Cluster Development Programme of UNIDO have conducted sessions on cluster formation.

Strengths

- Indian Machine tool industry maintained a satisfactory 6% growth rate even in the subdued economic and industrial environment of 2001.

- Enhanced features, competitive pricing, and marketing focus has increased demand for Indian-made machine tools in overseas markets, particularly in Europe, United States, and East-Asian regions.
- The India offers both conventional as well as CNC metal-working machine tools.
- Indian-made machine tools are currently exported to over 50 countries; major ones being United States, Italy, Brazil, Germany, and the Middle East. Lathes and automats, presses, electro-discharge machines, and machining centres formed the bulk of export orders for Indian manufacturers. These machines from the Indian basket are generally favoured in overseas markets primarily due to their cost-competitiveness, as compared to that available elsewhere.

Weaknesses

- Machine tool industry in India is scattered all over the country. Bangalore is considered to be the hub of machine tool industry.
- Second hand capital goods import has attracted a modest 25% customs duty. This liberal policy has had an adverse impact on the quality of the products manufactured by them. It also causes losses to local capital goods industry.
- There has been low volume production due to low demand.
- There have been no major investments for a decade.
- There is lack of focused efforts for export.

Opportunities

- A reasonable order inflow from the automobile industry as well as adequate business from the Defence sector led to increased production by machine tool manufacturers. (Source: Indian Machine Tools Manufacturers Asscociation).
- There is an increasing need for low-cost automobile manufacturing solutions offering enhanced value added features, higher precision, greater reliability, reduced set-up and idle time. It has led users to seek CNC machines.
- Further, there has off late been a perceptible change in the image of the 'Made in India' brand in overseas markets for Indian-built machine tools.

Threats

- CNC machine are suffering from lower margins.
- Output by domestic metal-working machine tool manufacturers in 2001 calendar year declined by 14 per cent to Rs. 5,137 million marking the fourth year of decline, since 1997, for the Indian machine tool industry.
- Most of the Indian companies are hamstrung by excessive capacities, and investing in new capacity might not be feasible in the wake of increased competition.

Strategic Action Points

- Potential to penetrate overseas machine tools market should be exploited. Dependence on domestic market alone does not look viable enough.
- We also need to develop machines for cellular manufacturing.

5.8.10 Analysis - Indian Paper Industry

Introduction

There are nearly 406 mills and 6.2 million tonne installed Capacity. It contributes 4% of Manufacturing GDP (2001 data). There is no significant contribution to exports.

Strengths

- Presence of Raw Materials in the country.
- Industry is intrinsically very rich in product mix.
- Large Players have firmly established themselves (BILT, ITC Bhadrachalam, TNPL)

Weaknesses

- Availability of Raw Materials - Though raw material is produced (bagasse etc.) the system to collect it is inefficient. As a result companies are importing waste paper. (At 5% customs duty)
- Presence of SMEs - SMEs cannot reach the economies of scale in operation or make the necessary capital investments for complying with pollution norms.
- Technology is yet to become world class. There is a very high energy consumption level.

Opportunities

- Export Markets are untapped and untried. Alternate fuels and substitutions can also be explored. Demand is dependent on a multiple factors like GDP Growth, IIP, Population increase etc. Provided the outlook on these factors are positive, opportunities to grow definitely exist.
- Restructuring and modernization is possible by large companies.
- Asia accounts for a fourth of the world production. While the European and American output is largely based on soft pulp and is of better quality, the Asian production is a mix of hard wood pulp, soft wood pulp and other agri-residues.

Threats

- WTO Norms and Environmental Regulations can cause problems for the SMEs.
- Energy dependency of this sector is a major handicap.
- Multinational Companies - Asia Paper and Pulp (APP), Stora-Enso etc are strong competitiors.
- Faulty projection of macro-economic variables will find the companies with wrong inventories at the wrong time.

World Trends

- Global paper industry in on an acquisition binge. The effect will not spread to India.
- Abroad, companies stress on energy savings – In India, the stress is on reducing capital cost (Reason: High CoC).
- North America and Scandinavian countries such as Finland and Norway accounting for 55 per cent of the output.

Cross Country Comparison - China

- 20% of industrial pollution due to paper industry. Government took the initiative and closed down polluting factories.

- All factories less than 5,000 million tons capacity were shut down by June 1996. Factories below 20,000 million tons capacity in stages are recognized as losers and are on the way to closure.
- Allowed entry to foreign players as early as the 1980s.
- Companies are investing in plantations, which will solve the environmental problems as well as provide adequate raw material for the demand growth projected in future.

Strategic Action Points

- Since Paper is an energy intensive sector, special attention to the issue of pollution is necessary. China also suffered from industrial pollution due to paper industry (nearly 20% contributed by the paper industry). Government took initiatives and closed down polluting factories. All factories less than 5000 million tones were shut down in June 1996.
- SMEs should not be allowed in this sector - Reasons: Pollution, Inefficiency, Zero potential to become growth drivers.
- Power sector reforms will have a favorable impact since 20% of production cost is on power.
- Infrastructural reforms will facilitate raw material procurement.
- Duties on raw materials should be taken off to reduce production costs -Reasons: Cost competitiveness of firms.
- Companies should be encouraged to set up own feeder plantations - Reasons: Future raw material demand cannot be met with existing supply. {Paper products like envelopes; pads should be dereserved from under SME category. SMEs make cheap, low quality products. Larger companies can achieve better quality paper products that can be exported. The market for SSIs will not get affected, since a large number of the population will still go for cheaper items.

5.8.11 Analysis - Petrochemicals Industry

Introduction

It has an estimated turnover of Rs. 30,000 Crore, with a capacity of about 5 million tonnes. It contributes 7% of Manufacturing GDP (along with Rubber), It has no significant contribution to exports. Industrial raw materials like plastics, rubber are the outputs. The quality of petrochem output will directly affect the demand and supply.

- Demand is around 7.5 mt. The supply is 5 mt.
- The country imports about 25% of its domestic demand.
- Demand expected to double by 2010. Adequate capacity additions are underway.

Strengths

- Strong players in this sector are there in the Indian market.
- Huge plans are on the anvil for capacity addition by large firms.
- Impending entry of Shell, Dow Chemicals etc will strengthen India's place as a petrochemical manufacturing base.

Weaknesses

- It is dependent on Naphtha, whose prices fluctuate based on world Crude prices.

- The production costs are higher in India, compared to global majors due to the higher cost of feedstock.

Opportunities

- Investments in R&D will improve products.
- With the commissioning of new plants like the one at Haldia and the entry of MNCs like Shell and Dow Chemicals, India may become a net exporter in a couple of years.

Threats

- Rupee depreciation is a major threat, since feedstock has to be imported.
- Fluctuations of Naphtha prices affect the price of the products considerably.
- Globally, the industry is moving towards value added products but in India, due to poor investment in research, the industry is not able to capitalize on this score.
- 16% excise duty on naphtha, building blocks and fiber intermediaries and 24% on polymers are worrying the industry.

Strategic Action Points

- The industry's potential to export to small Asian and African countries on the east coast are very significant. The budget of 2002 has also provided for reductions in customs duty for polymers from 35% to 30%.
- Economies of scale are the most critical factor, which affects the bottom line of petrochemical plants. Hence it is better for the industry to consolidate (e.g. RIL-RPL-IPCL combine) rather than fragment. The initiatives of government to disinvestment are also helpful in further consolidation.
- The country is insulated partially from increased imports, due to its high Import duty. However, any reduction in import duty is likely to adversely affect the industry's fortunes.
- Government has resorted to applying anti-dumping provisions against East Asian imports of polymers in 2002. Strategic use of anti-dumping measures may provide short-term relief to firms in the country.

5.8.12 Analysis - Petroleum and Allied Industries

Introduction

Firms in this sector can be divided as: -

- Exploration and Production (E&P) Companies (ONGC and Oil India Limited).
- Refining and Distribution Companies (IOCL, BPCL, HPCL).

Its contribution is nearly 7% of Manufacturing GDP. There is no significant contribution to exports. It is important to any state and India also it assumes a lot of importance and government attention.

- India is the fourth largest oil consumer in the Asia-pacific after Japan, China and South Korea.
- Past fluctuations in global oil prices have had the single largest impact on Indian economy.
- Second only to Coal, Oil accounts for 32% of India's total energy requirements.
- India is considered to be amongst the fastest growing petroleum consuming markets.
- Gave employment to 138009 people in 1998; trend is growing.

- Demand Growth estimated at the rate of 7% a year - Will double by 2010.
- Demand present level of 96 million tonnes; Estimated demand is 195 million tonnes per annum by 2011-12.
- Supply for FY 2001-02 was 32.03 million tonnes against 32.42 mt in FY 2000-01, a drop of 1.2%.
- Demand is met by imports. Cumulative import of crude and petroleum products during FY 2001-02 was $14 billion (In 2000-01it was $ 15.7 billion) lower by 10.4%. The fall reflects lower international price of crude oil.
- There is huge supply gap and there is need for alternate fuels, because of shrinking reserves.

Policy Measures

- The oil pool account that used to make payment to the oil company out standings has been abolished. It has been done with the payment of oil bonds. The Government has issued sanction to the RBI to make payment of Rs 9,000 Crore to oil companies against a part of their dues.
- NELP, many private players are permitted for exploration and production of crude oil.
- In E&P, Indian oil and natural gas fields have opened up to the private sector as well as to foreign participation under production sharing contracts.
- The refining sector has been opened to the joint sector (public-private) partnerships as well as to the private sector for new refineries.
- According to Sengupta Committee Report, the stand-alone refineries should be merged with the oil PSUs.

Strengths

- Self-reliance in refining has been already attained.
- Strong players - ONGC and IOCL had already entered into collaborations to set up joint ventures in international markets. ONGC registered Rs. 6.2 billion profits last year, an Indian record.
- Hard for MNCs to enter since the tech-intensive upstream sector is under virtual monopoly and the distribution and logistics-intensive downstream sector requires huge resources and presence.
- New reserves of natural gas found by Reliance will also strengthen the sector.

Weaknesses

- India currently imports about 70% of its total crude requirement.
- Exploration costs are one of the highest in the world.
- Technology is rusty, needs urgent updating.
- India has no bargaining power with cartels like OPEC.
- Crude procurement: Only RPL has established risk hedging for crude oil and products, whereas for PSUs, IOCL acts as sole canalizing agent, which at present is not hedging.

Opportunities

- Golden Quadrilateral: The entire stretch of 5,952 Km is expected to account for approx. 55% of the total diesel market.

- OPEC is becoming businesslike instead of the typical cartel it was.
- "Petroleum Industry world over is shifting its emphasis to redefine growth strategies through emerging technological excellence, incorporating Portfolio, Resource and Knowledge Management, harmonizing - technology, people and process with economic, social and environmental system - integration, from cost-reduction to more diverse asset-management practices, improving productivity with environmental and social responsibilities" - ONGC Chairman.

Threats

- Sector will become susceptible to volatility in the international market, after complete deregulation.
- Energy dependency of the nation is also a factor for concern.
- Competition (private/ international)
- India has one of the highest reserve/ production ratio which is expected to be sustained in the coming decades. This may lead to a depletion of resources in the future.

Strategic Action Points

- Single regulatory authority for both natural gas and downstream petroleum activities is required.
- A 'Petroleum Planning and Analysis Cell' is already functioning.
- Energy is strategically the most important sector. India needs a policy on energy security. India's share to world oil and gas production is less than 1% with petro-product consumption of less than 3%. Demand is projected to grow at 7% compared to a world-estimate of 2%.
- Ownership of individual companies should follow a transparent and efficient set of disinvestment guidelines that do not change from bid to bid.

5.8.13 Analysis - Indian Rubber Industry

Introduction

At present about 47 per cent of the natural rubber consumed in the country goes into the production of auto tyres and tubes, 13 percent in cycle tyres and tubes, 10 per cent in footwear, 7 per cent in belts and hoses, 6 per cent in camel black, 5 per cent in latex foam, 4 per cent in dipped rubber goods and 8 per cent in other rubber products. The size of the industry can be gauged by the following data: -

- 32 tyre factories.
- 250 medium scale units.
- 6,000 small scale units (of which about 600 are well-established units)

The industry has an annual turnover of Rs. 13,500 crores

- Exports goods worth Rs. 1,300 crores
- Contributes Rs 3,000 crores to the exchequer in terms of excise.
- Gives direct employment to 3.5 lakh workers
- Manufactures every conceivable rubber product from toy balloons to aero tyres.

It contributes nearly 7% of Manufacturing GDP together with Petro-industry. There is no significant contribution to Exports. The rubber board with an annual outlay of Rs. 68.55 crores in 1998-99 is a supporting agency.

Strengths

- Indian climatic conditions are very good for rubber.
- Large players have firmly established themselves.

Weaknesses

- Dependence on auto industry is a problem.
- Although India is the fourth largest producer of rubber in the world, the exports of rubber goods from India are very low at $ 338 million in 1996-97 against a world trade in rubber products of $ 21,000 million.
- India has a huge potential for export of rubber products but lack of quality products and formulation technology are major deterrents.

Opportunities

- Road rubberisation can be undertaken in a large scale.
- Healthy projections of demand can also help.
- R&D - RRI etc are developing high yield clones, which can improve raw material production.

Threats

- Environmental Regulations are a problem for the industry.
- Price of Solvent Oil and other raw materials are high and are inhibiting factors.
- Excise and Custom duty hikes raise the cost of product.

Strategies

- There is a need to set up export oriented rubber industries in India. India's exports of non-tyre rubber products rose from Rs. 404.07 crores in 1996-97 to Rs. 410.00 crores in 1997-98 and those of auto tyres and tubes from Rs. 794.42 crores to Rs. 888.00 crores.
- Countries like Malaysia, Indonesia, Thailand do not depend on Auto sector alone. Farmers are more organized and thus are able to show greater productivity at lesser cost.
- The tyre industry has 16 major producers of tyres with 7 of them accounting for 95 per cent of the total turnover. The total installed capacity for tyre manufacture is about 37 millon nos. per annum. The demand emanates from three segments, the replacement market, purchase by new vehicle manufacturers (OEMs) and export markets.
- Of these, the replacement market is the largest with a share of about 54 percent and OEMs market 39 per cent in 1997-98. At present the industry exports tyres to some 50 countries. About 30 per cent of the exports are to the US. Export is roughly 7%. In 1998-99, the exports of tyres in value to Americas went up to Rs. 292.05 crores (from Rs. 270.68 crores in 1997-98), but fell in other destinations. The exports to Pakistan fell to Rs. 31.89 crores (Rs. 41.30 crores), West Asia Rs. 137.61 crores (Rs. 179.68 crores), other Asian countries Rs. 198.46 crores (Rs 257.82 crores) and Africa Rs. 69.29 crores (Rs. 67.35 crores). In recent years South Korea has started giving tough competition to Indian tyre manufacturers/ exporters in exports of cross-ply tyres by quoting lower prices.

5.8.14 Analysis - Steel Industry

Introduction

In 2001 the Global production of Steel was 846.9 million metric tonnes of crude steel. India produced 27.3 million metric tones. In India, a major part of steel is consumed in engineering applications, followed by automobiles and construction. The Indian steel industry has invested about Rs 30,000 crores in expansion and modernisation during the recent past raising the total investment to about Rs 90,000 crore so far.

Strengths

- India is the eighth largest producer of steel in the world (27.3 million metric tones of crude steel).
- India has emerged as the largest exporter of stainless steel to China for the first quarter of FY02.
- India is gaining in terms of market share and production of steel.
- India has good quality of Iron ore.
- The balance life of Iron ore at current level of production is 135 years, according to Geographical Survey of India.

Weaknesses

- Per capita consumption is one of the lowest in World.
- The industry has been facing over-capacity, poor demand growth and declining tariffs.
- Exports that had been have been adversely affected due to the imposition of anti-dumping duties by the US
- High cost of power and low quality of coal.

Opportunities

- Trend towards flat steel HR and CR steel demand is high.
- Low per capita consumption of steel offers opportunities for domestic demand increase.
- Of late there has been some revival in the automobile sector (two-wheelers, passenger cars and HCV). This recovery in the automobile sector is expected to benefit the steel industry.
- Step up in government expenditure on infrastructure projects is also likely to benefit the steel sector.
- The Indian steel industry does have opportunities in markets such as Iran, Iraq, West Asia, the Philippines and Bangladesh.

Threats

- With domestic demand being weak and international prices touching twenty-year lows, the price of steel products in the country has also been weak.
- USA, India's largest destination for long products has changed their style of construction. There is over capacity in beam and joists, which are the products India exports to it.
- In 2002 Chinese steel production rose 25.3% or 3.0 million tons versus September 2001.
- Today, in India, the problem is one of the human attitude and not one of the absence of raw materials or of technology.
- The particular crisis, which the Indian steel producers face is, the rising excise duties and the lowering of the customs duties. Such a fiscal imbalance takes away most of the competitive

advantages of the steel producers. The FDI inflows would finance the trade imbalance of the government rather than finance real investments in this case.

- Rising cases of dumping against the Indian steel exporters in the USA, EU and now in South East Asia is a problem.

Strategic Action Points

- Given the limitations arising from trade actions, it is important to shift the export focus to value-added, downstream products - galvanized, cold rolled, blanks etc - which do not attract trade actions like hot-rolled (HR) coils.
- It is also important to protect the domestic steel industry from imports as trade actions of the developed countries might result in diversion of steel into India by the affected countries. Diversifying steel exports in terms of basket of products as well as countries can do this, so that any trade action can be countered by a shift in markets or in products.
- India is one of the major steel producers in the country; it has a very low per capita steel consumption. Therefore, the best prospects for the Indian steel producer is to concentrate in increasing the steel consumption at home. Effort should be on producing flat steel, whose end use applications are automotive body panels and domestic 'white goods' products. This is because of an important structural change within the Indian industry where consumer durables have become a more important sector than either civil construction or the village industries - both being very important consumers of steel.
- India has a very good source of iron ore but poor quality of coal and expensive power. Hence we should have technologies, which can compromise on the quality of coal and on power. Corex, Romelt and such technologies should be preferred in the future.
- Integrating all the processes of steel-making - pellets plants, mini blast furnaces, electric arc furnaces, induction furnaces, rolling mills and coating lines rather than leaving them stand alone should be taken as a policy.
- Government initiatives to disinvest loss-making plants will reduce government control and also bring about a cultural change in the industry. Some of the components of the industrial culture are industrial relations, attitudes of the labour, the responsiveness of the management, the sense of purpose and of mission, dedication, responsibility and accountability. If these attitude components are missing then it is very difficult to become competitive even with the best of technologies. Today, in India, the problem is one of the human attitude and not one of absence of raw materials or of technology.
- There is a need to increase the component of R&D towards the development of new products to suit the changing needs of the end-users. UTS rails by Bhilai, Loco rails by Durgapur and Ultralight steel sheets for autobodies by TISCO are such examples.
- The rising cases of dumping against the Indian steel exporters in the USA, EU and very recently in South East Asia shows that slowly and steadily the Indian steel producers are becoming important. Good relationship with ASEAN countries will strategically help India not only in steel but also in various other products.

The Indian steel industry does have opportunities in markets such as Iran, Iraq, West Asia, the Philippines and Bangladesh. Middle east and rest of Asia (except China) which consume 17.6% of crude steel (International Iron and Steel Institute, 2001). This offers a good market closer to India, but the South East Asian countries are stuck with over capacities. Thus, Middle East should be focused upon.

6

CHAPTER SIX

Small and Medium Enterprises

THERE has been an increasing realization of a need to introduce the concept of Small and Medium Enterprises (SMEs) in place of Small Scale Industry (SSI). SMEs represent over 80% of the industrial base of most of the developed countries and so most of these countries have a concept of SMEs rather than SSIs.

There is a growing recognition worldwide that SMEs have an important role to play in the present context given their greater resource-use efficiency, capacity for employment generation, technological innovation, promoting inter-sectoral linkages, raising exports and developing entrepreneurial skills. Their locational flexibility is an important advantage in reducing regional imbalances. The future of SMEs is of major policy concern given their strategic importance in any discussion of reshaping the industrial sector. This is more so in the case of India, which has one of the longest histories of government support to the small-scale industrial sector since independence compared to most developing countries.

In the Indian context what is neither small nor large is being loosely defined as medium. Further, enterprise encompasses businesses, services and industries. In the broadband of 'small', the discussion extends to medium as well. Another possible connotation for the SMEs is the small manufacturing enterprises.

6.1 Definition

Countries do not use the same definition for classifying their SME sector. Nor does universal definition appear to be necessary. The definition in use, depends on the purposes these definitions are required to serve and the policies which govern the SME sector thus defined. The three parameters generally applied by most countries, singly or in combination are:

- Capital investment in plant and machinery
- Number of workers employed
- Volume of production (turnover of business).

Despite the lack of universal quantitative norms, the SMEs as a class are clearly distinguishable in any country, developed or developing. The factors that set them apart are essentially qualitative and comparative. On the qualitative side are their internal management structures, decision-making processes, financial practices, trading styles, attendant risks factors etc. Most SMEs are one person shows or are run by two or three individuals, usually relatives, friends or business partners, who take most of the decisions. There is no distinction between private and business assets and subjective and personal factors play a large role in decision-making. The personal stakes SMEs have in their businesses are much higher than those of corporate executives in their companies. This enhances the attendant risk and commits entrepreneurs more strongly to the success of their ventures. They are small and medium sized in comparison with the large entities with which they share a given economic space. SMEs therefore come in varying sizes and SMEs in one country may well be larger than the "big" companies in another. Not withstanding their absolute sizes, the problems confronting SMEs appears to be similar in most countries whether developing or developed. It is these features, which set them apart as a distinct group and it is these factors and not the quantitative definition, which are common and have universal applicability. A quantitative definition in each national context is, however, advantageous, as it makes it easier to target macro-policies to a specific group of enterprises. It facilitates the provision of technical assistance and the channeling of benefits under various policies, fiscal and financial concessions, and other incentives.

The definition used by the Indian authorities is based on the level of investment in plant, machinery or other fixed assets whether held on an ownership, lease or hire purchase basis. It seeks to keep in view the socio-economic environment in India, where capital is scarce and labour is abundant. However, a definition exists only for tiny and small units, medium sized enterprises are not defined either technically or legally.

The definition as recently revised places an investment limit on plant and machinery of Rupees 30 million for a small-scale unit. Units with investment not exceeding Rupees 2.5 million are classified as tiny units.

TABLE 16

Varying Definition of SMEs in Indian and Other Nations

	Country	Employees	Fixed Assets (US $ mn)
Asian countries	India		# < 2.55
	Bangladesh	< 10	< 2.20
	Sri Lanka		# < 0.33
	Myanmar	<100	< 0.80
	Pakistan		# < 2.25
	Philippines		< 0.49
	Vietnam	<300	# < 0.07
Industrialized Countries	Canada	<500	
	European Union	<250	< 24.57
	Japan	<301	< 0.77
	*Korea	<300	
	**Malaysia	><200	
	Singapore	<100	< 7.16
	Taiwan		< 1.23
	United States	<500	

 # Not an official definition but commonly used
Sources: Asia Pacific Tech Monitor, Vol. 13. No.2 (1996)
Technology Adoption by Small and Medium Sized enterprises in Malaysia, IDRC Publication (1988)

Small and medium enterprises, both in size and shape, are not uniform across the globe. This asymmetry comes in the way of any effort of their integration. The way they are defined depends on the stage of economic development and the broad policy purposes for which the definition is used. According to a World Bank study, there are said to be more than 60 definitions of small and medium industries used in 75 countries surveyed (cited in Kim Seung Jin and Suh Jang-Won, 1992). The most commonly used definitions relate to either size of employment and/or quantum of capital investment /fixed assets. A comparison of the definition of SMEs in India with those of other selected countries is given in Table 16. An interesting observation is that SMEs in most industrialized countries are defined by employee size and not by the fixed asset investment.

Indian Definition

> *An Industrial undertaking in which the investment in fixed assets in Plant and Machinery, (whether held on ownership term or on lease or hire purchase) does not exceed Rs.3 Crore is defined as Small Scale Industry (SSI). This ceiling is also applicable for ancillary industrial undertakings.*

6.2 SSIs IN INDIA

In India, small-scale industries were given due importance in the process of industrialization as far back as 1951. The Industries Development and Regulation Act legislated by the Centre in that year, became the framework for the small-scale industrial sector's development. The "reservation policy" was introduced in 1967 in an attempt to protect SSIs from competition. 44 goods were reserved for SSIs. Large corporations were allowed to enter this sector on condition that 50 per cent of their produce would be exported. As a result, SSIs dominated readymade garments, leather goods, auto components, electrical appliances and the hand tools industries. Over the years, the number of these goods kept growing and it stood at over 800 in 1989. Abid Hussain Committee, set up by the government, observed in January 1997,

> *"This policy is now inconsistent with the current trade reforms which allow free import of a large majority of the goods and most of the remaining can be imported under the Special Import License."*

6.2.1 Why SME?

In India, the definition of small-scale industries is mainly in terms of investment ceilings, which have changed over the years to keep pace with economic development. Though employment and turnover are also used to define small industries, as these indicators are implicit in the requirement for registration under the Factories Act, the core definition of SSIs in India remains based on investment limits - "historical costs of plant and machinery."

Owing to the concessions/exemptions granted to the sector in India, the SSIs opted for horizontal expansion to take advantage of the extant policy provisions. This inhibited expansion in capacity and modernization. With increasing liberalization, domestic industry needs to adapt superior quality standards for domestic and international markets. With the introduction of the concept of MSE, SSIs would have opportunities for vertical expansion without loosing the concessions available for them. The present policy discourages the small units to grow into bigger ones because of the low investment limit in plant and machinery and the artificial props such as excise duty exemptions.

Creation of SME category would provide the fillip in providing the much needed technological advancement and upgradation, optimum sales of economy, and vertical growth. The change in size and level of operation would lead to corporatization of SMEs and it would also have integration with broader markets. All these would obviously facilitate seamless growth of small to medium and eventually even to large-scale units. This would also attract more foreign investment in the sector.

The Planning Commission Study Group on Development of Small Enterprises, which submitted its final report in May 2001, has suggested that tiny, small and medium establishments could be redefined in terms of investment limits of Rs 25 lakh, Rs 5 Crore and Rs 10 Crore respectively. The Government should think of creating a separate category of medium-scale units with investment limits of Rs 10-15 Crore and encourage them to raise equity capital, including foreign equity to supplement institutional finance. While the broad policies should be the same for the SMEs, the tiny units could be given a much higher level of institutional support to promote self-employment.

6.3 Current Policy

The small-scale units can get registered with the Directorate of Industries/District Industries Centre in the State Government concerned. Such units can manufacture any item including those notified as exclusively reserved for manufacture in the small-scale sector. Small-scale units are also free from locational restrictions. However, a small-scale unit is not permitted more than 24 per cent equity in its paid up capital from any industrial undertaking either foreign or domestic. As of May 2002, 749 items are reserved for exclusive manufacture in the small-scale sector.

Manufacture of items reserved for the small-scale sector can also be taken up by non- small-scale units, if they apply for and obtain an industrial license. In such cases, it is mandatory for the non-small scale unit to undertake minimum export obligation of 50 per cent. This will not apply to non-small scale EOUs that are engaged in the manufacture of items reserved for the SSI sector, as they already have a minimum export obligation of 66 per cent of their production. In addition, if the equity holding from another company (including foreign equity) exceeds 24 per cent, even if the investment in plant and machinery in the unit does not exceed Rs 10 million, the unit loses its small-scale status. An IEM is required to be filed in such a case for de-licensed industries, and an industrial license is to be obtained in the case of items of manufacture covered under compulsory licensing.

Table 17

SSI Reserved Items

Food and allied industries	Other chemicals and chemical products
Textile products including hosiery	Glass and ceramics
Art silk/man-made fibre hosiery	Roofing tiles
Wood and wood products	Flooring tiles
Paper products	Ceramic table wares and allied items in stone wares semi
vitreous ware and Earthen wares	
Leather and leather products Including footwear	Mechanical engineering excluding Transport equipment
Rubber products	Electrical machines, appliances and Apparatus including
electronics	
Plastic products	Electric appliances
Injection molding thermo plastic Products (1)	Electronic equipments and components
Injection molding thermo plastic Products (2)	Transport equipment boats and Truck body building
Chemicals and chemical products Laboratory chemicals	Auto parts components and ancillaries And garage eqpt and
reagents	
Water soluble wood preservative Based on copper	Bicycle parts, tricycles and Perambulators
chrome, arsenic Boric compounds	
Basic dyes	Miscellaneous transport equipment
Azo dyes	Sports goods
Acid dyes	Stationery items
Napthols	Organic chemicals, drugs and drug Intermediates
Reactive dyes	Others
Fast color bases	Natural essential oils

Source: Handbook of Industrial Policy

A small-scale unit manufacturing small scale reserved item(s), on exceeding the small-scale investment ceiling in plant and machinery by virtue of natural growth, needs to apply for and obtain a Carry-on-Business (COB) License. No export obligation is fixed on the capacity for which the COB license is granted. However, if the unit expands its capacity for the small scale reserved item(s) further, it needs to apply for and obtain a separate industrial license.

6.4 Cross-Country Comparison on SMEs:

- Since 1977, the number of SMEs in Sri Lanka has increased 3 folds. Liberalization enabled the higher utilization of capacities due to the free availability of equipment, plant and machinery, tools and raw materials. This in turn contributed towards setting up of new businesses, expansion of existing enterprises and also improvement of quality standards of goods and services provided by SMEs. However, just like in our case, there has never been any targeted development of this sector.
- Almost 73% of SMEs in Sri Lanka are involved in manufacturing. At present about 50,000 registered and 125,000 unregistered manufacturing units are in operation in comparison to the approximately 1000 large industry units. In the manufacturing sector of the country, SMEs constitute 86% of the industrial establishments. SMEs contribute almost 18% of industrial output, 17% of value added and account for about 24% of industrial sector employment.
- In the sixties, Singapore had granted tariff protection to labour-intensive, export-oriented manufacturing industries and others that generated jobs. In the earliest years of its reforms in 1978, China promoted the 'township and village enterprises'. The result was that the contribution of this sector to her GDP rose from 12 to 39% by 1993: "China has adopted policies in favour of labour-intensive goods." Korea and Malaysia followed similar policies.
- The governments of Japan and Korea "promoted labour-intensive, small-scale enterprises through special credit facilities and protective government regulations" in their period of rapid economic growth. Only after they have achieved high levels of employment and growth on the basis of such strategies have they had gone after foreign investment.

6.4.1 SMEs in China

Since 1992, China's small and medium enterprises (SMEs) in all sectors of the state, townships and village, and private sector have conducted a series of radical and successful reforms of their ownership and governance structures.

Some of the innovative arrangements are: -

- Joint stock co-operative (or joint-stock ownership)
- Selling, leasing, taking over, merging
- Restructuring through Sino-foreign joint venture ('grafted' joint venture) and
- Corporatization (or equitization).

6.4.2 Structure of Chinese SMEs

The category of SMEs in China includes almost all TVEs (Town and village enterprises), private enterprises, household enterprises, which together produced more than one-third of Chinese GDP.

The large-and medium-size enterprises are mainly state-owned enterprises (SOEs). In this sector the control has now been handed to local governments - mainly counties and cities.

Private and Household-run Enterprises - These have shown the most rapid growth since 1992, due to improvements in the policy treatment. Two features of these enterprises are: -

- These enterprises have operated in tandem with community authorities such as village governments and with community-run TVEs.
- An increasing number of production firms in this sector have adopted joint-stock partnership and some have even become limited liability companies.

"Township and village enterprise" (TVE)

It is a specific term to identify those firms that are either collectively established by or initially based on and closely associated with rural communities such as townships and villages. It does not include State Owned Enterprises (SOEs) and enterprises owned by urban collectives, although it does include enterprises owned jointly by urban and rural enterprises.

6.4.3 Restructuring methods used by China

Joint-stock co-operatives

The features of such an arrangement is: -

- Managers and employees own a majority of the total share of the firm and share-holdings among them differ on the basis of paid subscriptions.
- The firm is closely held, implying that ownership shares are typically not freely marketable, although subscribed shares can be transferred within the community.
- One-man one-vote is usually employed.
- The firm is small or medium-sized.

Corporatization

By this method a considerable number of small enterprises have been converted into limited liability share-holding companies. During this process, a certain number of state-owned shares is offered for placement and thereby reduces government holdings. This phenomenon emerged because of inefficiency of the institutional arrangements existing in China for SOEs.

Grafting

In China, corporatization with foreign participation is often denoted by a horticulture term 'grafting'. There are two ways of forming these so-called 'grafted joint venture' (GJVs). Chinese SOEs convert their existing fixed assets (including production buildings, equipments, lands, etc.) into shares in the joint venture, while the foreign partner invests in capital, technology or new-equipment. Chinese SOEs first restructure themselves into share-holding companies, and then transfer their shares to foreign investors.

Sales

There are numerous ways for SOE sales, i.e., state-owned assets of SOEs are sold to legal persons or individuals. Normally, a package deal for selling off all assets and liabilities of a SOE to one or two buyers is welcome by local governments. If there are more buyers an auction approach is taken.

Leasing

Leasing is a transitional arrangement that does not involve any property rights re-structuring. The ownership of those companies in which state assets are leased to non-state business-persons for a definite period remains unchanged. They are used for those enterprises that bear heavy liabilities, which nobody would like to buy or take over.

Bankruptcy

For those enterprises that have too many debts where none of the above methods is suitable to their restructuring, bankruptcy becomes that last resort.

6.4.4 Factors enabling growth of TVEs

This phenomenon is unique in the sense that the emergence of rural entrepreneurs and enterprises has not been so large in any country. It is not an outcome of a carefully defined plan. The government policy changed from tolerance to encouragement. And the TVEs used a large amount of surplus rural workforce, which was a problem for Chinese government. The good environment for TVEs is explained by the following factors:

- Village governments wanted revenue and were constrained by budget deficits. So they looked for some revenue earning measures.
- Community governments supported the initiatives and encouraged the process.
- China also loosened state control and thus TVEs entered the market to capture goods market shares.
- Flexibility of TVEs due to smallness of scale and a loose accounting system.
- Comparative advantage in the international market because of the overseas Chinese investors. The TVEs located in coastal areas are attractive, simply because the provinces of Guangdong, Fujian, Zhejiang, and Jiangsu have had close links with the overseas Chinese, historically.
- Support from SOEs came because the TVEs provided a way for the SOEs to diversify their sources of supply, provide cheap costs of labour and land use, and the flexibility to escape rigid controls of the state sector.

6.4.5 China's TVE Experience - Conclusion

China is a good example to follow because it was able to overcome a rigid mindset of state-control enterprises and become an all encouraging state. In India the Panchayati Raj system already provides lots of governance power to the villages. Such a district control is essential for easy development of local firms and enterprises. As in the case of China, government will have to be proactive and incubate the model of local firms with global tie-ups.

6.5 Present Status of SSIs in India

The Small-scale Industries sector, over the years, has grown steadily and occupied an important place in the economy. Contribution of the sector in terms of generation of employment, output and exports is quite significant.

- The number of registered units in the SSI sector has increased from 0.42 million as at end-March 1974 to 3.37 million at end-March 2001. These numbers are exclusive of the units in the unregistered sector. (Unregistered manufacturing accounts for 23 per cent of the total capital employed and 84 per cent of the workforce - Accenture report, 2001)
- The importance of the Small Scale Industries sector is well recognized from its significant

contribution to the socio-economic objectives of growth in generation of employment, output, exports and fostering entrepreneurship. Currently, the sector accounts for around 95 per cent of the industrial units in the country contributing 40 per cent of the manufacturing sector output and approximately one-third of the nation's exports. At the end of March 2001, there were 3.37 million modern SSI units providing direct employment to around 1.86 million persons.

- It has been estimated that a Rs. One million investment in fixed assets in small-scale sector produces 4.62 million worth of goods and services with about 10% value addition. Also, Rs. 1,00,000 of investment in the fixed assets in small-scale sector generates employment for four persons.
- While the food products sector leads in employment by employing 0.48 million, non-metallic mineral products and metal products follow with 0.45 million and 0.37 million respectively.

Under the planned economy regime, the important role assigned to the SSIs was largely because of their capacity to generate employment. However, it is interesting to note that under the new policy dispersion too, the SSIs are expected to play an important role. This is because they have access to cheap labour, the use of which makes the economy more cost-effective. In spite of the lip service, SSIs stand neglected both by policy-makers and researchers.

6.5.1 Need for Streamlining Institutional Support for SSIs

Many of the State Associations and the State Secretaries/Commissioners (Industries and Commerce) question the efficacy of most of these institutions. The general view is that several of them either lost their utility or require a thorough revamp in tune with the emerging aspirations of the SSIs. Thus it is very clear that though the intentions are no doubt for the development of the small-scale industries of India, there are a lot of gaps as far as implementation is concerned.

Most of the institutions are not clear about their exact functions and have ended up in providing common services rather than the specific functions for which they were created. Almost all the institutions, except a few, which provide technical knowledge or are supposed to be the knowledge base for this sector have proven to be inefficient and in some cases even irrelevant for the present era of liberalization. Even the institutions under the state government, which can be expected to perform better, have shown a lot of short-comings due to lack of technology resources and lack of follow up and supervision. It is high time now that the central and state governments realize the inefficiencies of these institutions and go for a revamp of strategy and streamlining of these institutions. Sri Lanka, adopted a similar strategy as early as 1993.

The suggestions made by the Nayak committee, Kapoor committee and the Abid Hussain committee have to be considered seriously if the government is earnest about providing assistance to this sector. Several State Governments, recognizing the need for entrepreneur education, have set up separate Centers or Institutions for entrepreneurship development, notable among them are: Gujarat, Orissa, Andhra Pradesh and Maharashtra.

The Nayak Committee and Kapoor Committee have brought out several inadequacies on the part most of the afore-mentioned institutions that need to be addressed with seriousness. 90 per cent of the institutions/ industries contacted and the Small Industries' Associations have complained of inadequate and untimely credit flow and onerous burdens of collaterals/guarantees as the major obstacles for their sustenance and growth. Thus schemes like the CGTSI were very much welcome and the government must look to work on similar lines, to simplify the procedures, to encourage the working in this sector.

6.5.2 WTO-Compatible Policy for SSI

Despite numerous policy measures during the past four decades, Indian small-scale industries have remained mostly tiny, technologically backward and lacking in competitive strength. Notwithstanding their lack of competitive strength, small-scale industrial units in India could survive due to product and geographical market segmentation and policy protection. From 1985, the economy started moving towards market-oriented systems, and this process accelerated after the crisis of 1991. These changes have reduced the importance of the above three factors.

- First, there has been a shift, during the past two decades in the economic policies of nations, especially developing nations, from 'policy regulation' to 'market orientation' exposing their industrial units to greater market competition.
- Second, globalization in the sense of increasing integration of world economies is taking place, resulting in intensifying the market competition.
- Thirdly, the past few decades have experienced rapid technological developments in numerous areas.

The post-liberalization business environment has become harsh for the small-scale industries (SSI) sector because of increased internal and external competition. In addition, the far-reaching impact of the various WTO norms are now threatening to further affect the fortunes of small and medium enterprises. Unfortunately, despite sufficient notice and the growing awareness of the impending threats, the SSI sector does not appear to be adequately prepared for the new challenges. While a number of units in the sector have been striving hard to obtain ISO or BIS certifications and compete against cheaper imports, the overall picture appears gloomy for want of proper policy support. Small-scale units are the most vulnerable as these units do not have the required resources and capabilities either for innovation or for purchase of technologies.

The creation of a new Ministry of Small-scale Industry and Agro and Rural Industries in 1999 did not make any difference to the sector's plight. Soon after its creation, the new Ministry decided to do a detailed sector-wise study of the impact of various WTO agreements on the SSI sector but nothing seems to have happened since. Not surprisingly, though the sector has grown at a rapid pace post-Independence, the incidence of sickness is on the rise.

While the official figures show only about 10 per cent of the over 32 lakh SSI units as sick, the unofficial figures put this figure at over 40 per cent. Given the crucial importance of the SSI sector to the economy with 40 per cent share in the total industrial output, 35 per cent in exports and over 80 per cent in industrial employment, it deserves all the policy support the Government can offer. Though various organisations including FASII have been advocating more protection for SSIs to meet the challenge from imported goods, experts feel that over-protection from the government has only led to arrested development of the small-scale sector.

6.6 STRATEGIES FOR REVIVAL AND GROWTH OF SSIs

Development of Small-Scale Industries is necessary and vital for industrial development. However, it must be recognized in all fairness, that many small entrepreneurs in the country are facing genuine problems:

- Poor quality infrastructure.
- Inadequate access to institutional credit.
- Delayed payments by large industries.
- Procedural delays in getting government clearances.

- Harassment by inspecting officers.
- Rigid labour laws.
- Technological obsolescence.
- Non-availability of skilled manpower.
- Lack of marketing facilities and
- Difficulty in competing with well-established foreign and national brand names.

One crucial argument for legal reforms in the SSI sector relates to competitiveness is that if laws are simplified, SSI products will become quality-upgraded and increasingly competitive in the international market. Some of the critical success strategies for SSI revival are:

6.6.1 Public-Private Partnership (PPP)

Public-Private Partnership (PPP) between public institutions and private business could be one innovative strategy to tackle the issues pertaining to SSIs. This would entail contribution of resources by both as also participation in the planning and decision-making processes and leveraging the competencies of each for collaborative, promotional efforts. The new SSI policy of the Indian government provides some scope to forge such partnerships. Partnerships may be for financial assistance, for manpower development, creating testing and research facilities or even for support services for promotion and development of SSIs.

Lack of adequate and timely institutional credit is a frequent complaint of the small-scale industry.

- Banks insist on collaterals, which the small entrepreneurs, with limited resources, are unable to furnish.
- Bank officials frequently point out to the high level of NPAs and the poor culture of repayments to justify their cautious lending policies.
- The problem is that there is no trust and understanding between bank officials and industrialists in India.
- During recession, the bankers often cut back on lending, impose heavy penalties for defaults and try to recover as many advances as possible. This makes it nearly impossible for the industry, and especially the small-scale industry, to survive through a recession.

Contrast this with Germany where a close partnership exists between the local banks and the small-scale industries.

- The banks often have their representatives on the boards of the borrowing companies and closely monitor the companies' operations.
- In times of recession, they reschedule the loans, cut interest rates, and even extend additional loans to finance high inventories.
- In recognition of this support, the companies in boom periods compensate the banks for the sacrifices made by repaying the rescheduled loans with higher interests.
- It would be in the interest of both the banks and the industry in India to adopt such a collaborative relationship.

The non-availability of skilled manpower is an important constraint, especially for hi-tech industries.

- Government training institutes are often ill-equipped, follow out-dated curricula and are theoretical in approach.

- Specialized training institutes can be started on the PPP concept wherein infrastructure can be provided by the government while the management can be left to either an NGO or an industry association.
- The Centre for Research and Staff Performance (CRISP) in Bhopal set up by the government of Madhya Pradesh and Germany in collaboration with the local industry is a good example of such a partnership in manpower development.

A network of testing and certification facilities needs to be put in place for the use of small-scale industries, especially for those who wish to enter the export market.

- Specialised facilities often require expensive and sophisticated instruments.
- Government has neither the funds nor the expertise to provide these. The PPP concept can work well here.
- The new SSI policy of government of India provides a good opportunity to SSI associations to set up and operate testing laboratories by availing the capital subsidy of 50 per cent being offered by the government.

Providing basic infrastructure to SSI clusters through industrial estates has been the traditional strategy of most governments for promoting small-scale industries. However, increasingly small-scale units, especially in hi-tech areas, are in need of specialized infrastructure and services of international standards.

- Government is not in a position to provide and maintain such facilities.
- Pure private initiatives in infrastructure are often not feasible because of low returns and long gestation periods.

The PPP approach can provide several possible solutions.

- Specialized facilities can be jointly funded by government and the user industries and thereafter, handed over to expert private agencies for maintenance.
- Industrial estates set up by the government can be handed over to associations of local industries for maintenance along with the tax revenues.
- Infrastructure projects can be structured to make them attractive to private investors by offering them as 'packages' - combining commercially viable projects with the not-so viable ones or by offering subsidies and tax concessions for the unviable segments.

The PPP approach can succeed only if there is an attitudinal change on the part of government officials, small-scale entrepreneurs and industry associations.

- Instead of a regulatory and paternalistic approach to SSI development, the government, especially at the state and local levels, will have to adopt a more democratic, participative and power-sharing approach.
- The small entrepreneurs will have to focus more on group development rather than individual goal fulfillment and subject themselves to community discipline.
- So far the role of SSI associations has been largely that of representing their grievances before the government and of seeking greater concessions, especially in the matter of taxes.
- In future, the associations will have to play a more positive role and take on responsibilities in implementing development programmes and enforcing regulations as in the case of environment protection, safety of workers and so on.

- With PPP, the government can gain from enhanced efficiency, sustainability of schemes, and higher outreach.
- The small entrepreneurs can overcome their scale-related disadvantages, reduce transaction costs and influence decision-making.
- Finally the industry associations can expand their activities, participate in policy formulation and attract more members.
- Thus PPP can create a win-win situation for all parties.

6.6.2 De-reservation of Items for SSIs

Small-scale industries enjoy certain inherent strengths such as lower over-head costs, flexibility in production, informality in labour relations, exploitation of local resources and skills, capacity to execute small orders and to offer customized services.

The policy of reservation of certain items for the exclusive production by the small-scale sector has little relevance today with most of the reserved items, placed on Open General license. In fact, it is only serving to prevent the Indian industry from benefiting from economies of scale and from becoming globally competitive. This is particularly so in the case of labour-intensive and resource-based industries like the manufacture of garments, leather goods, toys and agro-processing. The Abid Hussain Committee had forcefully argued in favour of scrapping it as far back as in 1997.

Fears that dereservation will lead to the extinction of the small-scale units is based on wrong presumptions. For example, countries such as China, Bangladesh, Malaysia, Philippines, Sri Lanka have no reservation for the garment industry and yet the small-scale units, are thriving.

- The large units specialize in executing large orders of institutional garments.
- While the small units take on smaller orders of high fashion, seasonal garments.
- Collaborative relations between the large, medium and small units through sub-contracting arrangements are well established.
- This can happen in India too. Government's policies should seek to strengthen and promote such collaborative efforts rather than keep them apart through protectionism.

6.6.3 Manufacturing Flexibility

Periodic recessions are going to be a regular feature of a liberalized and globalized economy. Therefore, flexibility in operations, especially for the small-scale units, will be critical for survival.

- A lean and mean structure alone will see them through the bad times.
- However, the present rigid labour laws and complicated procedures for getting government's prior approvals for lay-offs and retrenchments are contributing substantially to the spread of industrial sickness.
- They are forcing industrialists to adopt illegal methods to downsize their operations and encouraging them to opt for capital-intensive technologies rather than labour-intensive technologies.
- The present labour laws, in trying to protect existing jobs, are in effect, hampering the creation of new jobs.
- Flexibility in operations for the industry combined with adequate compensation for the temporary loss of jobs for the workers could easily be achieved by judiciously amending the existing laws and rules.

- A majority of the labour laws are central laws, though the subject falls in the concurrent list.
- The state governments need to be given the freedom to bring about necessary amendments to meet the local situation.

At present, the states need the prior consent of the Government of India and the presidential assent for any proposed amendment, a cumbersome procedure which is needlessly putting a brake on some reform-oriented states who wish to make labour laws more in tune with the changing economic environment.

Industrial firms in India are widely perceived to lack the necessary flexibility to respond to changes in markets and technology. The regime of state-directed industrialisation and labour policy is to be held responsible for having imposed rigid rules on firm management decisions. Labour laws and job security regulations have made labour use by large industrial firms inflexible [Chaudhary 1996].

- Industrial regulatory policy has been restrictive
- On the choice of capacity expansion, product mix and location decisions of large firms.
- Technology policy worked to limit foreign technology purchase, duration of technology agreements and royalty payments.

During the period 1963-64 to 1978-79, it is observed that domestic competition sharpened with an increase in the number of firms in both homogeneous and differentiated markets.

With the relaxation of capacity licensing and import restrictions in the 1980s, industrial markets experienced greater domestic and import competition often with many players with foreign technological collaboration [Jacobsson 1991; World Bank 1996]. The growth of organised manufacturing employment decelerated and employment elasticity of output turned negative. The unorganised sector accounted for most of the manufacturing employment.

TABLE 18

Industries With High Share of Contract Labour

Industries with High Share of Contract Labour (Percentage of Total Employment in Census Sector)

Industry Group	1986-87	1980-81
Beverages and tobacco	50.9	9.1
Non-metallic mineral products	23.9	14.6
Food products	19	16.5
Basic metals	16.9	8.5
Wool and silk	12.5	5.5
Chemical products	10.2	6.8
All industries	10.1	4.6

Source: ASI 1980-81 and 1986-87 (census), Labour Bureau, ministry of labour.

Micro-economic studies have documented the widespread use of market mediated work practices and outsourcing of products. It includes greater use of contract, temporary and casual workers, on-site labour, only sub-contracting and production sub-contracting among others. Aggregate data shows that the share of casual workers in large factory employment rose from 4.6 per cent in 1980-81 to more than 12 per cent in 1993-94. It is primarily the strategic response of management to changes in both product and factor markets.

Use of Sub-contracting for Flexibility

Sub-contracting can be defined to mean the manufacture of goods by one firm for another based on latter's specifications [Lazerson 1990]. Such orders may include "the treatment, processing or finishing of materials or parts by the subcontractor at the principals request" (cited in Productivity, vol 35, 1995). Contract manufacturing is included.

The two main forms of sub-contracting are commercial and industrial sub-contracting. In commercial sub-contracting the role of the principal firm is limited to marketing and distributing the sub-contractors' finished product. The principal firm itself may not be engaged in the manufacture of that product.

Two large (even rival) firms are found to engage in sub-contracting arrangements. For example, Videocon supplying Power Wash machines to Godrej-GE. Suppliers of refrigerators, electric fans, thermo-ware steel furniture and domestic electric appliances constitute the numerous instances of horizontal sub-contracting between large and small firms. It is possible to distinguish a number of sub-species of sub-contracting arrangements between firms. In Japan sub-contractors of the various automobile manufacturers are organised into tiers, with lower tier sub-contractors supplying higher tier sub-contractors. It is argued that in response to changing market conditions sub-contracting has emerged as an alternative strategy to vertically integrated production in developed countries [Lazerson 1990]. What are the determinants of sub-contracting arrangements?

Large industrial firms in India, till recently, worked in an environment of control and regulation. They were subject to capacity restrictions and labour market regulations. Sub-contracting of final products, components, services and labour by large firms are known to be widespread and are believed to be rapidly growing. The practice of sub-contracting is an indicator of the search efforts of firms to gain new margins of flexibility in competitive markets. It was shown that sub-contracting intensity has risen in aggregate manufacturing.

- Putting-out production or traditional sub-contracting, based on material supply to other factories, has not been a major source of flexibility.
- The industry group of consumer non-durables is found to have the higher sub-contracting intensity compared to basic and intermediate goods. The latter represent process technology and continuous flow methods of production in contrast to consumer non-durables in which batch production methods seems to be widespread.
- Sub-contracting intensity and labour intensity are found to be positively associated. Both technology and labour market regulation are identified as sources of higher sub-contracting activity of factories using more labour per unit of output.
- The hypothesis that sub-contracting intensity and factory size are non-linearly related to cost and output relationship in the theory of production was supported by regression estimates. It showed that sub-contracting first falls and then rises with employment size of factories. This is attributed to the organizational dis-economies of large factories.

Indian industry is in the process of restructuring to improve competitiveness. In this context sub-contracting would be an important strategy of large firms. It is also a major source of flexibility. In course of time there would be qualitative shift in the nature of sub-contracting perhaps with an emphasis on the use of skilled labour and specialization rather than on taking advantage of poor labour standards in the small-scale factories. To the extent that sub-contracting is a response to the constraint of labour regulation, the method and the process of labour market reforms need greater attention.

Upgrading Technologies

Small-scale units play a critical role in the Indian economy in the creation of employment opportunities and wider dispersal of industrial production across regions and individuals. However, the general observation has been that the majority of these units are not competitive (Ayyar Report of the Expert committee on SSE). Notwithstanding their lack of competitive strength, the small-scale units of India have survived so far due to product and geographical market segmentation and policy protection. The importance of all three factors is decreasing, as expected, as the economy becomes more and more liberalized. Ultimately, in a market economy, small-scale units have to sustain themselves on their own competitive strength by successfully facing competition from large-scale units, including multinationals. Even to provide employment (an argument that has been widely articulated in favour of protection to small-scale units in India) in a sustainable way and at higher wages, these units have to be competitive and commercially viable.

Commercial viability and competitive strength of any industrial unit primarily depend on the technology they use. A study by T.A Bhawani on the impact of technology on the competitiveness of the Indian small-scale automotive component units show that technology, as represented by the level of mechanization, materials, labour, and the plant and machinery, has a significant positive impact on sales turnover. A higher degree of mechanization in terms of the use of NC machines has a significant positive impact on the sales turnover, followed by materials, labour and capital. Technology helps to improve the competitive strength of any industrial unit either through cost reductions, productivity/quality improvements, or a combination of all.

The small-scale industry in India needs to upgrade technologies on a continuous basis to remain competitive. Adopting information technology to increase productivity and leveraging their advantage of flexible operating systems, the small enterprises in the US have performed exceedingly well during the last decade. There is no reason why small enterprises cannot do the same in India. There are very effective software packages available in the country for manufacturing units that can eliminate waste, compress production cycles, and improve the quality of the products. If they are expensive and beyond the reach of individual small-scale units, the packages can be purchased by groups of units and used jointly on a time share basis, paying hourly fees. The government can provide the infrastructure and train the managers in the use of the software.

6.6.4 Consortium of Small and Medium Scale Industries

Difficulties in marketing their products has always been a major problem for the small manufacturers.

- Government agencies have often failed in providing assistance since officials have little understanding of market intricacies.
- Instead of competing severely against each other through undercutting in prices, the small manufacturers could try a collaborative approach by setting up marketing consortia.
- The small and medium-scale bulk drug manufacturers based in Hyderabad have jointly set up a company to facilitate execution of large export orders which they were not able to do individually.
- Four small leather goods manufacturers in Calcutta have taken a similar step.
- Government could provide infrastructure support for such consortia (showrooms, godowns, trade fairs and convention centres) and promote trade delegations.
- E-commerce has, of late, opened up tremendous opportunities for cost-effective marketing for small-scale industries which need to be exploited fully.

6.6.5 Clustering as a Strategy for SSI Development

The Concept

An industrial cluster (Porter 1990) is a set of industries related through buyer-supplier and supplier-buyer relationships, or by common technologies, common buyers or distribution channels, or common labour pools. Industry clusters are concentrations of competing and related firms, bringing together both the firms producing final goods and their suppliers and contractors.

Importance

The relevance of cluster development lies in the fact that in an increasingly globalized world, factors external to the business, but internal to the regional economic foundation, are increasingly important for the creation of competitive advantage. For nations, especially developing nations, competitiveness cannot be measured by the performance of a single firm alone.

Whether it is a nation, state, or region, a host location has a set of unique local conditions that strengthen the ability of dwelling companies to compete in an industry. The competitive advantage of a location does not normally arise in isolated companies but in clusters of companies. The active set of relationships among the members of the cluster enables firms for individual efficiency and competitiveness.

Cluster Structures

The presence of backward and forward linkages in an industry is one of the critical characteristics of clustering. The linkages can be buyer-seller relationships, common customers, or other relationships. Based on these aspects, cluster structures can be analyzed as below

- *Buyer-Supplier Relationships.* This cluster structure consists of core companies that produce goods and services that are sold to final consumers. It also encompasses companies at earlier stages in the value-adding chain that supply the inputs–intermediate goods, services, and raw materials that are used in the assembly of final goods and services. Distributors of final goods and services, where separate from the producers, may also be a part of these clusters.
- *Competitor and Collaborator Relationships.* This cluster structure consists of companies that produce the same or similar goods and services at a specific level in the value chain. Such clusters exist because competitors frequently share information (often unintentionally) about product and process innovations and market opportunities. In such a scenario, it makes more sense for these companies to formally collaborate to develop such innovations in pre-competitive or strategic alliances. While there is cooperation in some areas, there must be rivalry among firms in order to create global competitiveness, and encourage constant innovation.
- *Shared-Resource Relationships.* These relationships exist when firms rely on the same sources of raw materials, technology, human resources, and information even though they may use these resources to produce goods and services for very different markets.

Advantages

- Costs of research and development can be shared among the cluster members. This will be a much-needed alternative to importation of technology that has been practiced so far by Indian firms.
- With the adoption of WTO norms, environmental norms will become more stringent. Setting up pollution treatment plants is beyond the capability of many firms, especially in the small and medium sector. Provided the member firms are closely located, industrial units can set up a common pollution treatment plant. In accordance with the amount of effluents output by each firm, the capital and maintenance costs of such plants can be proportionately shared.
- Since a group of firms are involved information critical for competitive advantage is not confined

in bits and pieces to any single firm. This kind of sharing of information provides richer insights about opportunities and threats than is available from more traditional economic development strategies.

- Effective strategies can be developed for the collective benefit of the industry. Policy formulation and regional organization will become more focused and successful.

How to Make Clusters Successful

Lundequist and Power (2001) observe that successful clustering can occur in many forms as below:

- Industry-led initiatives to build competitiveness and competence within an existing base
- Top-down public policy exercises in brand-building
- Visionary projects to produce an industry cluster from scratch
- Small scale, geographically dispersed, natural resource based, temporal clusters that link into global rather than national and regional systems and sources of innovation, competitive advantage and strategic assets.

Characteristics of a successful cluster are as follows:

Institutionalization

The institutionalization of ongoing processes under the banner of clustering is decidedly better than starting an industry cluster from scratch. Rather, the existing sources of regional competitive advantages have to be leveraged on. Indian examples of such successes include Tirupur and Ludhiana.

- *Ties with educational institutions* - The key challenge is to create resources and capacities that facilitate the transfer of ideas and innovations into new commercial products.
- *A distinctive cluster vision* - A widely anchored vision will be an important platform for a successful cluster strategy. The vision should be flexible while being focused.
- *Handpicked network brokers* - Authorization should be given to competent people or organizations to act as cluster 'drivers'. The cluster driver will be the interface between policy makers and cluster members.

Marketing

In order to strengthen the competitiveness of the cluster a clear cluster brand needs to be developed. The importance of focused branding cannot be undermined, since,

- Brands strengthen the attraction of the cluster for investment, venture capital, skilled workers and new entrants;
- Brands help to unite actors in a shared purpose and identity;
- Brands often complement individual firms' marketing and collaborative-marketing activities.

Upgrading

As cluster development reaches stages where it needs a boost in capital investment and policy investment, the non-availability of these factors can be extremely harmful. In such instances, government has a crucial role to play.

Challenges in Execution

Critical Mass

For cluster formation to be successful, there should be a critical mass of Competitiveness Factors. The factors may be information, skills, relationships, and infrastructure in a particular field. Having a critical mass of these competitiveness factors provides the foundation for a growing cluster.

The Role of Government

The challenge for Indian government is that of facilitating cluster development and upgrading. Porter indicates that clusters must not be developed or created from 'scratch' within the context of declining markets and industries. This is one of the main pitfalls of cluster-based industrial policy. Hence the emphasis should be on reinforcing and building on established and emerging clusters, not creating entirely new ones.

Constraints like government regulations, basic infrastructure, education, training policies) are a result of government policy and therefore require its intervention. Essentially, any government policy that results in increased costs with no potential for long-term competitiveness or social benefits needs revision. This clearly highlights the need for government focus to go beyond improvements in the general business environment.

Cross Country Comparison - Sweden

From the experience of Sweden, it is clear that clustering should be addressed as an approach that represents a shift away from narrowly focused firm-based strategies to more holistic regional economic development approaches. In this sense, industry cluster is a pro-active policy tool and approach based on negotiated and, often, collaborative efforts to manufacture.

Building Institutions

On the policy front, Sweden started off 'Klustergruppen'; a Cluster Focus Group set up by NUTEK. (The Swedish Business Development Agency). This group is necessarily a partnership between regional and national policy practitioners that aims to link significant regional cluster initiatives together in order to learn from each others' experiences.

The group holds regular meetings and also invites academics, consultants and policy-makers. In accordance with their recommendations, the Swedish Ministry of Industry plans 'Regional Growth Programmes'. These programmes have gradually evolved into national policy instruments for implementing regional industrial policy.

Thus, as far as Swedish policy is concerned, cluster development is developing into a recurring area of priority. In the aluminium sector, for example, the Aluminiumriket started off as a private sector initiative. This was based on a need for skill development of labour. Sweden successfully integrated this beginning with its formalized cluster development initiative and achieved the following objectives

- Close cooperation between industry and educational institutions
- Development of R&D

Biotechnology was another domain where the cluster 'Medicon Valley' has been developed by successful policy initiatives into a huge centre for pharmaceuticals and biotechnology. Central to such policy moves has been the idea of turning the region into Europe's pre-eminent hub for life sciences R&D and production.

India's Experiences in Clustering

India has already exhibited some successful industrial clusters as shown in the Table below. The kind of rightful attention that these regions have managed to obtain has helped bring about major strides in economic and social well-being.

The table below is testimony to the immense possibilities of cluster-based development in India. Developed naturally, i.e. without any sort of government initiative to start with, these clusters have developed themselves quite well.

TABLE 19

Successful Industrial Clusters in India

Name	Statistics
Panipat	◆ 75% of total blankets production in India
Tirupur	◆ 80% of India's cotton hosiery exports
Ludhiana	◆ 95% of Indian woolen knitwear,
	◆ 85% of India's sewing machines
	◆ 60% of bicycle and bicycle parts.
Agra	◆ 800 registered and 6,000 unregistered small scale units making shoes
	◆ Daily production value of 1.3 million dollars
	◆ Exports worth US $ 60 million per year

TABLE 20

Sample Sectoral Distribution of Clusters in India

Sector	Clusters
Machinery and Parts except electrical	20
Cotton Textiles	15
Chemical and Chemical Products	14
Metal Products	13
Hosiery and Garments	10
Food Products	9
Non-metallic Mineral Products	9
Electrical Machinery and Parts	8
Wool, Silk and Synthetic Fibre Textiles	8
Transport Equipment and Parts	7
Others	25
TOTAL	138

The above table is not exhaustive and shows a sample of 138 clusters. There are estimated to be around 350 clusters presently existing. However, the distribution and performance statistics can be safely assumed to be representative.

TABLE 21

Regional Distribution of Clusters in India

State	Clusters
Maharashtra	25
Gujarat	20
Punjab	15
Rajasthan	14
Uttar Pradesh	13
Haryana	12
West Bengal	9
Tamil Nadu	8
Himachal Pradesh	5

Source: UNIDO

The findings from the UNIDO report are as follows:

- The industrially developed states of Maharashtra, Gujarat, Punjab, Rajasthan, Tamil Nadu and Haryana have the maximum number of clusters.
- The top six industrially developed states out of the 25 states in the country account for 71% of the clusters.
- Surprisingly, the state of Madhya Pradesh, which has the maximum number of small-scale enterprises (12.7%) as per the 2nd All India Survey of SSI, seems to have very few clusters (1.4%).

Following the international recognition of clustering as a major strategy for economies of scale, time and effort, clustering has started to pick up in our country as well. Several initiatives in India have attracted foreign assistance and interest.

Factors that can favour Cluster Development in India

- **Infrastructure**

Development of infrastructure-based clusters is a relatively new phenomenon and these are mainly induced clusters. Such clusters are in the field of electronics, software, floriculture and biotechnology. The reason why infrastructure based clusters are government induced is that infrastructure has been and still remains primarily the responsibility of the government.

- **Co-existence of Modern and Traditional Clusters**

India is one of the oldest civilizations and it is very common to find traditional skill based industrial clusters. In several of the traditional sectors, modern small units producing the same goods through mechanized means also exist. Any program or policy that has to be initiated for development of such clusters would have to take into account such diversities within the clusters.

- **Industrial Education and Labour Training**

India has a good infrastructure for industrial education. There are approximately 5,000 science and technology related institutions providing training to about eight lakh students every year. Even in the traditional sectors, large skilled work force is available. India has very large skilled and technical manpower. There is a well-diversified network of Industrial Training Institutes (ITIs) in almost every district providing training that does not relate to the need of specific aspects of the locally based industry.

- **Reservation Policy and Clusters**

In order to provide a protective shield to SSI in India, government has reserved certain items for their exclusive manufacture in the small sector. The medium and large-scale units are not permitted to make these items. In some of the industries, large and medium scale units had been set up before the reservation policy for that particular product was enforced. There are 74 clusters making products that are reserved for SSI and 64 clusters make products that are not reserved for SSI out of the sample of 138.

Policy Imperatives

The UNIDO study on SME clusters in India stresses on several policy imperatives, which are discussed below:

◆ **Specific Policy for the Clusters**

There is no special policy either of the central government or state governments for the promotion of the SME clusters in general. There have been, however, attempts to strengthen the clusters by providing support through the general schemes related to technology upgradation, infrastructure provision and training through various institutions at the national and state levels. Some of the industrial estates developed for particular industries have been developed, but only to a limited extent.

◆ **Organized Efforts from Government Machinery**

The governments at both the center and state level has been supportive of clusters. Government sponsored research institutions that are supposed to provide assistance directly to the small units are amply set up. The functioning of these institutions is regulated under a framework of centrally designed rules and regulations. State government institutions have tried to emulate the central structure leading to avoidable duplication in some cases. In this context, a central level institution for cluster development has to be developed. The Swedish model, analyzed earlier in this section may be relied upon.

◆ **Involvement of Industry Associations**

Little initiative has been taken to help develop the capabilities of the private sector to start self-help programs. The government wants to use the private associations to set up these delivery mechanisms; it finds only a select few that are worthwhile and strong enough to shoulder the responsibility. The major problems of the associations in or outside the clusters relate to lack of financial resources, poor democratic set-up, lack of vision, and involvement in petty politics rather than provision of services to its members with a long-term perspective to improve their competitiveness.

◆ **Skill Development**

Existing educational institutions like the ITIs, which provide practical training and skill development should be changed to private or quasi-private institutions. Till now the maintenance of these training institutions was the responsibility of the Government but now the industry is being looked at as a possible contributor.

◆ **Moving from Informal Cluster Structures to Formal Organizations**

Only the bigger end of small-scale units and medium scale units are set up in industrial estates with a formal infrastructure. The private industrial estates are very few and mainly limited to the industrially developed states of Maharashtra and Gujarat. There are several disadvantages associated with operations based in home and un-authorized locations. Not only such units have problems in expansion and modernization due to space constraints, they may also be denied registration as SSI units and thus denied access to credit and other support facilities.

◆ **Extending Savings and Credit Institutions**

In India, financial institutions do not have a special policy or a program for providing support to units in a cluster. The assistance is provided on the lines of a general policy sometimes even ignoring the special requirements of an industry or a cluster. For example, the working capital needs of the food industry fluctuate as the raw material is seasonal but the financial institutions because of their general norms find it difficult to support such demands.

Cluster specific credit supply is needed, as is prevalent in Italy. In such a situation, banks and other financial institutions extend cheap credit by charging interest in a varying manner according to the cyclical fluctuations in the business that the cluster members are involved in.

- **Marketing and Export Promotion**

Indian Government has established 'Export Promotion Councils' for the products with high potential of exports. These councils provide information and advise to exporters. An example is the handloom export promotion council, which provides manufacturers with precise information about foreign trade fairs, cloth specifications and other such vital information. In case of ancillaries and sub-contracting with large firms, the setting up of SIDBI in 1991 has been particularly helpful.

- **Building Institutions**

Some proactive initiatives in building the necessary support institutions for SME clusters has been taken as shown in the table.

TABLE 22

Institutions Providing Cluster Building Assistance

Name of Institution	Parent Organization
'BISNET' by Federation of Indian Chambers of Commerce and Industry (FICCI)	Industry Association
National Small Industries Corporation (NSIC)	Government
Asia Pacific Center for Technology Transfer (APCTT)	Government
Maharashtra Chamber of Commerce and Industry' (MCCI), Pune	Private Sector
TANSTIA-FNF Center, Madras	Private Sector

However, there are some significant issues in catering to the needs of smaller firms. The UNIDO report elaborates as below.

> *" The type of information usually needed by the tiny and very small enterprises is very specific to the state government rules affecting their operations. They may be interested in knowing about the suppliers of specific type of machinery suiting to their needs. Such grass root level and specific detailed information is not yet available on the information networks. Secondly, the information providers, in most of these clusters and local areas, do not virtually exist who can source and supply the necessary information. Thirdly, the sourcing of most information for the small entrepreneurs so far has been by informal channels where there may not be any payment involved. This leads to a tendency for the entrepreneurs to expect free information, which may not be possible in the current scenario. "*

- **Effectiveness of Upgradation Facilities**

Upgradation facilities are several but not very effective since they are more closer to academic and research environments and as a result are often lacking in industry perspective. An increasingly commercial approach has been taken recently in case of organisations like Council for Scientific and Industrial Research (CSIR).

- **New Product Development**

India has limited capabilities in the area of 'Product design and development'. Most of the clusters and the SME units work on the designs or samples given to them by the domestic or foreign buyers. Broadly, there seems to be little innovation in product design and new product development in the clusters. This has its implications on the profit margins. The institutions promoted by the Government for developing new products/ processes have not been able to fulfill the objectives with which they were started.

- **Pollution Norms**

Natural clusters in India like Food Processing, Leather and Textile Processing started off as a few small sized units using low energy intensive technology. Growth in these clusters took place along with analogous increase in pollution levels. The impending conformity to WTO enforcement of pollution norms are sounding the destiny bell for many of these firms. Government is yet to institutionalize common pollution control facilities. The Leather industry in Madras and Hosiery industry in Tirupur are a few pro-active examples where the industry itself has started these common facilities.

The Bevel Model of Cluster-Competitiveness

The Bevel Model can be used for analyzing the competence of a nation in establishing industry clusters. The model identifies four critical parameters of successful clustering as discussed below.

The overarching parameter in building clusters is policy support, which includes the following parameters. The components are:

Policy Support

- Supportive Legislation
- Potential and Efforts for Formal Organizing
- Building Support Institutions

FIGURE 21 - THE BEVEL MODEL FOR CLUSTER DEVELOPMENT

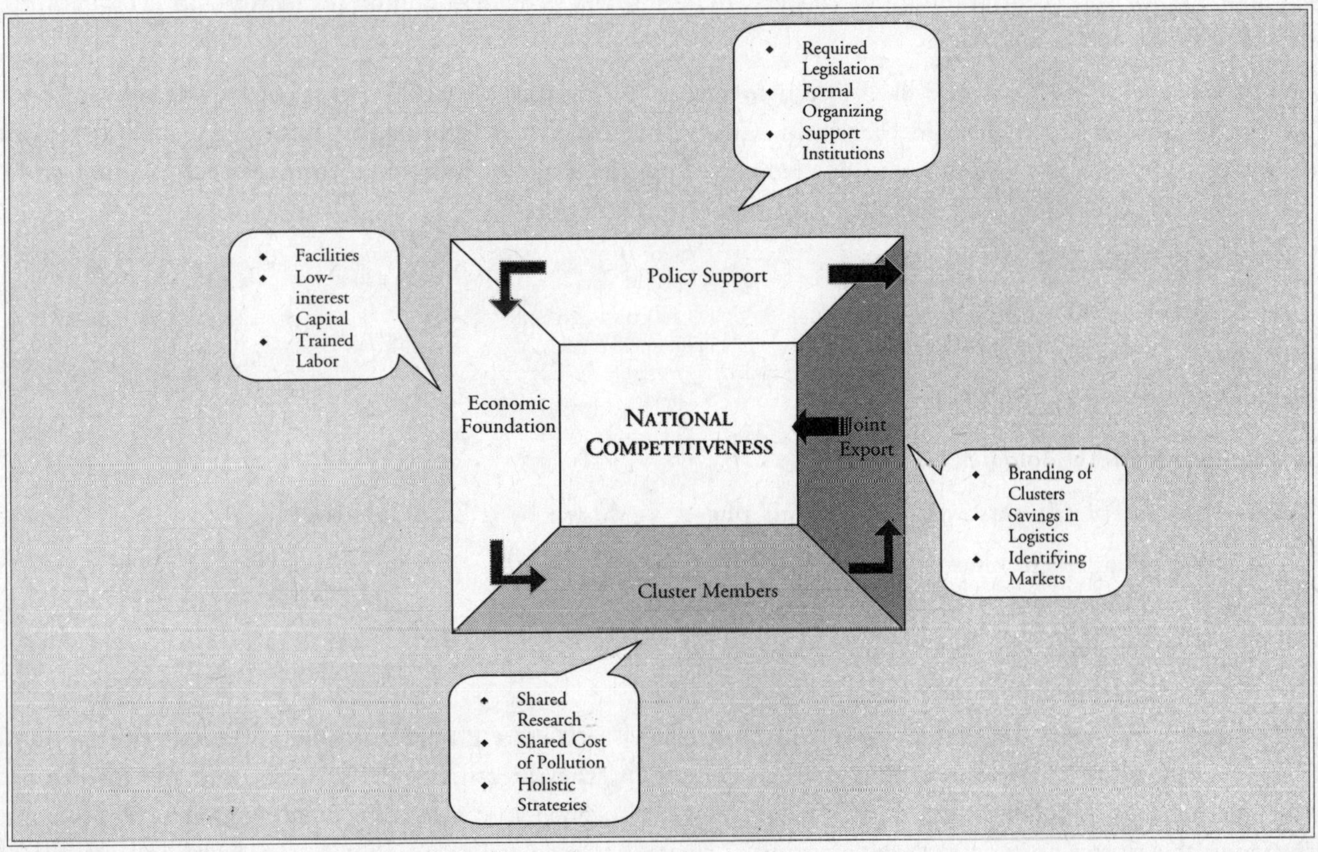

Economic Foundation

Long range planning with a successful vision will lead to economic sustainability of clusters. This kind of economic foundation will include

- Infrastructure Facilities
- Low-interest Capital
- Trained Labour

Cluster Members

Once the formal organizing of cluster building has been launched, the cluster members should commit themselves to the following parameters. Appointing a network broker or a mediating institution will make it easier for individual firms to work out collective strategies and details of cost sharing. The broad factors that come in to play are:

- Shared Research and Development
- Shared Cost of Pollution Norms
- Holistic Strategies for the Industry

Joint Export

Once a formal working system has been put into action, cluster members can look forward to true synergy. This is the step when clustering leads to strategic advantages for the cluster over foreign industries of similar attire. Continued joint efforts become very important in this stage. For instance, a good marketing acumen is very important for successful branding of clusters. Branding will provide a common identity and eventual brand equity for the cluster as a whole.

The concept of joint export will also lead to savings in logistical costs. The costs of logistic movements in India match some of the highest in the world. Costly investments in information technology such as logistics solutions etc. can be shared among cluster members. Logistic needs of individual firms in such a situation can prove to be complementary and generating further cost reductions.

- Branding of Clusters
- Savings in Logistics
- Identifying Markets

Suggested Policy Directions

Implementation Methodologies

Development of Clusters will involve four phases as shown by a UNIDO study:

- Diagnostic phase
- Pilot phase
- Strategic and institutional networking phase
- Self-management phase.

Diagnostic phase - A diagnostic study will be needed to generate crucial information about the economic and social conditions of the cluster analyzed, its potentials, the inter-firm relationships, and the institutional support mechanisms. If conducted in a participatory way, it also provides a means for building awareness, promoting trust among the cluster actors and developing initial cluster action-plans in which the local firms and institutions are closely involved. The diagnostic study helps to strengthen institutional mechanisms for the implementation of these plans, by

- Fostering cooperation among the cluster actors through the very process of assessment–formulation, and

- Identifying leaders and suitable counterparts who are willing and competent to assist ongoing implementation.

Pilot phase - Here groups of firms sharing similar constraints are formed and specific action plans and initiatives are prepared and implemented. During this phase co-operation should start bearing concrete results to the participating enterprises.

Strategic and institutional networking phase – In this phase, long term strategic projects are conceptualized, their feasibility studies are conducted and their implementation begins. These projects generally entail an increase in the degree of specialization by process and/or by product of the firms involved; the restructuring or creation of common service facilities; the launching of new product lines and common brands. During this phase the involvement of both technical and financial institutions becomes essential for the realization of the strategic projects.

Self-management phase - The networks gain greater autonomy and gear up for even keener achievements.

Some Crucial Points

- Since information and distinctive image building is a crucial ingredient in realizing the cluster's vision of being an internationally renowned marketplace, this sort of access and support is crucial.
- The leadership in a cluster project with many partnership members needs to have the members' confidence. It is necessary that there is someone who can act as network broker and represent an organization without taking part of any specific organization.

In instances where the clusters have failed, the chief reasons have been:-

- Lack of knowledge.
- Need of relevant experience.
- Paucity of rigour in the analysis of cluster strengths etc.

In conclusion, policymaking must be based on a vision that draws strengths from the existing clusters while keeping in mind the above factors of failures and continuously checking for directional progress.

6.7 Industrial Pollution Control: Importance and Strategies

Introduction

Sustained progress on pollution control in a developing country like India, depends on clear evidence that its benefits and costs compare favorably with those of other social investments. Regulating all pollutants under all conditions is neither economically nor politically sustainable. Regulators have limited skills and resources and they will rapidly lose political support if the public regards them as unfair or ill informed.

Agencies can avoid serious trouble if :

- They target a small group of serious polluters.
- Limit regulation to a few critical pollutants.
- Effectively measure these pollutants as well as regulatory compliance, and,
- Publicly document their activities.

Instead of relying only on inflexible stick wielding traditional regulation, a new flexible approach requiring interaction among various stakeholders would be more effective and useful. As the environmental agency

FIGURE 22- A MODEL FOR POLLUTION CONTROL

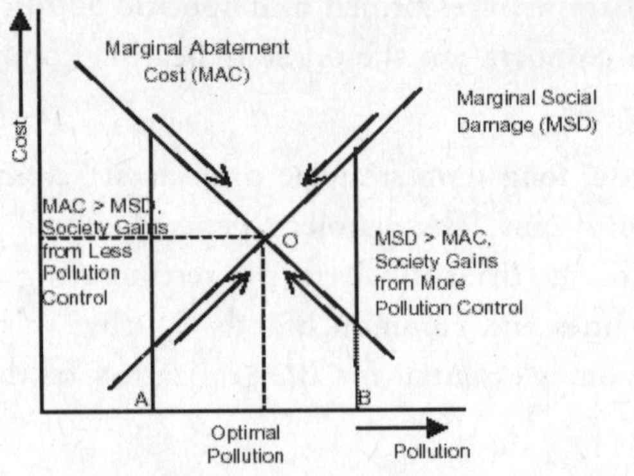

FIGURE 23- DECIDING ON PERMISSIBLE LEVELS OF POLLUTION

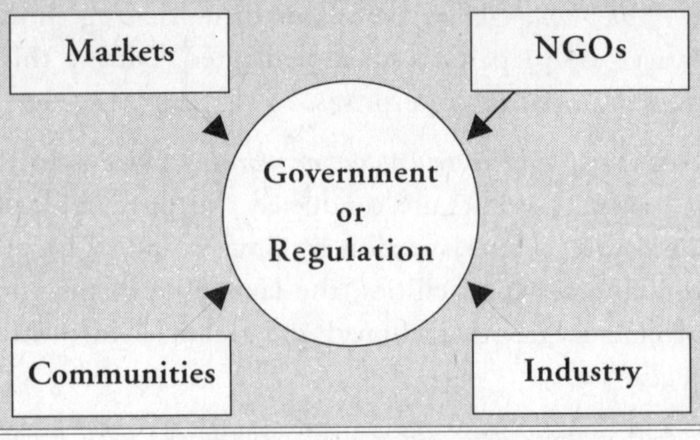

exerts influence through numerous channels, it becomes more a mediator and less a dictator. The structure of pollution control should have four pillars namely markets, NGOs, communities and industry. The regulator / government should be at the center.

The figures give the following insights:-

- Pollution abatement is subject to opposing effects.
- Each additional (or marginal) unit of pollution creates more damage than the previous unit (progressively more respiratory disease from air pollution, fewer fish in contaminated water, etc.). This is called the Marginal Social Damage (MSD).
- Each additional unit of pollution control costs more than the previous unit. This is called the Marginal Abatement Cost (MAC).

The optimal choice for the regulator is the level of pollution where MAC and MSD are equal (MAC = MSD) (point O figure1). On the other hand, at point A of figure 1, where MAC is greater than MSD, the society benefits if pollution control is reduced and vice versa is true for point B.

When managers reduce emissions, it also increases the plant's marginal abatement cost (MAC). Since the managers are usually cost minimisers, their preferred choice is that level of pollution intensity, where MAC and marginal expected price MEP are equal (figure 2). A firm having low MAC tends to be less pollution intensive for the same level of regulation (i.e. point A instead of point B in Figure 2).

The variation in compliance rate across plants can also be explained by the balancing of MAC vs MEP. A case of severe pollution occurs (point D) when a plant with high MAC faces a weak regulatory and information environment in which MEP is low. Such a plant will pollute more. Strengthening regulation will shift the MEP schedule (from MEP1 to MEP2) and motivate the plant manager to lower over all costs by reducing pollution intensity (movement from point D to point C).

Flexible Regulation Approaches

In recent years, a number of countries have shifted from traditional regulation based systems to more flexible systems.

- **Pollution charges:** they are based on "polluter pays" principle and it levels the economic playing field by confronting all managers with the same price for each unit of pollution of pollution generated. Under this

FIGURE 24 - MAC VS. MEP

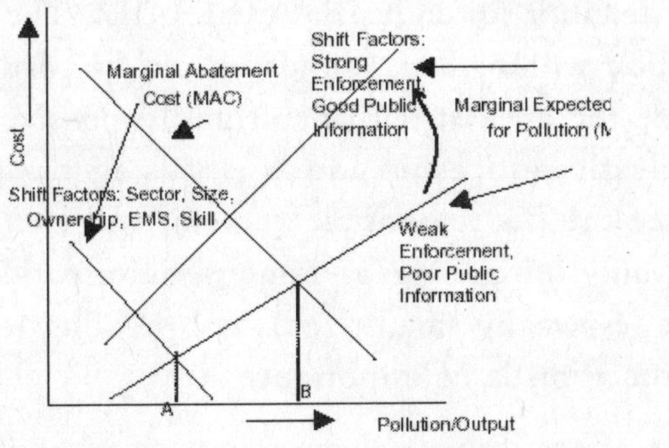

FIGURE 25 - POLLUTION: VARIED COMPLIANCE ACROSS UNITS

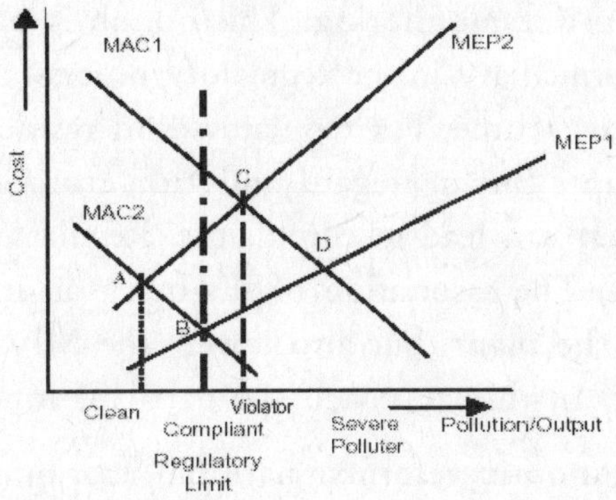

system managers are allowed to adjust their operations until they have minimized pollution-related costs, i.e. charges plus the abatement costs. Pollution charges necessitate knowing the pollution profile of each plant. So presumptive charges based on the engineering estimates of pollution plants of different kinds can be used. This helps the plants to either pay or ask for reduction in charges by proving that their pollution levels are lower than the estimates. The regulators can also outsource the fee collection so that they may be better focused.

- **Public Disclosure:** in this declaration the pollution contribution or environmental performance (i.e. rating) of firm is done. Firms are affected by it and will tend to shift their MAC schedule, i.e. move from point D to B (Figure 3) in anticipation of public disclosure even for the same level of regulation. Ratings or public disclosure has many benefits for different stakeholders. For instance, equipped with performance ratings, citizens are in much stronger position to negotiate pollution control agreements in neighbouring factories. With ratings stock market can more accurately value companies' environmental performance and banks can factor pollution-related liability into their lending decisions. With ratings, clean firms have an incentive to identify themselves, and the regulatory agency can concentrate on serious polluters and expose them in public.

- **Targeting Enforcement:** the regulatory agencies, which find it difficult to go for regulatory reforms can still rely on traditional approaches rather than in Market Based Intermediaries (MBI) and can flexibly target plants for monitoring and enforcement instead of regulating all factories. Empirical research in several countries e.g. Brazil has shown that a few large plants contribute the bulk of the pollution. Targeting large pollution sources with low abatement costs would raise the expected penalties (i.e. shift from MEP1 to MEP2) and this would approximate the results of a charge system. These plants may also react more than small and medium factories because they usually have more skills to draw on and have the ability to spread their administrative costs over many units of activity. In Brazil agencies have allotted A, B and C grade to factories and then largest A factories are targeted.

- **Technical assistance:** Government can also cut pollution by providing technical support for converting to clean production, thereby lowering MAC. This approach is a promising alternative to traditional regulation because it wields a "carrot" rather than a "stick". This arm's length regulatory model works effectively when well-informed managers can easily revamp their technology as incentives shift.

- **Environmental management system (EMS):** assistance from large companies to Small and Medium Enterprises (SMEs) can reduce pollution by lowering MAC rather than raising MEP as in conventional regulation. In Mexico 11 large companies many of them MNCs agreed to provide assistance to 22 small and medium scale suppliers who were interested in improving their environmental performance. At the end of nine months of implementation, 80 per cent of plants reported lower pollution and nearly 50 per cent reported improved compliance and waste handling.

- **Two-tier monitoring:** This option is particularly relevant for industries estates (IE). The IEs may be the useful intermediary in the regulatory process. The pollution within their boundaries can be partially thought of as non-point source. For regulators and residents outside the area, it is difficult and expensive to monitor individual plants but aggregate pollution from an IE is clearly noticeable and it is this aggregate pollution from an IE that can lead to complaints. Regulators can make the IEs responsible and this will form the second tier of control. The association of IEs being an insider not only bridges the asymmetric information by raising the MEP for the plant, but also lowers the MAC of plant (especially small-sized) by providing technical and other support. In this approach the participation of IE association is of importance.

- **Economic reforms:** national economic policies affect industrial emissions so strongly that they constitute "pollution's hidden half" (Wheeler 2000, "Greening Industry"). The cleaner production generally results from economic reforms - reducing barriers to international trade, privatizing state enterprises, developing new stock markets, eliminating subsidies for energy and raw materials and deregulating domestic industries. The efforts of the government to influence MAC or raising MEP can be made at different levels, national, sectoral or plant.

7

CHAPTER SEVEN

Supportive Base Sectors

THE supportive base sectors enabling manufacturing industry development are infrastructure, education and manpower skills. Each one of these is analyzed to have an understanding of the basic issues in these sectors that may foster or hamper the growth of the manufacturing sector.

7.1 LABOUR SKILLS

The labour cost in India is cheap by developed country standards, but the competency levels are far below.

7.1.1 Per Capita Cost of Labour

The cost of labour is one of the most important inputs to the manufacturing process. The following graph captures the cost of labour per employee in India, China, Indonesia and Thailand. It is evident that India does not enjoy any distinct advantage in per capita cost of labour. Thus, its manufacturing sector will have to face pressures from import of goods from neighbouring countries like China, which will be able to produce at a cheaper rate.

FIGURE 26 - PER CAPITA COST OF LABOUR - COMPARISONS

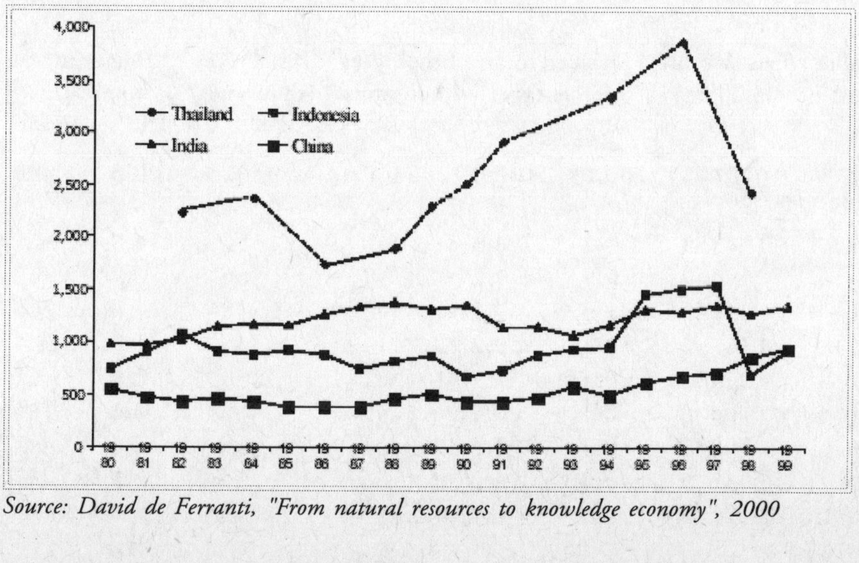

Source: David de Ferranti, "From natural resources to knowledge economy", 2000

7.1.2 Manufacturing Value Added (MVA) Per Capita

The following table shows the Manufacturing Value Added per capita at constant 1990 prices in US $ indicator. India's per capita MVA has always been less than all its competitors. Developing countries lag far behind the developed countries but India is further lower in the ranks among the developing countries.

TABLE 23

International Comparison of Manufacturing Value Added

MVA (At constant 1990 prices in US$)	Year	India	South and East Asia	Developing countries	Developed countries
MVA per capita (US $)	1980	36	84	161	3712
	1990	60	154	203	4430
	1997	81	267	290	4829
	1998	83	267	291	4880
Real average annual growth rates of MVA (percentage)	1970-1980	4.6	9.0	6.8	2.9
	1980-1990	7.4	9.0	5.3	2.8
	1990-1998	7.4	9.5	6.9	2.2
Real average annual growth rates of per-capita MVA (percentage)	1970-1980	2.3	6.7	4.5	2.1
	1980-1990	5.2	6.9	3.1	2.1
	1990-1998	5.5	7.9	5.2	1.5
Share of MVA in GDP (percentage)	1980	14.2	21.1	19.5	22.9
	1990	16.6	24.6	21.2	22.0
	1997	18.2	28.8	24.0	21.6
	1998	17.7	28.9	24.0	21.4

Source: UNIDO

7.1.3 *Employment and Value added: Manufacturing Sector Break-up*

This table shows the break-up of value added in manufacturing sector. The value added is measured in terms of Rupees and is shown for each industry within the manufacturing sector for 1985 and 1999. The value added changes can only be observed if the employee number in each sector is also taken into consideration.

TABLE 24

Employment and Value Added

Value Added (at current prices, in national currency) Employment by Manufacturing Branch (ISIC)	Value Added (currency=Rupee)				Employment			
	Factor value (millions) 1985	**Factor value (millions)** 1999*	**Share in total (percentage)** 1985	**Share in total (percentage)** 1999*	**Employees (thousands)** 1985	**Employees (thousands)** 1999*	**Share in total(%)** 1985	**Share in total(%)** 1999*
TOTAL MANUFACTURING (300)	192044	1670143	100.0	100.0	6469	9108	100.0	100.0
Food products (311)	17764	138955	9.2	8.3	957	1356	14.8	14.9
Beverages (313)	1666	19546	0.9	1.2	47	66	0.7	0.7
Tobacco (314)	2851	24243	1.5	1.5	303	538	4.7	5.9
Textiles (321)	26414	136752	13.8	8.2	1414	1526	21.9	16.7
Wearing apparel, except footwear (322)	1073	25465	0.6	1.5	61	286	0.9	3.1
Leather products (323)	641	6800	0.3	0.4	39	57	0.6	0.6
Footwear, except rubber or plastic (324)	638	5851	0.3	0.4	31	59	0.5	0.7
Wood products, except furniture(331)	899	3399	0.5	0.2	66	66	1.0	0.7

Contd..

Furniture, except metal (332)	81	829	0.0	0.0	8	10	0.1	0.1
Paper and products (341)	2877	24073	1.5	1.4	125	197	1.9	2.2
Printing and publishing (342)	3464	24249	1.8	1.5	152	157	2.3	1.7
Industrial chemicals (351)	14838	167707	7.7	10.0	207	381	3.2	4.2
Other chemicals (352)	14172	164168	7.4	9.8	329	499	5.1	5.5
Petroleum refineries (353)	4259	79200	2.2	4.7	16	37	0.2	0.4
Misc. petroleum and coal products (354)	1885	15064	1.0	0.9	34	42	0.5	0.5
Rubber products (355)	4487	34500	2.3	2.1	83	149	1.3	1.6
Plastic products (356)	2057	26206	1.1	1.6	54	137	0.8	1.5
Pottery, china, earthenware (361)	334	3973	0.2	0.2	30	26	0.5	0.3
Glass and products (362)	1255	9371	0.7	0.6	56	47	0.9	0.5
Other non-metallic mineral prods. (369)	9587	67247	5.0	4.0	340	411	5.3	4.5
Iron and steel (371)	22137	188054	11.5	11.3	541	462	8.4	5.1
Non-ferrous metals (372)	1424	37498	0.7	2.2	55	214	0.9	2.4
Fabricated metal products (381)	5260	39498	2.7	2.4	187	299	2.9	3.3
Machinery, except electrical (382)	18627	116267	9.7	7.0	440	514	6.8	5.6
Machinery electric (383)	14859	118617	7.7	7.1	350	439	5.4	4.8
Transport equipment (384)	15226	159656	7.9	9.6	473	985	7.3	10.8
Professional and scientific equip.(385)	1463	9070	0.8	0.5	41	60	0.6	0.7
Other manufactured products (390)	1806	23885	0.9	1.4	30	90	0.5	1.0

Source: UNIDO

7.1.4 Average Annual Growth of Value Added and Employment

The growth rate is a better indicator of the demographic profile of the manufacturing industry in terms of the Value added and the employee growth rate in each one of the industry.

TABLE 25

Labour Productivity and Wage Rates

(Current prices, in national currency) and selected real growth rates by Manufacturing Branch Branch (ISIC)	Value added per employee at current prices (Currency=Rupee)		Wages and Salaries per employee at current prices		Average annual growth rate (%) of: Value added at 1990 prices	Employment	Value added per employee at 1990 prices
	1985	1999*	1985	1999*	1985 -1999	1985 -1999	1985 -1999
TOTAL MANUFACTURING(300)	29687	183365	14284	57681	...	...	...
Food products(311)	18569	102450	7657	33970	4.99	2.85	2.09
Beverages(313)	35795	295932	12827	51159	4.99	2.76	2.17
Tobacco(314)	9409	45087	5135	15226	1.84	3.55	-1.64
Textiles(321)	18680	89636	12817	43891	6.27	0.84	5.38
Wearing apparel,except footwear(322)	17590	89114	8098	31787	-3.78	13.73	-15.39
Leather products(323)	16443	119491	8902	36438	0.46	2.62	-2.10

Contd...

Footwear,except rubber or plastic(324)	20581	98768	13194	33405	-3.78	5.33	-8.64
Wood products,except furniture(331)	13687	51526	6608	24073	1.37	0.53	0.83
Furniture,except metal(332)	10125	87263	9250	39368	1.37	0.81	0.56
Paper and products(341)	22997	122086	14116	55894	7.33	3.50	3.70
Printing and publishing(342)	22789	154844	15033	85207	7.33	0.70	6.59
Industrial chemicals(351)	71804	440684	25527	104483	7.94	4.85	2.95
Other chemicals(352)	43041	328972	17247	65957	7.94	3.15	4.64
Petroleum refineries(353)	267862	2164#	41950	155716	3.15	6.53	-3.17
Misc. petroleum and coal products(354)	55441	361003	18676	74628	3.59	1.27	2.28
Rubber products(355)	54049	232250	16707	60466	4.47	4.43	0.04
Plastic products(356)	38093	191147	11870	52123	8.32	7.37	0.88
Pottery, china, earthenware(361)	11133	151232	10133	50216	8.96	-1.34	10.45
Glass and products(362)	22411	199751	10679	64567	8.96	-0.78	9.82
Other non-metallic mineral prod.(369)	28157	163521	9721	48171	8.96	1.21	7.66
Iron and steel(371)	40919	406839	18993	89113	7.83	-1.17	9.11
Non-ferrous metals(372)	25891	175171	20582	59727	7.83	11.06	-2.91
Fabricated metal products(381)	28162	132029	14001	57236	2.13	3.86	-1.67
Machinery,except electrical(382)	42335	226277	19119	86816	6.92	1.50	5.35
Machinery, electrical (383)	42455	270285	20646	88569	10.47	1.83	8.49
Transport equipment(384)	32158	162104	20603	76581	7.53	5.78	1.66
Professional and scientific equipm.(385)	35534	150113	16856	74180	7.51	2.85	4.53
Other manufactured products(390)	60724	265389	13080	56100	4.51	9.80	-4.82

Workers in the advanced countries do more marginal value addition than their counterparts in India, China or other developing countries. India has not been able to bring the cost of MVA per capita below that of its neighbouring countries. Moreover, the marginal value added by the workers in India is also lower than other major players in Asia. Thus Indian firms that are into labour intensive work will have to incur higher costs and lower profit margins.

7.2 Education Sector

Education is an essential tool for achieving a sustainable future. Simply providing more education, however, is not the answer for creating a sustainable society. Current global consumption patterns show that the most educated societies leave the deepest ecological footprints. An appropriate basic education, therefore, should be re-oriented to include more knowledge, skills, perspectives, and values related to sustainability than are currently included in schools.

7.2.1 *Experiences from Other Countries*

The United States of America

- Till the beginning of the Second World War, even in the most advanced of them, the US for example, the rate of enrolment at college was not more than 5 to 10 per cent. Joining a university was something of an exception.
- The situation started changing in the US after the passage of the veterans act in 1946. A large number of young people who had joined the armed forces were demobilized when the war ended.
- In order to enable them to pursue their studies that had been interrupted, new enabling

legislation was passed and financial support was extended to these students. This phenomenon soon picked momentum.

- By the 1960s or so, the phenomenon of mass higher education had become noticeable, enough to be identified and written about. For more than a quarter century now, that country has had approximately half the population opting for higher education.

Certain structural changes also accompanied this development in higher education.

- Till the 1940s, higher education was imparted either in the universities or what were called four-year colleges.
- In the changing situation, two crucial changes were made. One was to establish a large number of two-year colleges. They existed either as independent units or were combined with four-year colleges.
- Much more significant, however was the second innovation which was introduced. A substantial proportion of what was done at the two-year college level was vocational in character and focus.

As a result of these changes, two things happened.

- The first was that, despite the growing size of the student body, the character of the universities did not undergo any change for the worse as in other countries.
- Their mandate was higher education and that mandate did not have to be diluted; the new channel of education provided the necessary safety valve, so to speak.

Universities too were of two kinds.

- There were private universities and
- State-run universities.

The two-year colleges, even when they were a part of the four-year college programme, maintained their focus on vocational education. Over the years, their number has grown.

Canada

- The educational system in the country is more or less modeled after the American system.
- The numbers involved were small but the kind of academic stratification, which had taken place in the US, came to be introduced in Canada also.
- Today, the percentage of those who go in for higher education in that country is approximately one-third of the total as against one-half in the US. It is the gross domestic product (GDP) of a country that eventually determines these variations.

Japan

- Japan is the second strongest economy in the world.
- The curve of development in the country followed the American model and even the proportion of young people who go in for higher education is nearer the American than the Canadian figure.

UK and Others

- The proportion of students opting for higher education is below 30 per cent in the UK.
- In Germany and France, it is nearer 25 than 30.
- Italy too is moving in that direction.

- In former USSR developments that were taking place were interrupted a decade ago and a trend cannot be seen in current date.

Conclusions

The experience of these advanced countries makes a few things clear.

- The first one is that secondary education, is so rigorous particularly in the UK and western Europe, that even if students do not go on for higher education, they are capable enough to cope with the demands of the market.
- The second feature is seen best in the case of the UK. The term used for post-secondary education in that country is 'further education'. In essence, further education is vocational in character and is provided to young people after the age of 16. Education at the school level is mandatory till the age of 16. Those who pass out have two options at the stage. One is to join the world of work. Approximately half the young population does that. The other half however takes up further education - whole time or part-time. But what is more significant is that even though the university system has expanded a great deal in the UK during the past few decades, expansion in respect of further education has been higher in terms of scale as well as coverage than at the university level. Even in respect of outlay, more and more funding is being made available for further education.

Further, the emphasis on further education did not mean that the higher reaches of knowledge and advanced research were diluted in any way. If anything, the decades at the end of the war are known for having thrown up and defined a certain number of universities in the US as 'research universities'.

- In India, which boasted of at least a few good universities not so long ago, things have been deteriorating over the years. One important contributory factor has been the thoughtless multiplication of under-graduate colleges. In certain cases, colleges over-shadow, if not determine, how the universities function.
- India did not choose to alter or modify what began to be done in colleges from the mid-19th century onwards. The duration of schooling was certainly extended but the kind of restructuring required at the college level was neglected. Our colleges have been multiplying in numbers in order to take care of the growing student population but their academic thrust has remained more or less unchanged.
- Even though the duration of schooling has been extended by two years, the quality of education at the school level has not improved significantly. Students spend much more time at school than they did before; an additional two years, to be precise. But when they finish, they are only marginally better than their predecessors some decades earlier.
- Secondly, when students come to college, a curious kind of situation prevails. Co-ordination between school education boards and the universities is weak as well as faulty. Of late, a considerable proportion began to opt for commerce instead of the humanities. Nonetheless, it is the liberal arts tradition that has been growing apace.
- This has created a severe imbalance in the training of the right kind of manpower required by the economy.

7.2.2 Recommendations

- The present system of education prepares children only for entry into the modernized sector, which is not growing fast enough with the result that there are problems of congestion and mis-management apart from lack of employment.

FIGURE 27 - NET ENROLMENT RATIO IN EDUCATION, 1997

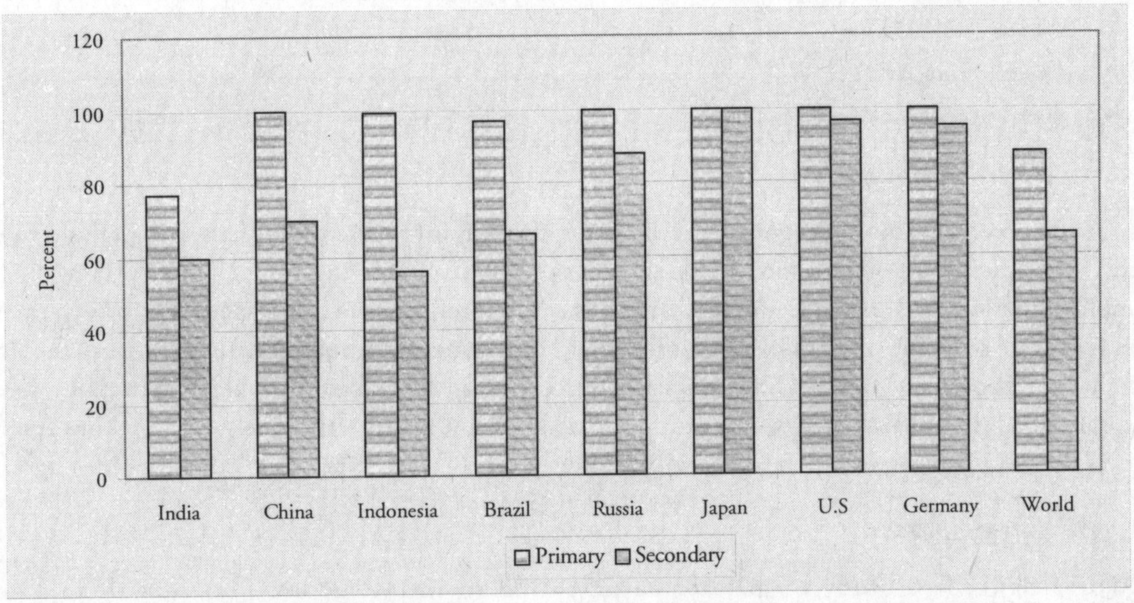

- To the extent that the informal sector of the economy could have grown, we have not utilized our system of education to act as a facilitator and a lever for progress.
- Had we vocationalised our system of education so as to keep in step with what was happening on the economic front, things would have been certainly better.
- Diversification is necessary not only in the urban sector, it is equally necessary in the rural sector. Such a step alone will ensure progress in the country-side.
- Even this step cannot be taken without some local and regional modifications.
- It needs to be recognized that while a certain proportion of students will enter the world of work after school, it also needs to be recognized that they are not really equipped for work either in the rural or in the urban sector.
- Every level of education has to be aligned and attuned to the stage of physical growth and intellectual development at which children happen to find themselves and the needs of the economy. It is only after young people have crossed the age of 20 or so that they are mature enough to do things on their own. Therefore, a system that looks after and trains the 16 to 19-year old youth will have to be devised and maintained.

7.3 INFRASTRUCTURE - INTRODUCTION

One of the primary roles of Government in any nation is creating the right ambience for utilizing its factors of production effectively, so that existing comparative advantages can be translated to national competitiveness. A study into how far has India as a nation been successful in doing this in the past will make for interesting insights. Of the numerous studies, debates and analysis about India's developmental issues that have been undertaken in the past, one common and serious weakness that has been repeatedly pointed out was poor infrastructure. Here in this section we will analyze this common ground of weakness and furnish recommendations for policy formulation and implementation.

For the purpose of drafting strategies for manufacturing competitiveness, the sectors that are short listed and that are analyzed are as follows:

- Power
- Ports
- Roads and Bridges
- Telecommunications

7.3.1 International Comparison

Developed nations have long recognized the need for clearing infrastructure pitfalls as a prelude to economic development, while many developing countries missed the bus by following either faulty policy visions or by incorrect implementation. For example, in case of roads, the 'autobahns' or high-speed motorways in Germany were instrumental in reducing the time frame of logistics movement for manufacturers within the country to less than 24 hours. Providing for the differences in size, USA can be looked at where inter-state highways did the same function of simplifying road transport. India as a nation is still tentatively undertaking huge projects such as the Golden Quadrilateral, a measure which should have neared its completion decades ago.

7.3.2 Private Participation

The World Bank, in a study on private participation in infrastructure, provides three main trends happening in developing countries.

- Across 130 developing countries, private participation has grown from $ 17 billions to $ 128 billions in a span of 7 years (1990 - 1997).
- Private funding is increasingly recognized as the one viable solution to capital scarcity in infrastructure, with almost all of the countries showing some form of participation.
- Latin American countries and East Asia received comparatively larger amount of private sector investment.

In this context it is useful to analyze exactly where India stands as far as attracting private funding to infrastructure is concerned. A country specific report for India from the same agency, World Bank, says that:

> *"The full potential of the private sector to meet India's pressing infrastructure needs is largely untapped. With few exceptions (principally in the power sector in the state of Orissa), there has been little divestiture of existing assets. The private sector has built new infrastructure, such as independent power projects and new port sites, and established new companies that compete with public operators, particularly in the telecommunications sector. However, the productivity and efficiency improvements that private management and ownership could introduce to existing public sector service providers under an appropriate regulatory regime and with competition when possible would help to relieve some of the current infrastructure constraints."*

7.3.3 Indian Scenario - Overview

Reports on India's infrastructure development in the past year conclusively say that the expediency of reforms is slowly getting recognized. Off-shoots of progress in information technology are slowly being co-opted while the fallacies of the past are being corrected. The visible slowdown and gap in infrastructure projects after the Dabhol fiasco is being filled in.

Recently, the telecom sector and road sector have attracted greater policy action and hence witnessed comparatively higher growth rate. Growth has been excellent in the telecom sector, proof to the benefits of rightful policies. The road sector has been another area of considerable attention, with a number of projects happening under the National Highways Development Project (NHDP). The Golden Quadrilateral projects are progressing with contract awards getting under way. Power sector reforms, especially in the area of distribution are going on, although the sector is so mired in troubles that apparent benefits of reforms will take time to trickle down. One example of pro-active policies happening is the recent handing over of Delhi Vidyut Board's (DVB) distribution functions to private companies-BSES and Tata Power. Ports and shipping as well as railways maintain as laggards in the reform program, still vying for the rightfully needed policy attention.

Another aspect of infrastructure projects is the large scale financial support, the need of which has begun to be recognized in the policy circles. Three large funds has been created as follows

- Infrastructure Equity Fund.
- India Power Fund and,
- City Challenge Fund.

Internationally, India figures in the list of top 10 developing countries by money value of infrastructure investment. (Table)

TABLE 26

Infrastructure Projects 1990-2000 in Developing Countries

Top 10 Countries	Total Investment (in billion $)	Number of projects
Brazil	123	195
Argentina	78	163
Mexico	55	128
China	51	253
Malaysia	34	59
Philippines	32	65
South Korea	30	24
Indonesia	27	60
India	25	110
Thailand	21	68

Source: World Bank, India Infrastructure

However, a closer analysis of the above information reveals the following details

- Latin American Countries (LAC) are way ahead of Asian countries in infrastructure investment
- China and India has a relatively larger number of projects with regard to the total investment. In particular, for India, the 25 billion dollars of investment is spread across 110 projects.
- When one considers the break up of private investment in the above table none of the Asian countries rank in the first 10.

This categorically asserts the importance of private sector participation in infrastructure, since that is the only factor that enables LAC to move so much ahead in investment rankings.

Secondly, the large number of projects that India has managed to undertake might be perceived to be the distribution efficiency of development. But in the real sense, it also implies the failure to undertake large scale projects successfully.

7.4 Power Sector

Power is the primary sector in infrastructure, being absolutely essential to the entire manufacturing sector and also to emerging industries like software. Given its importance, power generation and distribution in India is woefully below required standards.

7.4.1 Main Issues

Cost effect of power is very pronounced in the manufacturing sector. To illustrate, 40% of the production cost of textiles are power tariff payouts. Other energy intensive sectors in manufacturing such as paper and paper products, cement, fertilizer, glass, iron and steel etc also heavily depend on the dynamics of the power sector. As such, power sector reforms form the basis for trans-national competitiveness of firms in these sectors.

Across the world, success in manufacturing sector has been achieved with a congruent development in infrastructural facilities, most importantly power. There is a need for a revamp in power sector as far as our nation is concerned. The analysis of India's position in power sector with reference to the quality of power supply reveals the following anomalies:

Reliability

The unreliability of electricity supplies in India is severe enough to constitute a constraint on the country's overall economic development. Aluminium plants, for instance, are so sensitive to power fluctuations that a short fluctuation in power will cause extensive plant damage. The risk factor in our nation with regard to power is so high that 60% of Indian firms find it better to install their own power plant, compared to 30% in China (IMF- CII, 2002).

Cost Efficiency

Cheap availability of power will make manufacturing cost competitive. Per unit cost of power from the national pool sometimes run as high as 450 paise for high-tension users. Understandably, companies venture for captive power generation. But again, this calls for additional investments, adding to the cost curve. Cost of power in China is almost 1/3rd of Indian levels (IMF- CII, 2002).

Ease of Connection

For a new business, especially in the SME sector, getting a power connection often calls for 'baksheesh' expenses. For SMEs in particular, cost of power is a direct deterrent. Being small, investments in the form of power plants are totally ruled out.

7.4.2 Cross-Country Comparison

The break-up of power generation in India and the US at present is as follows:

One of the strengths of our nation in power sector is the presence of natural resources. Capabilities do exist in hydel power, nuclear power and non-conventional energy. The potential of setting up of smaller sized hydroelectric plants should be fully explored. China followed a strategy of small sized hydel power plants very successfully. Smaller power plants contribute to smaller towns and rural areas. This eventually contributed to a large extent to the development of TVEs, the Town and Village Enterprises which are the Chinese model of SMEs. The TVEs account for nearly 40% of total.

As a second step, nuclear power can be explored. The uneventful functioning of nuclear power plants in

FIGURE 28 - STRUCTURE OF POWER GENERATION INDUSTRY - INDIA VS. US

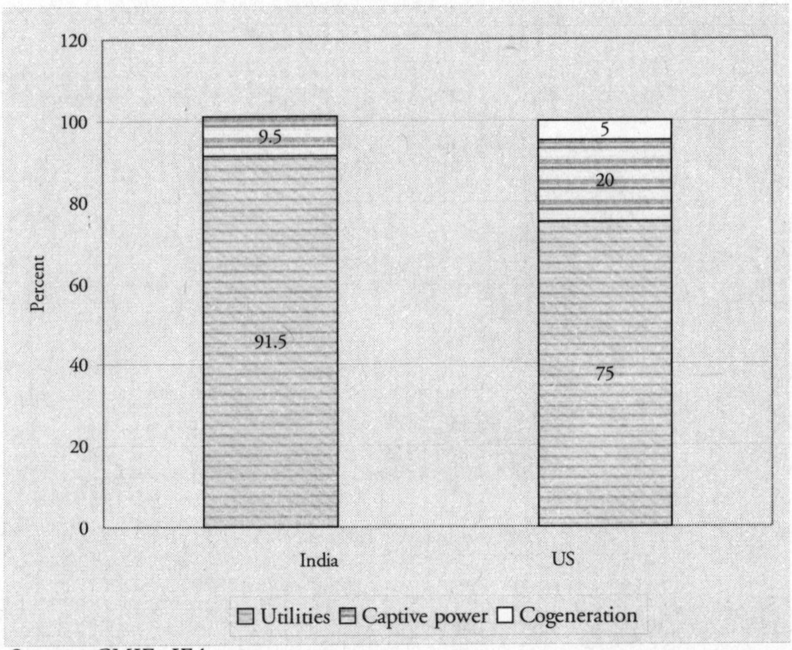

Source : CMIE, IEA

several parts of the world shows that the safety of nuclear power plants is not all that suspect. For example, 15% of total power generation capacity in US is nuclear-centric. Capability of successfully handling nuclear arms can easily be extended to nuclear power as well. As an externality, the technology benefits and know-how from nuclear power generation might be used in other areas. Besides, as the rest of the industry is privatized, nuclear power generation can be one area in which the government can still retain strategic control.

7.4.3 Demand - Supply Scenario

According to the Ministry of Power, the elasticity ratio of growth rate of power requirement vs. GDP growth rate is currently projected to be around 1.5%. Thus, given a GDP growth rate of 7%, the growth rate of power supply has to be around 10% in the years to come. The current power generation is around 100,000MW. Thus an additional 10,000 MW has to be added every year for at least 10 years.

What is missing in all these projections is a holistic, long-term view. Pegging the power needs to a downsized vision of GDP growth is faulty. If we plan to achieve a GDP growth rate of 10 percent, obviously the power needs will be much greater, and the lack of efficient supply will become much more pronounced.

7.4.4 SWOT Analysis

Strengths

- Well-established and vast Transmission and Distribution network.
- Regulatory mechanism for tariff setting is getting established.
- Independent judiciary with well established legal system for dispute settlement.

Weaknesses

- Distribution of power follows rotten traditions.
- Systemic problems in SEBs, mainly excess manpower.
- Shortage of capital and consequent problems in achieving plan targets.

Opportunities

- Potential for private investment in future.
- Highly qualified Engineering and Technical personnel in the country.
- Availability of highly trained I.T. personnel and expertise.
- Rising awareness of power as the prime mover of economy.

Threats

- Political resistance against tariff revisions.
- Uncertainties of privatization in particular and reform programs in general.

7.4.5 Cost Dynamics

Changing the existing industry orientation from predominantly coal based to a balanced break up (Chart) makes tremendous strategic sense.Economically, relying on costly fossil fuels will turn out to be a negative sum game in the long term. As shown in the table, hydel power is the cheapest to operate, with per unit costs in the range of 30 paise. Similarly, nuclear power also ranks quite well when compared to coal and oil based fuel costs. Hence there ought to be a strategic shift towards smaller hydel plants and more nuclear power generation.

FIGURE 29 - PLANT TYPES AND BREAK UP OF POWER GENERATION IN INDIA

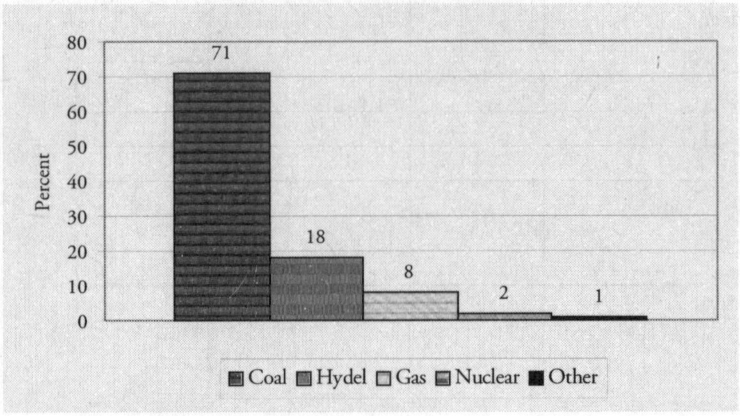

Some among the manifold advantages of such a strategy are listed below:

- Lesser costs when compared to fossil fuel based plants.
- Better distribution of power across the geography.
- Reduced T&D losses in cases where local distribution of power from small power plants is undertaken.
- Conservation of India's coal reserves which will otherwise be used up in power generation.
- Moving away from high ash content coal will reduce pollution.

FIGURE 30 - DESIRED PLANT MIX IN FUTURE

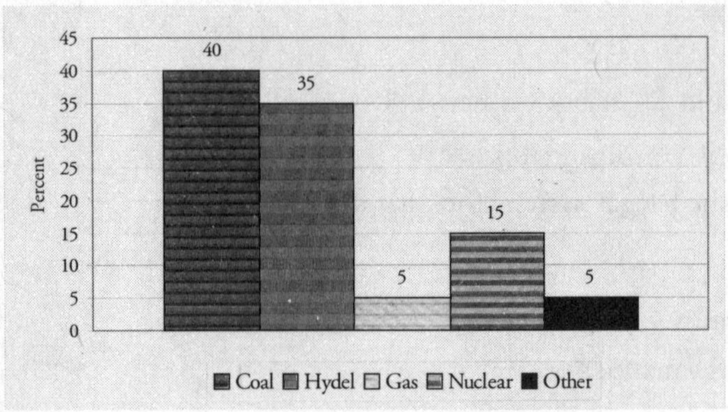

TABLE 27

Cost Dynamics of Various Types of Power Plants

Type	Capital Cost (Approx.)	Operating Cost (Paise/Unit)
Thermal - Coal Based	5 Crore	80
Thermal - Oil Based	5 Crore	200
Hydel	7 Crore	30
Nuclear	7 Crore	50

Source: Estimates by Capitaline

7.4.6 Analysis of Investments Required

The Udesh Kohli Committee which was set up by the government to analyze the fund requirement in the power sector has reported that the country requires a staggering Rs 8,00,000 crore in the next ten years to increase the power capacity by 1,00,000 to 2,06,266 MW by March 2012.

The planning commission report says that India need to spend Rs 5,66,000 crore, mostly on cement, steel and aluminium, in the next five years to 2007, to add 47,939 MW of power and to improve transmission and distribution. Of the capacity additions planned, 17,311 MW is expected from hydro, 28,323 MW from thermal and 1,300 MW from nuclear plants.

One glance at the balance sheets of SEBs is enough to understand that such huge investments are beyond any of them. Infusion of private sector thus becomes a dire necessity, rather than a pro-active reform measure.

7.4.7 Policy Imperatives

The externalities of power sector reforms are perhaps the most significant among the potential benefits of various policy modifications trickling down to the manufacturing sector.

Unbundling

Indian power sector is a good illustration to the inefficiency and potential market distortions of an integrated model of power supply. Having the various components of power supply viz. generation, transmission, supply and distribution under the control of single entities created inefficient monopolies. To move towards a more efficient model, the first step to be taken is to unbundle the various functions and systematize them under stand-alone entities. This will pave the way for more competition and lower the costs of power supply.

Privatization

In the case of generation, the all India average of total factor productivity (TFP) is about 34% when indexed to US standards (McKinsey, 2001). This figure dramatically improves to 80%, while benchmarked against Indian best practices. As much of the productivity gap is attributed to excess manpower and its inefficient deployment, private sector entry into generation can bring about dramatic improvements. Hence it is strongly recommended that policy amendments be made to facilitate private sector participation in generation.

Sorting Out Distribution Problems

Power distribution scenario in India is pathetic at best, with the T&D productivity of SEBs as low as 4% when compared to US. It should be kept in mind that SEBs accounts for nearly 52% of the total ownership in our nation. The measures to be taken on an immediate basis are as follows:

- Curtail T&D losses to the international benchmark of 15% in five years.
- 100% metering of all connections

Exorbitant losses in State Electricity Boards

Over the years, SEBs have demonstrated several ways in which business should never be done. The cumulative losses of SEBs in 1998 alone was over 22,650 crores.

Low Labour Productivity

The single most important factor causing low productivity is the excess manpower in SEBs. As shown in the Table, the all India average for employees per million units of electricity is 3.6. Much of this manpower is concentrated in non-core activities like administration, accounts and so on. In addition, because there is such a huge workforce, many SEBs have built on ample manpower in their HR departments as well. Lay offs and merit based promotion being out of question, the little revenue streams of the SEBs is squandered in salaries and benefits.

Lack of Rational Tariff Structure

The tariff structure followed in our nation is not conforming to economic sensibility. The method of implementation forces the manufacturing sector to unfairly bear them for the sake of household and agricultural consumption. Ours is perhaps the only country in which industries bear the costs of irrigation and irregular monsoons. The need for rational and uniform tariff cannot be understated in this state of affairs.

TABLE 28

SEB Manpower Statictics

State	No of Consumers	Employees per 000 Consumers	Total Sales (Mn Units)	Employees per Mn Units Sold
Andhra Pradesh	10,070,992	7.4	23,766	3.1
Assam	762,533	26.3	2,023	9.9
Bihar	2,150,000	17.1	7,512	4.9
Delhi	2,282,547	11.4	7,636	3.4
Gujarat	6,379,563	7.8	26,780	1.8
Haryana	3,390,245	16	8,947	6
Himachal Pradesh	1,286,812	22.3	2,262	10.8
Jammu and Kashmir*	690,988	26.8	1,789	10.3
Karnataka*	8,110,033	5.7	15,194	3
Kerala*	4,913,041	5	7,021	3.5
Madhya Pradesh	7,626,632	12.8	24,428	4
Maharashtra	11,959,395	9.3	43,984	2.5
Meghalaya	114,744	38	541	8.1
Orissa*	1,218,678	25.3	4,876	5.9
Punjab	4,778,805	18	18,808	4.6
Rajasthan*	4,612,902	12.2	14,172	4
Tamil Nadu	11,906,988	8.1	26,943	3.6
Uttar Pradesh	6,708,798	13.8	25,224	3.7
West Bengal	2,762,593	13.6	8,751	4.3
All SEBs	91,726,593	10.8	270,657	3.6

Source: Planning Commission 1998

Transmission and Distribution Losses

Theft accounts for nearly 22% of productivity loss, along with poor organisation of functions and projects. In a drive to give free power for agricultural activities, the need to monitor the outflow of such subsidized power was ignored. As a result, power connections to all the corners of India, envisioned vociferously in political statements, did not include metering of power in all connections. Only recently the CERC has recognized the need for 100% metering, an operational objective that should have been dictated by sheer economic sense quite earlier.

State Policies - Direction and Uniformity of Progress

The India Infrastructure Report - 2002 has compared the performance of individual states regarding power reform and the report shows the urgency of need in SEB restructuring is yet to be grasped by many states. Seven states alone have been keen on SEB restructuring. This is not enough to make inroads in power sector reform, since state barriers do not apply to power grids. Moreover, uniformity of power sector reforms will ensure that distribution of industries will also be uniform, thus avoiding possibilities of regional disparities in the future.This clearly shows that the Central Government should take charge in enforcement of reforms. This can be done through the CERC, which will impart more power and autonomy to the regulatory body as well.

While taking such measures, it is important that a time constraint is set up in reform process. This will ensure meeting of targets and goals and the accompanying spillover of benefits to other sectors.

Determination of Tariff Rates

Ahluwalia (2000) investigated the various models of cost transfer to consumers followed in power industry globally. Thus, the tariff structure of a power company can be arrived at by:

- Cost of Service method.
- Performance Based Ratemaking, and,
- Revenue or Price Cap method.

In the cost of service method, a utility has to periodically disclose the expected costs, its sales projections and its investment plans to the regulator. The regulator checks these data and arrives at a prudent rate of tariff. In Performance Based Ratemaking (PBR) the regulator prescribes a set of cost and investment norms that are also linked to performance standards. Hence the recovery of costs through tariffs is linked to the achievement of explicit, pre-determined targets and performance benchmarks. In the Revenue Cap method, the total available revenues for the electric company are fixed. The tariff per unit is variable, depending on the sales realization.

Of the above three methods, the first two are more intrusive in nature, with the regulator enjoying tremendous flexibility to fix the tariff norm. Overall, Ahluwalia concludes both the PBR and the RC/PC approach is superior to the COS approach as they assure tariff stability and reasonable tariff levels for consumers on two counts. Firstly, it provides incentives for efficiency enhancement and hence ensures results. Secondly, it forces government to commit to subsidy outlays in advance and to budget for such outlays over a medium term perspective.

What to Privatise?

While there is strong reason to invite private sector power generation, what lies in doubt is the attractiveness of such open offers to private companies. Again McKinsey argues that it will not be possible to attract investment in the supply side unless and until the demand side has been purged of inefficiencies. Practically, this means that market distortions in distribution have to be sorted out as a pre-requisite to the privatization of power generation as a segment.

Upadhyay (2000) argues that a disintegrated structure with predominantly private ownership will be ideal for our nation. But there are issues involved while looking at T&D segment. The kind of investment required and the extent of returns available in such investment are simply not attractive for a private player. To illustrate, no private company can singly own the national grid and gain a profitable toll out of it. The grid, however, is necessary for compensating fluctuations and differentiations in availability of power. Issues such as this call for monopoly or consortium ownership as opposed to competitive bidding of T&D. Taking such factors into consideration, a viable model of unbundling power sector might be as shown in the figure below.

Summary of Policy Recommendations

A clear vision statement for power will give direction to policies regarding power sector reform. Towards this end the government should come up with a Comprehensive Reform Programme that encompasses all areas of power supply. It should have time bound objectives in line with the policy imperatives. Three year planning and review periods will be viable. To sum up, power sector reforms are preconditions for success in manufacturing. Quality power is the first step to quality manufacturing.

FIGURE 31 - MODEL FOR UNBUNDLING POWER SECTOR

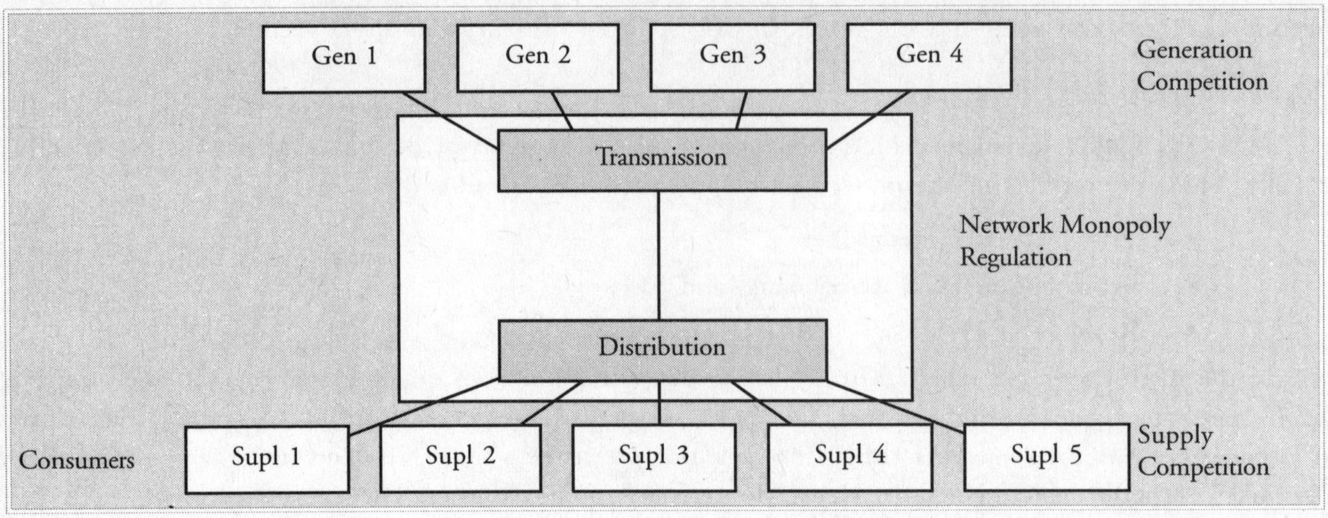

Source: EPW, 2000

7.5 TELECOMMUNICATION SECTOR

Introduction

The average tele-density in India is 2.94 basic telephones and 0.37 cellular telephones per 100 people. In advanced countries the tele-density is averaging in 50-60 range. The production of telecom equipments in India increased from USD 1.3 billion in 1993-94 to USD 2.48 billion in 1997-98, and is expected to reach USD 5 billion in 2002. The requirement of telecom equipment by various users during the five-year period from 1997-2002 is estimated at USD 22.3 billion, of which equipments worth USD 18.5 billion will be produced in India.

The target for the year 2000-01 was 5.58 million lines but only 5 million connections were provided. Waiting list in India was 1.9 million in March 1999, which has only grown further to more than 3.0 million now, which still validate the demand supply mis-match.

The changing lifestyle of human beings enhanced by Internet, facilitated by mobile communications and enriched by e-commerce would give a real boost to this industry. As trade and industry grows, telecom services also have to expand commensurately because it is the life-blood for the modern trade and commerce. In developed countries telecom industry is viewed as the prime mover of GDP growth and India is now increasingly getting ready to follow the same trend.

Telecom equipment manufacturers are the other category of players in the industry, which was exclusively reserved for government enterprises until 1984. Till that time it was dominated by Indian Telephone Industries (switching, transmission and terminal equipments), Hindustan Cables (cable products) and Hindustan Teleprinters (telex machines/ modems). Thereafter, private entry was allowed in the manufacture of telephone instruments, cables, transmission equipment, small switching exchanges developed by C-DoT and manufacture of large exchanges.

Strengths

- India's telephone network with 27.4 million direct exchange lines (DEL), as of July 2000, is one of the largest in the world and the third largest among emerging economies (after China and Republic of Korea).

- It is expected to bring in an investment of $ 37 billion (Rs1,72,400 crore) by 2005 and $ 69 billion (Rs 3,21,500 crore) by 2010.
- The country's telecom network consists of 25,000 exchanges with 26.6mn working connections as on April 2000.
- There is significant advancement on the disinvestment front.

Weaknesses

- Although the sector was opened to private competition almost ten years back, the same is not reflected in either telephone penetration or investment in this sector.
- The most important factor is the abysmally poor telephone penetration of 2.94 lines per 100 population as against the world average of close to 15 and a very high 50-60 in some of the advanced countries.
- Monopolized by the government, the sector was a big victim of huge operational inefficiency and customer apathy till recently.
- At present, private sector is allowed to manufacture the entire range of telecom equipments. Now several new private players like Himachal Futuristic, Global Telecommunication, Bharati Telecom, Tata Telecom, Shyam Telecom, etc. are serious players in the field.

Opportunities

- The industry grew at 20% per annum over the last four years.
- Change-over from fixed license fees to revenue-sharing-cum-entry fee regime has taken away one of the most important obstacles to telecom growth in India as the high license fees quoted by the players under the earlier regime had made projects of most of them totally unviable.
- India has set a tele-density target of 7 by 2005 and 15 by 2010. This translates to an addition of approximately 75 million lines by 2005 and 175 million lines by 2010. The rural tele-density has been targeted at 4 by 2010 with reliable telecom facilities in all villages by 2002.
- Government's massive investment plans amounting to Rs 832.50 billion for infrastructure is another factor. As the government opens up long distance telephony, a dozen corporations have plans for Rs 36,000 crore investment in the sector.

Threats

- The value-added services of recent origin include cellular, radio paging, public mobile radio paging, trunking, global mobile positioning communication services, VSAT services, electronic mail, voice mail, internet services etc. So the development has still to take place. Technology penetration will be difficult given the basic infrastructure of India.
- Private players are now flocking to this sector with unexpected enthusiasm and the scenario is set to witness fierce competition, both in basic telephony, cellular and other value-added services.

Strategic Action Points

- The sector being capital intensive, government was the only player so far. Though private players are now permitted, only large players with deep pockets can enter and operate successfully in this field. Smaller players are likely to sell out or merge with larger players.

- Again, the industry is very technology-intensive and hence the chances of obsolescence are immense for different modes of communication. In advanced countries this has already changed the dynamics and the structure of the industry. The abolition of 10% surcharge on the import duty is a positive step for the telecom industry.
- The buzz word is broadband and integrated services encompassing basic, ISP, STD/ISD, mobile and other value-added services under one roof or through consortium and alliances. Concessional rate of 5% customs duty admissible to specified equipments and parts thereof for basic telephone service, cellular service, pager services, V-SAT, PMRTS and Internet services extended up to 31 March 2002.
- A strategic research by Frost and Sullivan estimates that the Indian basic telecommunications equipment market will reach a level of $ 8 billion by 2003, which was just $ 0.9 billion during 1996-97.
- The transmission equipment industry and, more so, the WLL, though currently it is not being manufactured in India, is expected to generate revenues of $1.423 billion during 2002-2003. The convergence of entertainment, media and telecom would again only increase the need for more telecommunication facilities.
- The government proposes to introduce a Convergence Bill to cover telecommunications, information technology, and information and broadcasting sectors in an integrated manner.

7.6 Ports

One of the chief imperatives for globalization is easy movement of raw materials and finished products across national borders. This directly implies that any country that wishes to be in the global map of trade has to figure out ways to reach foreign markets in a competitive manner.

Ports are among the most important components that create the critical ambience for foreign trade. Policy-makers need to realize that the benefits of information technology as heralded vociferously cannot be materialized without strategic investments in ports.

7.6.1 Cross-Country Comparison

To understand how vital ports are to the manufacturing sector we need to look at some of the successful East Asian countries.

CHINA

According to an UK Government report, the barriers caused by differing business cultures no longer present the same difficulties to foreign investors in China. In 1999 Chinese ports handled 1.3bn Tonnes of goods and this figure will continue to rise as China's economy grows. Quote unquote, 'This is particularly true of 'major' ports since these facilities tend to possess foreign affairs divisions whose members have been trained specifically in western business practices.'

In the past the Chinese have proved to be extremely capable of initiating port development without external investment. However they recognize that the next stage of infrastructure expansion will require foreign input and have identified container terminals as being the area most in need of such attention.

INDONESIA

In Indonesia, as early as the 1980s, maritime transportation became the focus of a major investment program. A series of regulatory reforms followed. In the beginning, like many areas of government policy, shipping and port policies had become increasingly restrictive and bureaucratic during the 1970s and early 1980s. For instance, foreign vessels were by then permitted to dock at about 100 of Indonesia's approximately 300 registered ports only if they had a local Indonesian agent. Credit goes to the policy makers to say that such policies were dramatically liberalized during the mid and late 1980s.

In 1985 a major deregulation was conceived. It included

- Suspension of local customs operations.
- Permission of market-determined freight charges.
- Abrogation of the special license for foreign vessels.

In 1988 the five licensing categories were simplified into two--oceangoing and regional shipping and inter-island shipping. New entrants, including foreign joint ventures, were permitted with no restrictions on size of fleets, and some categories of commercial businesses were permitted to operate their own fleets with no additional license. The investment program to expand the domestic fleet, which since 1984 had mandated the elimination of vessels older than thirty years, was suspended indefinitely, and the gateway hierarchy was effectively undermined by more liberal route permits, although investment in port infrastructure still centered on the four gateway ports.

7.6.2 Existing Scenario

The 6000-Km long coastline of India is studded with 11 major and 139 state Ports. The 11 major ports handling over 90% of the country's seaborne cargo, at 227 million tonnes during 1996-97, are under regulatory control of the Central Government, while the minor ports are administered by the State Governments concerned.

FIGURE 32 - CAPACITY AND TRAFFIC AT MAJOR PORTS (MILLION TONNES)

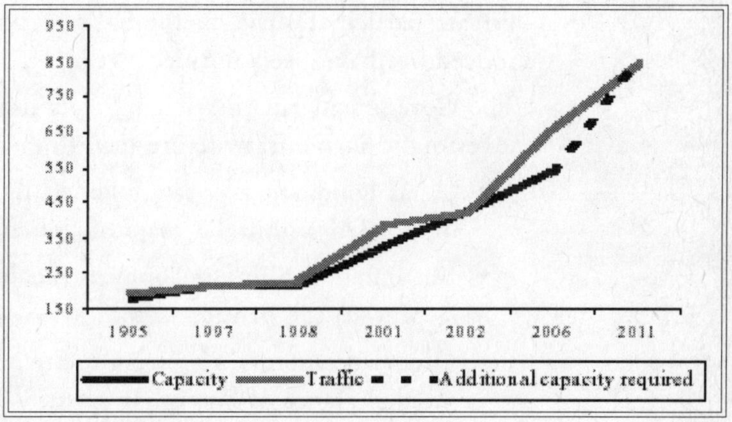

Source: EPW

On completion of 8th plan spill over schemes, now under execution, the cargo handling capacity will go up to 252 million tonnes, requiring additional capacity of 173 million tones to be created by 2002 AD. The financial requirement is estimated at Rs. 160 billion during 1996-2001. Internal resources of the ports, external assistance and Government Budgetary support will generate about 60% of the funds required.

7.6.3 Investment Gap

Several initiatives have been taken by the Central and State Governments to attract private sector investment. For the major ports, the areas identified include:

- Leasing out existing assets of the ports.
- Construction/creation of additional assets such as container terminals.
- Specialized cargo berths.
- Warehousing and storage facilities.
- Cranage/handling equipment.

- Setting up of captive power plants and dry docking/ship repair facilities, and
- Leasing of equipment for port handling.

FIGURE 33 - PLAN OUTLAY AND EXPENDITURE IN PORT SECTOR, INDIA

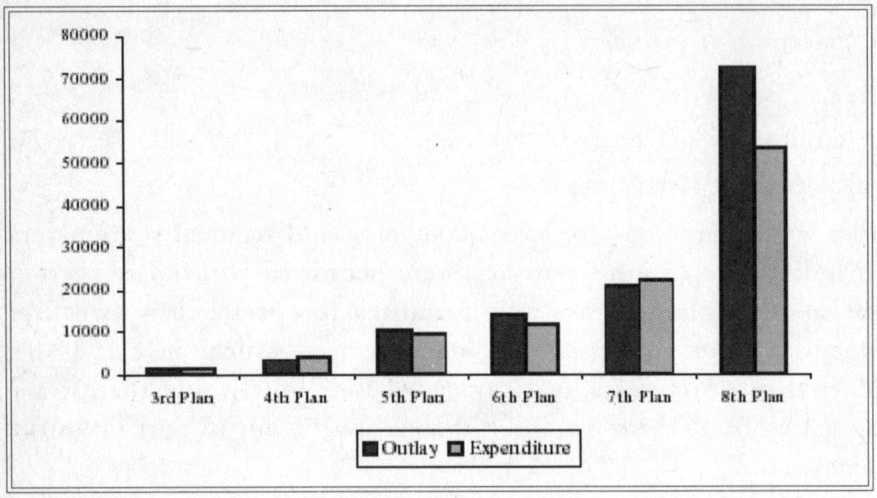

To realize these objectives, the government needs to analyze the sources and viability of investment on a long-term basis.

As there has been increased recognition of the need for better investments in the port sector, a greater plan outlay has been provided in recent five-year plans (Graph). According to a McKinsey study, the current level of investment in Indian ports is around 0.1 per cent of GDP.

7.6.4 Past Policy Measures

- State Governments have also undertaken development and regulation of state ports. Most of them have thrown open the development of such ports to the private sector.
- Private participation is on the basis of open competitive bidding. The tender conditions and procedures have been notified.
- The Government has since changed its fiscal policies to offer following incentives to the private investor in port infrastructure development.
  - A tax holiday for 5 years followed by 30% rebate on the earnings in the next five years to be availed within 12 years of the commissioning of the project.
  - An Infrastructure Development Finance Company has been floated to provide concessional credits to the private sector.
  - Automatic approval to foreign equity investment up to 74%. Proposals for 100% foreign equity need clearance of Foreign Investment Promotional Board (FIPB).
  - Tax concession of 40% to Financial Institutions on income from financing infrastructure.
  - Infrastructure funds have been exempted from Income Tax under certain conditions.
    - Subscription to debentures and equity of the companies engaged in infrastructure will qualify for Income Tax relief for the investor.

7.6.5 Policy Imperatives

Like many other sectors, India still does not have a National Policy on Ports. Such a policy has to be formulated, incorporating the following dimensions:-

- Throughput targets for ports need to be set according to world benchmarks.
- Export infrastructure at ports need to be improved.
- Indian Ports Act should be made applicable to major as well as minor ports. Out-dated clauses in the Indian Ports Act and Major Ports Act need to be amended.

- One of the chief reasons for increase in the gestation periods of port infrastructure projects has been the random checking and verifications imposed on companies. To put an end to such practices, clearances should be obtained from the relevant authorities before a project is handed over for development. The concession period should start after obtaining statutory clearances.
- A strong and independent regulatory authority is needed. At present the regulatory structure is independent port trusts. This creates vagueness regarding detailed aspects of berthing of cargo, as far as a multi-national shipping company is concerned. The regulatory functions of independent ports needed to be overseen by a 'National Port Trust Authority'
- Government expenditure, at the present levels of 0.1 percent of GDP is insufficient and needs to go up.
- The option of Public-Private Partnerships, such as those recommended in the power sector should be explored for closing the investment gap.
- Market dynamics should determine tariff, rather than government perspectives.
- A vision statement is long required. In tandem with the institution of the National Port Trust Authority, an integrated and long-term strategic master plan for port development in the country should be authored.
- External commercial borrowings from agencies like World Bank, ADB, OECF, etc can be sought after.
- Private sector participation is expected to reduce the gestation period for setting up new facilities and help bring in the latest technology and improved management techniques.
- According to McKinsey, investment in ports must be better targeted. Less focus on building new berths and terminals and more attention to removing bottlenecks in existing capacity is vouched for.
- Reducing red tape and bureaucracy in customs will make possible better utilization of existing capacity. This will contribute to faster ship turnaround, which is dismally below the world standards. Since the turnaround time is low, many ships prefer the Colombo port for unloading cargo for the region. Compared with China, where ship turnaround time is nearer to 48 hours, turnaround periods runs into weeks in India.

7.7 Roads

Developed countries have allowed development of network of arterial roads. They have allotted funds generously towards it. The U.S has its mega-network of Inter State Highways, Parkways and Beltways that blend well with the environment and enable the trade and commerce to flourish. It is the same with the Motorways of the U.K., Autobahns of Germany and Autoroutes of France. In Malaysia, North-South Highway is the shining example of how a well-designed national road network can trigger development phenomenally. These are worthy emulations for India.

7.7.1 Indian Scenario

The importance of roads and bridges to manufacturing cannot be undermined. At the time of independence, more than 70 per cent of freight traffic was through railways and roads shared the rest of the pie. Within a span of four decades itself, the situation has reversed. With 70 per cent of cargo traffic through roads alone, this mode of transport has evolved as a vital lifeline for manufacturing.

TABLE 29

Road Sector in India

Type	Name	Length
Primary	National Highways	58112
Secondary	State Highways	137119
	Major District Roads	470000
Tertiary	Rural Roads	2650000

The above data shows that the length of road network is quite big, especially when taking into account rural roads. But in comparative situations in other nations, developing 'bigger roads better' got more policy attention. The thrust in our case should be to improve the quality of existing roads. Merely to 'promote' as many roads as possible into highways without achieving the true quality that is expected from such a road is wastage in the long run. Planning of roads also leaves a lot to be desired, with a huge number of roads being rendered redundant to compensate loss of land to interested individuals. In fact, one of the off-shoots of the Urban Land Ceiling Act that kept real estate prices irrationally high was to impede road development. Development authorities have to face litigations in order to reclaim land from private parties for road development.

Government Spending

Government spending on roads kept on decreasing through the five year plans. First five year plan allocation was 6.7 per cent of the total plan outlay. By the time the Eighth five year plan was brought out from the womb, the plan allocation had diminished to 3 per cent. This is but one testimony of how such an important infrastructure sector like roads was ignored.

However, in the recent years, the Government has undertaken important initiatives like the National Highway Development Project implemented by the National Highways Development Authority of India. The project, slated at Rs. 540 billion. is aptly ambitious.

7.7.2 Key Facts of National Highway Development Project (NHDP)

- Three phases–the Golden Quadrilateral, the North-South East-West (NSEW) corridors and the Port Connectivity projects.
- The Golden Quadrilateral is 5,843 Km of National Highway connecting Delhi, Kolkata, Chennai and Mumbai. It is to be completed by mid 2004.
- North-South East-West corridors is 7,413 Km long, Kashmir to Kanyakumari - 3,948 Km Silchar to Porbandar - 3,465 Km, scheduled for completion by December 31, 2007 against earlier date of 2009.
- In April 2000, the Government approved under "Other projects" along with NHDP, the network of National Highways of a length of 400 Km connecting 11 major ports to highways of the country with an outlay of Rs. 4,000 crores. This is essential for fast movement of export/ import cargo from the hinterland.

Innovative Approach of National Highways Development Authority of India (NHAI)

National Highways Authority of India (NHAI) has adopted some innovative and practical approach for implementation of NHDP, the salient features of it include: -

- Timely completion of project through:
    1. Prompt decision-making and faster payment to suppliers.
    2. Reward for every completion and penalty for every delay.
- NHAI is achieving better quality roads through high quality work specifications and use of state-of-art technology.

- The authority has in place a lean officer-oriented organization tapping private sector efficiencies and capabilities by out-sourcing project preparation, work supervision and maintenance.
- Procurement and information dissemination by own website/highway patrol.

To sum up on the performance of NHDP, there has been considerable progress. All six NHDP projects awarded under annuity-based tolling have reported financial closure.

7.7.3 Investment incentives

Some of the policy incentives to attract foreign and domestic investments:

- Private sector has been allowed to retain toll money.
- 100% tax exemption for 5 years and 30 % relief for next 5 years, of the choice of the contractor, within a period of 20 years.
- Concession allowed up to 30 years.
- Duty-free modern high-capacity equipment for highway construction for highway construction can be imported.
- NHAI has been permitted to participate in the equity of BOT projects up to 30% of the total equity.
- Government to provide land at no cost and free from all encumbrances.
- NHAI/Government is to provide capital grant upto 40% of the project cost to enhance viability on a case-to-case basis.
- Housing and real estate development, which are integral parts of the highway projects, are being treated as infrastructure and entitled to the same tax benefits.
- Strengthening the Central Road Fund.

7.7.4 Central Road Fund

- The Central Road Fund Act 2000 gives statutory power to establish a Central Road Fund for development and maintenance of national highways and state roads, development of rural roads, construction of under or over-head railway bridges, erection of safety works at unmanned railroad crossings.
- It is hoped that CRF will ensure funds raised through the cess will be used for road development. The cess on petrol and high-speed diesel (HSD) will be levied and collected at the rate of rupee one per litre.
- The proceeds of the cess shall first be credited to the Consolidated Fund of India; the central government may credit proceeds to the fund.
- Any fund provided by the central government for the development and maintenance of state roads is also to be credited to the fund.
- The balance to the credit of the fund is not to lapse at the end of the financial year.
- The fund is to be administered by the central government which is to allocate and disburse money to concerned departments.
- Projects of state roads, approved by the central government using set criteria to be financed out of the share for state roads shall be monitored by the central government.

7.7.5 Rural Roads

- The Pradhanmantri Gram Sadak Yojna was launched in August 2000. The Prime Minister has reiterated the government's commitment to this programme. This 600 billion dollar project aims to bring together the 140,000 villages, thus refining the rural market to the needed maturity for Indian firms to exploit.
- The scheme expected to connect 1.4 lakh habitations with the highway network, has a total outlay of Rs. 58,200 Crore over the seven-year period.
- The approval of the planning commission has been waived for it.
- The roads constructed under this scheme will be maintained under Panchayati Raj system.

7.7.6 Problems

- The recent initiatives by the GOI are to be appreciated, but political influence leads to lots of complexities at state and national level. The local contractors in nexus with the local politician siphon the funds allotted for road development by ensuring that the favoured contractor gets the contract. The contractor does not construct the road as per directed standards.
- Corruption is the root cause of inefficiency. Land acquisition can sometimes be a highly corrupt process.
- Time and cost overruns are another problem which vexes the construction of roads. The reasons for this can be classified as:
  - Geographic barriers and topography of the country, which will require better road building technology
  - Important landmarks associated with the local population that causes needless delays in implementation.
  - Fund scarcity, which leads to relying on petty projects for local development rather than all-encompassing road projects that brings new dimension of connectivity.

7.7.7 Policy Suggestions

- The umbrella of implementation, the NHAI is an excellent policy initiative and should be given a centralized vision. Funds for carrying out projects within time bounds should be allowed. Mere provision of more funds for the authority will only help to bureaucratise the authority.

7.8 An Integrated Model for Infrastructure Development

What is currently needed is a single, over-arching infrastructure policy instead of having separate vision statements for individual ministries. This comprehensive framework of policy directions should contain specific recommendations on each infrastructure sector.

It gives the nation perspective as to what direction should be taken in future and what should be the individual policies to achieve them. In addition, comprehensive infrastructural policies will layout in specific steps the goals and achievement to be expected out of each sector. There are excellent synergies possible and needed between a manufacturing vision statement and an infrastructural vision. The advantage is that threats and weaknesses can be perceived as opportunities and addressed holistically.

The advantages of having such a holistic infrastructure policy are as follows:

State Level

An important factor to note is that in many states, infrastructure development will be led by any one sector. For example, Gujarat has more chances of a port-led development than a state like Kerala, where telecommunication will probably be the thrust area. Statewise detailed diagnostic studies have to be conducted to identify those industries that each individual state can excel in. Accordingly infrastructure development can take place. For example, a state that wishes to pin its hopes more on information technology and which accords less priority to manufacturing, will have to put initial priorities on telecommunication and power sector development. Thus for each state, development of some sectors will figure in the long range plans while the immediate imperatives will be addressed faster. Such an approach makes sense since capital raising and subsequent financial closure of infrastructural projects are not easy, considering the level of state finances and difficulty in attracting private capital due to the long gestation periods.

Central Level

An infrastructure vision 2020 needs to be written that will serve to give direction to infrastructure development. The imperative for a central level policy is that it will help integrate all the state level policies. Again, for undertaking huge projects that span multiple states also such a policy is an imperative.

A central level infrastructure vision should contain the following factors:

- A transparent set of guidelines for privatization with an emphasis on competitive bidding rather than limited tenders.
- Key thrust areas in which projects have to be undertaken to give an immediate respite to existing infrastructural bottlenecks.
- A 'National Infrastructure Mission' which reviews the progress and implementation in all the individual sectors needs to be constituted. This authority will have representatives from the individual sector policy-makers.

Implementability

The infrastructural vision will manifest itself in a set of projects that the Government plans to undertake in a fixed time frame. As discussed earlier a good precedent has been set in formulating the National Highway Development Plan in case of the road sector. The plan listed out some major projects that enumerate the requirements of the road sector.

In a similar way, the National Infrastructure Mission will have in its annual plans a set of key projects to be undertaken. Its performance will be measured by the extent to which it is able to co-ordinate and complete the sector developments.

In conclusion, a successful manufacturing strategy cannot happen unless and until the distortions in the infrastructure sector is cleared. The benefit of aggressive reforms in infrastructure is perhaps the most important factor that Indian manufacturing sector can count on, in its quest for global leadership.

8

CHAPTER EIGHT

Critical Issues Impeding the Manufacturing Sector

INDIAN Manufacturing sector has enormous potential for full capacity utilization and satisfying domestic demand. However, some of the key factors of production are mired in defects even now. A list of such factors that limit corporate India's growth potential is summarized below.

8.1 Availability, Cost and Reliability of Power

In the energy intensive sectors of manufacturing like cement, aluminium, steel and paper Indian companies find it much cheaper and reliable to build power plants of their own, instead of depending on state power. While in China manufacturing units get power at 50% lower rates, in our country companies pay as high as Rs. 5 per unit for drawing power from the grid. Power reform is perhaps the most urgent measure required to boost the chances of Indian manufacturing in a competitive world.

8.2 High-Cost Economy

India has achieved much, but our legacy of growth and development has left the country with a high-cost economy. It is this aspect that has been worrying most Indians as it affects their standards of living and, after the progressive opening-up of the national economy (as signatory to the WTO), even threatens their livelihood. India continues to be a high-cost economy even after 10 years of liberalization. It needs to correct itself, which requires efforts both from the government and the industry.

In order to change the high-cost aspects of the economy and streamline it with others in the world, Indians must change the way they do things. It applies as much to agriculture as to manufacturing - and to the government as to the producers. For example in 'India-The Growth Imperative', a report published by McKinsey Global Institute, states: "In wheat farming, which is the most important of India's crops constituting 3 per cent of its GDP and involving over 15 million farming households, the productivity is just 1.3 per cent of [that in] the US."

8.3 No "Leapfrog" Way to Sustainable Growth

Until recently, after the phenomenal success of the software industry, it was believed that unlike most other developed and emerging economies before it, India could leapfrog straight into a "knowledge economy", bypassing the manufacturing-led growth phase. Now, it is clear that our low education levels do not support "Knowledge-based industries". Nearly 75 per cent of the workforce is educated below middle-school levels. This means that even if some of the service segments continue to grow impressively, their overall contribution to growth, employment, and incomes will not be significant for a country of India's size.

8.4 Unattractive Destination for FDI

India has not been able to attract Foreign Direct Investment (FDI) compared to other Asian countries. The Accenture report states: "China has been remarkably successful in attracting foreign direct investment (FDI), with nearly half the share of the total annual flows to developing countries going to it." China has successfully targeted FDI in labour-intensive manufacturing exports, which is the ultimate test of efficiency of any economy. While over 60 per cent of FDI in China has been in manufacturing, over 50 per cent of it has been in the labour-intensive segments of manufacturing. On the flip side, there is another half that has been going to non-labour-intensive Chinese manufacturing. This implies that there may still be opportunities for India to catch up.

8.5 Migration of Labour

The worker-to- population ratio in manufacturing has declined from 418 per 1,000 in 1993-94 to 396 in 1999-2000. Thus, a historically observed phenomenon of migration of labour from agriculture to manufacturing has remained unnoticeable in India. While some migration of labour from agriculture to closely related services has happened, this has not been substantial enough to drain the former of its excess workforce. A general impact of this has been to keep over 60 per cent of the national workforce tied up in a low wage trap, thus curbing rapid growth of markets for manufactured products in rural areas -- barring, of course, a few FMCG categories. It has meant that in several important product categories where market size is crucial for achieving economies of scale in manufacturing, Indian companies have been cramped by smaller markets.

8.6 Lack of Global Vision

Indian manufacturers, on the other hand, have generally focused narrowly on domestic markets. They have lacked the vision to see global opportunities for their business and have operated at sub-optimal levels. In order to safeguard their entrenched positions they spare no efforts in lobbying for protection with the government even after 10 years of liberalization.

There is ample evidence that some of our manufacturers have been surviving under protection of trade barriers, which are still among the highest in the world. Despite tariff reductions throughout the 1990s, India remains among the world's most protected economies. For example, although import duties have significantly come down in India over the last 10 years, they are still the second highest in the world, next only to Chile. In China it is just 15 per cent, and in Pakistan, Sri Lanka, and Bangladesh it ranges between 20 and 22 per cent. Moreover, until 2000, India had the highest number of products covered under non-tariff barriers among the developing countries.

Various government policies that are a carry-over from the past also continue to weigh down on Indian manufacturing. For example, past policies of import substitution backed by protection to industries from overseas competition have left behind a highly diversified Indian manufacturing base. Besides, Indian manufacturers were so busy in the past with diversification that many of them could not see the need for acquiring global competence in any of their business areas, although they had a wide choice.

8.7 Inefficient Logistics

Product supply chains in India tend to be inefficient by global standards. There are too many transactions and invariably there are losses and inventory buildups. Slackness in upstream supply chains is primarily because of manufacturers' transaction-oriented approach towards suppliers'. This hinders commitment, as suppliers don't see their relationship with manufacturers as strategic. The low-scale operations of most Indian manufacturers also impose heavy penalties, making specialization difficult for key suppliers.

The logistics costs of Indian companies are many times that of the East Asian economies, as shown elsewhere in this report. Poor road infrastructure, a railway system that has lost 40 per cent of the freight market to an inefficient road transport sector and still not doing anything about it, high turnaround times in ports, lack of containerization in ports etc, are huge disadvantages to Indian manufacturers when compared to their foreign counterparts. A poor transport infrastructure makes forecasting of demand even more complex and the average inventory levels higher, negating the cost effectiveness all the while.

8.8 Indirect Tax

A cumbersome indirect tax administration is something else that effects costs. Indian indirect tax rates are high by international standards. At one time India's corporate tax was considered too high to attract FDI (an ultimate barometer of the country's overall competitiveness). However, though direct tax rates have come down here to global levels unlike China, India has not been quite successful in attracting FDI. Investments don't always chase tax havens -- infrastructure, political risk, bureaucratic interference, freedom to do business, and several such things also matter.

8.9 Poor Infrastructure

It has, in fact, been well recognized that India's poor infrastructure poses substantial barriers to the cost-competitiveness of its various sectors. World Economic Forum rankings of Indian infrastructure, barring airlines, are in almost all cases over 50 in a survey of 59 countries. The Rakesh Mohan committee had submitted a comprehensive report on what needs to be done to improve India's infrastructure in 1996. Except telecom and, recently, national highways, little progress has been made in other areas. In power, several initiatives that were taken during the mid-1990s misfired.

Worldwide, power is supplied at 5-6 cents per unit. In fact South Africa and Australia give power at only 2-3 cents to bulk users, whereas in India the cost of power to industry is as high as 10-12 cents. This is largely due to cross-subsidies to agricultural and residential users. While the cost of power is important, its quality is of equal significance. Frequency variations and voltage fluctuations damage manufacturing equipment and swell maintenance costs.

Turnaround times in most Indian ports vary from one to five days depending on the cargo. In developed countries the maximum is two days. It is far more logical to trans-ship cargo from Dubai than from here, as vessel frequency is much higher there. In Dubai volumes are 10 shiploads per day, as opposed to half or one here. In China this figure shoots up to 20. Freight by sea to the U.S. or Europe is far more expensive than from China or Southeast Asia. Logistical inefficiencies thus burden Indian importers and exporters with extra costs. In fact, many ships don't touch Indian ports because basic facilities just don't exist. They dump India-bound container cargo at Colombo port, where the vessel turnaround time is eight hours.

One reason why the progress on most infrastructure projects has been slow is the long gestation periods of many of these projects. Most investment bankers prefer short to medium-term funding, which makes it difficult to raise capital for infrastructure projects. However, with the opening-up of insurance and pension funds, things should improve. These funds are universally acknowledged as ideal investors in long-gestation projects. But this is for the future.

8.10 High Cost of Capital

The cost of capital in India is also high by global standards, upward of 13 per cent in comparison to 4-7 per cent in the US and Europe and 0.5-1 percent in Japan. Even the cost of water is high here by global comparison. For example, water supply to industry is charged at the rate of Rs30 a kilolitre as against Rs15-20 in China and other East Asian countries. Similarly, stamp duties on property are 10-12 per cent in India as against 2-3 per cent elsewhere. Thus, when one adds up many small things, the overall effect on costs is quite devastating.

8.11 Lack of Strategic Concerns

At the policy level Indian companies are influenced more by specific government incentives than strategic concerns. This leads to insufficient rigor in project selection. The sophisticated understanding of customer needs and robust marketing processes, firms can improve their top-lines. Similarly, by ironing out inefficiencies in supply-chain processes Indian companies can reduce their operating costs.

8.12 Under-Investment in IT

One of the major findings of the Accenture report is that Indian companies under-invest in information technologies (IT). The presence of information technologies of different vintage in companies constrains their ability to integrate information, and thus deprives them from realizing full value of their investments in it. Input costs are the other factor that affects cost-competitiveness of Indian firms. Moreover, the inadequate pressure on management performance owing to lack of transparency, lack of consumer awareness of their rights, and inadequate intellectual property rights protection combines to contribute to management complacency.

8.13 Labour Laws

It is generally believed that India's economy would grow faster were it not for inflexible labour laws. Opinions are divided. For example, while this ranks low in McKinsey's assessment, Accenture considers it a major barrier in its report. But the former admits that the current labour laws do inhibit productivity in labour-intensive and export-oriented manufacturing segments such as textiles and clothing, by making it difficult to shed workers when market conditions change.

In all the sectors studied, availability of cheap labour surfaced as one of the chief strengths that can be relied upon. In spite of this, labour culture and labour laws are not allowing companies to fully utilize this advantage. This is especially true in post-reform companies, which have grown a mellowed workforce over the years. Modern companies such as Hyundai Motor India Limited have been able to successfully counter labour woes. This is proof of the fact that sound and supportive policies especially from the part of the central and state governments as well as the trade unions is the starting point for leveraging India's labour power.

8.14 Government Ownership

The benefits of privatization has long been recognized internationally with very relevant examples for India such as UK, China *et al.* successfully reducing government exposore in factor markets. It has been widely criticized that the pace of privatization is slow enough in India to negate whatever benefits it might bring in. It is an urgent imperative for government to get out of mental blocks regarding disinvestment. There are many successes like the automobile industry, telecom etc to draw on. To capitalize on these, a matching political aggression is required.

8.15 Poor Support from State Governments

Indian constitution has some items on the concurrent list, which is supposed to give more autonomy to states while making policy decisions. The fact is that some aspects of this concurrent list create huge disparities in policies across States, often increasing the hidden costs of doing business in India, especially for multi-national companies. The lack of uniformity in political ideology across states, which is perfectly natural in a modern democracy, should not be reflecting on economic ideology.

8.16 Entry Barriers for Entrepreneurship

A culture that facilitates innovation is lacking in the country. New products and new markets come up as a result of innovation. As a nation, India cannot be continuously depending on foreign players to bring in innovative products that build entire markets. Across the world, innovation as an enterprise grew because of entrepreneurship. Apart from problems of the mindset, the atmosphere for entrepreneurship in our nation is dismal at best. It takes an average of 90 days to start a company in India, whereas in China it takes 30 days and lesser. Incentives for entrepreneurs is also very low.

8.17 Reservation of Products for SSIs

There are 836 products currently reserved for SSIs. The presence of capital investment limits specifying SSI sector are a major limiting factor for growth for these products. To enhance competitiveness in these sectors more capital infusion is necessary. Hence product reservation for SMEs should be called off.

8.18 Restriction on Foreign Direct Investment

China attracts the biggest portion of FDI inflow to developing countries. Chinese non-residents bring in a large portion, nearly 60 percent of the total inflow. A similar kind of FDI boom will be highly beneficial for India. This is because the restructuring programmes in sectors like infrastructure need phenomenal amounts of resources, which cannot be provided anymore from the kitty of government. Moreover, the benefits of successful international joint ventures in a technology poor country such as ours cannot be ignored. Hence restrictions on FDI have to be categorically removed.

8.19 Lack of Industry Focused Trade Zones

Countries that are known for specific products provide a consistent policy support to corresponding industries. Such a focus based on comparative advantages is lacking in Indian policy. One way of rendering strength to sectors of export potential is to provide them with trade zones which offer facilities like infrastuctural support, tax holidays, start-up assistance and single-window mechanism of procedures. Such clusters and areas will carry specific brand tags that will help in marketing them in international markets.

8.20 Uneven Enforcement of Rules

Companies, especially MNCs wanting to set up business in India, find it extremely difficult and complex to understand and adapt to the existing business systems. This is one of the chief reasons why they try to rope in Indian companies to gain an initial foothold and consolidate positions later on. The reasons for the uneven enforcement of rules are the disparities in policy-making across individual states. This has to be corrected by enacting over-arching laws in factor and product markets and by having holistic national vision and guidelines regarding supporting sectors such as infrastructure, education etc.

8.21 Lop-sided Industrial Growth

Industries in India are largely concentrating on establishing themselves in the following five states namely, Maharashtra, Karnataka, Tamil Nadu, Gujarat, and Andhra Pradesh. Flow of FDI, bank credit, and government support is cornered by these states. The other states have been left years behind in terms of industrial progress. This leads to a lot of social problems like, disparity in employment opportunities, lop-sided infrastructural growth, and bitterness among state relations. This in turn affects the long run development of the firms. Investors perception about a state's behavior largely guides their decisions to set up firms. Kerala for example has all the natural endowments and necessary infrastructure to host a number of industries, but the investor perception of its labour force is unfavourable. Local government too, do not help change this perception and hence little industrial activity is seen in the state. Clustering of industries in a few states is firstly a big drain on the natural resources of the state and secondly, progressively renders the other states less competitive and poor.

8.22 Regulatory Hassles

The time spent by the management of firms in dealing with government officials and regulatory authorities is very high and this is the basic cause for the reluctance of Multi-nationals and NRI investment in setting up manufacturing units in India. This is a great entry barrier for new entrants into the industry and particularly for first generation entrepreneurs. Moreover, this affects the progress and growth of the industry as productive time is lost in wasteful activity.

8.23 Lack of an Academic Liaison Model for Manufacturing

India has been able to produce a world recognized pool of Software professionals. In late 1980s computers attracted a lot of young students. The youth of the country could see a lot of opportunities in software services; better emoluments, excellent working conditions, foreign assignments and a fast growing career profile. At the same time India was liberalizing its markets and the scope of Information technology applications gained importance. Private schools in the country also included computer science as a main stream subject in their curriculum. But manufacturing firms in India were caught napping. The industry was set for a revolution in terms of increased computerization and automation of operations in the coming decade, but there was no initiative from the industry or the government to orient the students towards learning computers as a tool with specific application in manufacturing. Thus, though the industry was entering an era of automated supply chain management, inventory control, quality control, resource allocation, vendor management and e-business, they never attracted a focused student group who could provide such solutions to the industry with an Indian perspective.

8.24 Limited Talents Attracted to Manufacturing Sector

Engineering education in the country has seen an increase every year both in terms of number of institutes and the number of students enrolled. From the early 1990s the top ranked students in engineering qualifying exams preferred to join Computer Science, Telecommunication and IT, Electronics stream rather than the traditional Mechanical, Civil and Electrical branch. The former branches are heavily into software services and most of the graduates were absorbed by the software industry. Premier engineering institutes in India like the Indian Institutes of Technology also saw the best students enrolling for software and telecommunication courses. Civil engineering was the first to lose its halo-status followed by Electrical and Mechanical engineering disciplines. Top 100 students in the IITs consistently opted to enroll into Computer Science, Telecommunication and IT, and Electronics. Such a brain drain in the core sector went unnoticed. India needed software as a tool to leverage its industrial production, while the manufacturing industry lost its 'would be solution providers' at the entry level itself. This served as a body-blow to manufacturing sector in India with only limited capable talents available for nurturing the growth and sustainability of manufacturing sector. Moreover, the manufacturing industries took a very limited initiative to attract and retain rare talents shifting to the services sector.

8.25 Talents with IT Capabilities Required for Manufacturing

The opening of the Indian markets attracted foreign companies and investments to India. The companies in the manufacturing sector were typically handicapped with low level of process automation and huge cost of importing better technology. Even the imported technologies and software solutions for manufacturing were not customized for Indian firms. The cost of customization of the software is very exorbitant which adds to the production fixed costs. Moreover, the maintenance cost of imported technologies is also large and time consuming. NHPC's Kahiri plant in Himachal Pradesh imported 3 transformers from USA but the plant now has to rely on service engineers to come all the way from US in case of a breakdown. There have been instances when the concerned service engineers have refused to come to India and the delay in servicing costed the plant nearly Rs. 3 per unit of power. Though software courses are taught in several reputed institutes, they are mostly oriented towards a programming job rather than equipping students with applications relevant to modern manufacturers such as Autocad, SPSS, SASS, Pro-E, etc.

9

CHAPTER NINE

Turnaround Strategies for Manufacturing Sector

- Equalize sales tax and excise duties for all categories of players in sector and strict enforcement.
- Remove all licensing and quasi-licensing restrictions that limit number of players in affected industries - Banking, dairy processing.
- Reduce import duties on all goods to the levels of South Asian countries (10%) over five years.
- Allow unrestricted foreign direct investment in all sectors in manufacturing.
- Privatize all companies owned by government sector with a priority list- Banking and Insurance, manufacturing and mining, power, telecommunications, - begin with large companies - task should include the creation of pressures for enterprises to be competitive, relax rules and give more managerial autonomy, create conditions that enable PSUs to compete in equal terms with others.
- Competition policy should maintain and encourage competition in order to foster greater efficiency in resource allocation and consumer welfare.
- Reform labour laws by repealing section 5-B of the Industrial Disputes Act , by introducing standard retrenchment-compensation norms and by allowing full flexibility in the use of contract labour.
- Full Utilisation of funding options for development - efficient taxation, savings from investment in unprofitable government enterprises, FDI.
- Encourage entrepreneurship to convert the potential of skillful persons opting for VRS to productive use.
- Emphasis on strict enforcement of environment standards.
- Strict enforcement of labour laws protecting child labour and penalizing factories using child labour with heavy fines and stringent norms discouraging them to do so again
- Provide incentives and create institutional support for R&D and strategic alliances, grouping/ clustering for R&D.
- Create institutional structures for Public-private partnership, Public-educational sector partnership and Private sector-educational sector partnerships for development of applied and developmental research with a focus on basic research by government.

- Eliminate reservation of products for small scale industries and re-define investment limits for SSIs, tiny units and encourage development of TVEs and SMEs.
- Develop infra-structure support for SMEs and refine institutional support structures through closure of duplicating agencies and redefining the role in terms of developmental activities instead of regulatory activities
- Refine Education Policies to raise the skills and capabilities to suit the manufacturing sector requirement and give thrust on industry focused research in higher level educational institutions.
- Refine Education Policy to inculcate the culture of thinking at the school level to develop the young talents towards creative thinking and innovation for better technology innovation.
- Formulate an Infrastructure Policy to encompass all sectors and evolve plan of action with an Infrastructure Mission 2020.

9.1 Concentrate on Labour-intensive Sectors

Focussing on specific sectors such as the consumer-centric and labour/skill-intensive sectors, where Indian manufacturers are placed at medium to high-level on competitiveness quotient would have positive impact on aggregate manufacturing productivity. In order to speed up the process, flexible labour legislation and de-reservation of the small-scale industry is an immediate necessity. The re-direction of resources towards sectors that are likely to sustain international competitiveness of Indian manufacturing would also enable their transition towards more trade-oriented manufacturing.

9.2 Re-engineer Core Processes of the Company

Indian companies will have to show far greater urgency than in the past in addressing the internal aspects, in the wake of rapidly emerging global competition. They will have to recognize shareholder value creation as their principal objective and use this as a yardstick to measure management performance. They will have to build nimble-footed organizations that have accountability built in at all levels, in order to harness human energies to the optimum. The Indian companies, which have not yet re-engineered their core processes will have to do so soon so as to dramatically, improve their efficiencies and drive business value. They will have to consolidate and scale up their operations to global standards.

9.3 Professional Management Style Required

For manufacturing in India to be globally competitive, Indian firms must upgrade all the time to acquire and retain global competitiveness. Accenture lists management alertness to external stimuli, resource allocation across firms within each manufacturing sector, historical baggage that sectors carry, and resource (capital and labour) utilization levels as main drivers of manufacturing growth. It observes that by professionalizing management style and by focusing on leveraging core capabilities, Indian companies can achieve a 20-50 per cent rise in productivity. It reports as high as 30-50 per cent over-manning in Indian companies and suggests that by revamping organizational structure, properly leveraging human resources, focusing on team-building, and putting in place performance management systems that reward superior performance, Indian companies can substantially change their cost structures.

9.4 Identify the Research Areas

The manufacturers will have to estimate their technological bottlenecks and identify what makes their product less competitive and more costly in the world market. Mostly, Indian firms will be using low end technology into their production processes. Government companies are especially notorious in not identifying or communicating problems in the plants. A pro-active measure is required to document the needs of the company and then through conferences and symposiums communicate it across to the technologist. Pro-actively identifying and documenting technical requirements is the first step towards being competitive.

9.5 Marketing the Industry to Attract Students:

The manufacturing companies will have to market themselves to technical students. The charm of working in a top manufacturing firm is no more an attraction among students. Thus, it is necessary to rejuvenate the students by involving them in identified research areas or tasks of mutual interest through short term projects and provide the opportunity to create awareness and confidence on the scope for manufacturing industry.

The software firms have thrown up a number of cases in which young highly talented professionals leave their jobs, simply because of the routine stressful nature of work and some of them also develop an aversion to computers. Moreover, in India the standards under which the professionals work vary from firm to firm and there is lot of dis-satisfaction among the employees. Personal interviews with a number of colleagues in top software firms reveal that family life and free time is being missed. Manufacturing firms typically have a more traditional work schedule and this can be leveraged. Highlighting the problems of a software professional vs. the comforts in a typical plant manager's job is required during institute visits, seminars and other technical functions.

9.6 Better Working Conditions

Today, students are very concerned about the working conditions in which they will have to perform. The manufacturing firms have till now been unmindful about projecting their working conditions to the students. Typically a plant is considered to be hot, noisy and unpleasantly dirty. But in reality many top manufacturing firms from different sections like automobiles, electrical equipments, rubber, food processing etc. have taken up a lot of initiatives to improve the working atmosphere of the employees. Air-conditioned shops and state-of-art automatic machines help workers feel comfortable while working. The need is to follow a two-pronged policy. Firstly, highlight such state-of-art facilities to the students and seek feedback in terms of what more will they want. This will bring the initiatives in public notice. Secondly, continuously improve the working conditions.

9.7 Manufacturing-Incubation Centers

The firms from various sections of manufacturing industry will have to pool resources and finance long-term high-technology research programs in collaboration with IITs, RECs and research centers. The established need of the research will make it a good investment and the good incentives for such a project will attract attention among the student community. Long term association with the engineering colleges has to be revamped by bringing in new streams of research. For example, Computational Fluid Dynamics (CFD) is one field which is being hotly pursued by most of the top manufacturing countries. Indian students have gone abroad and contributed to research work in this field. They are in demand and have done very good work too. Similarly, Indian firms need to attract the best students to lead break-through technology projects.

9.8 Quality Managers will Yield Best Results

Competitive advantage of foreign firms over Indian firms is also due to the capabilities of mangers of the former. The professional approach of the MNCs is due to the hard work done by the managers at supervisory and top level. The culture of the company is affected by the firms' managerial quality. Manufacturing firms in India have to seek the best managers. As of now many manufacturing firms rely on their engineers to be a marketing manger or a quality manager. Increased presence in B-schools campus recruitment and internships is required. Sharing of information about the company and the industry of operation will help the would-be-managers to be interested in taking up jobs within the manufacturing industry.

9.9 Evolve a Technology Policy

Development of R&D and Innovation needs a separate technology policy to be formulated which will deal with the development of innovation and R&D across all sectors of manufacturing industry. By supporting basic research, applied research and industrial research, it should assist in the construction and expansion of an innovation system. It should consciously try to exert influence on the development of technology, in order to secure the competitiveness of industrial enterprises. A Technology policy should attempt to achieve a synergetic focusing of endogenous resources by government support, by active organisational measures and by taking on the role of a moderator in the dialogue between science and industry.

The government needs to develop broad based and varied research institutes to ensure that different organisations contribute their particular capability to industrial R&D and outline the structure for their contribution to research. The institutional support may include higher education institutes, government research establishments, industry labs, national research centers, and joint R&D clusters. Higher education institutes, especially technical higher education institutes will carry out independent and industry/ government sponsored research. The government research establishments undertake basic research and defence related research. The national research centres perform R&D which cannot be carried out either by the higher education institutes, the industrially oriented research institutes or by the industry itself such as CLRI. They focus on activities with a long-term orientation which appear to have a high degree of risk, require relatively large research teams and incur high costs. Industry laboratories do research for their own requirements which is basically applied and experimental research. Joint R&D clusters is a form of research in which several partners co-operate in order to tackle the topics by dividing the work up among themselves. Possible forms of co-operation range from an agreed exchange of results to a combination of innovative projects in a common laboratory.

An innovation supportive infrastructure can be started by government to provide services such as :

- Contact offices, situated in "trustworthy institutions" (such as chambers of industry and commerce) offering initial orientation assistance;
- Advisory offices with sufficient diagnostic capabilities to mediate specialist advisers and monitor the innovation process;
- Information brokers who carry out data bank searches for clients;
- Patent information centres which, based on the great volume of world-wide patent literature, pinpoint specific technological and business information for firms, thus supporting their R&D activities;
- Advisory centres, with teams of advisers that are able to make detailed problem diagnoses, perform specialised consulting in selected areas of technology, and act as qualified consulting mediators;
- Transfer centres as R&D establishments which perform further development and adaptation contracts in individual specialist areas, and can offer reliable expert advice;
- Demonstration centres, which provide neutral information (i.e. independent of manufacturers) on possibilities for application and use, advise employees of enterprises - especially SMEs - on purchasing decisions, and help them to acquire qualifications by running seminars on new technologies;

- Technology centres at R&D establishments which are primarily concerned with the development of new technologies, and whose specialists can cover the whole range of consultation from initial inquiries to in-depth consulting;
- Technology and incubator centres as a local community group usually consisting of new technology-based firms with the purpose of developing, producing and marketing technologically innovative products, services and processes.

The emphasis of technology policy should be on the support of existing innovation potentials and on the attraction of innovative enterprises by:

- Increasing and focusing the innovation potential by motivating enterprises to engage in innovative activities or intensifying innovation activities;
- Stimulating the founding of new firms derived from existing firms (spin-off's);
- Improving the conditions favouring innovation, particularly by innovation services and relevant infrastructure (R&D institutions, information and advisory institutions, supply of capital and venture capital, professional training/further education) in order to attract companies from other regions or from abroad to establish businesses;
- Dissemination of technologies.

9.10 Technology Assessment Board

Competitiveness of India Inc. depends to a large extent on the capability of Indian firms to move up the productivity ladder. How exactly do Indian firms address this issue? Approaches must differ in case of large industries and for SMEs. Large firms can get access to superior technology by either importing technology via royalty agreements or by entering strategic tie ups with foreign companies. In this scramble for technology, SMEs are left out since they do not have the required funds as well as the connectivity to improve or improvise on their existing technology.

Figure 34 - The Vicious Circle of Value Erosion

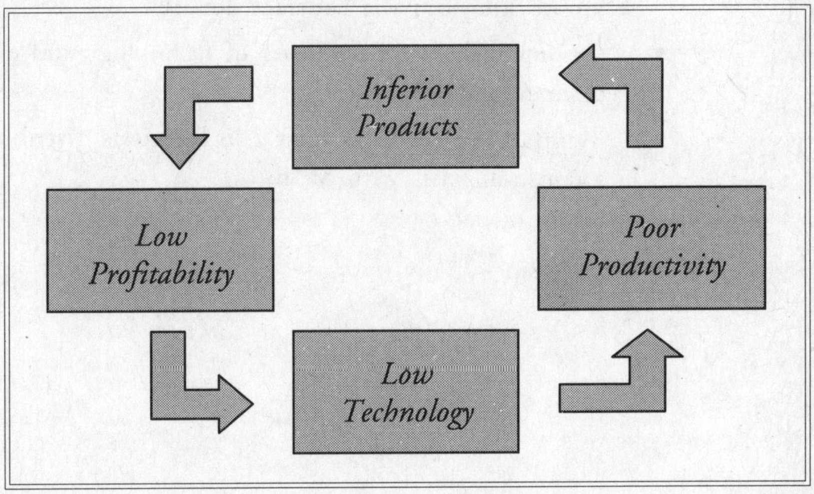

There are two issues in this regard

- ⊙ Firstly, technology importation is costly for SMEs.
- ⊙ Secondly, often only second hand technology is available for import.

There is a dire need for technology assistance for SMEs from the part of the policy makers. As analysed in the SSI section, there are a huge number of research and technology institutions that pursue research programmes with out proper commercialization.

The vision of such an institution should be formulated keeping two broad areas into consideration

- Firstly more funds for the research institutions to carry on their research.
- Secondly cheaper availability of technology to the small-scale sector.

FIGURE 35 - BENEFITS OF TECHNOLOGY FACILITATION

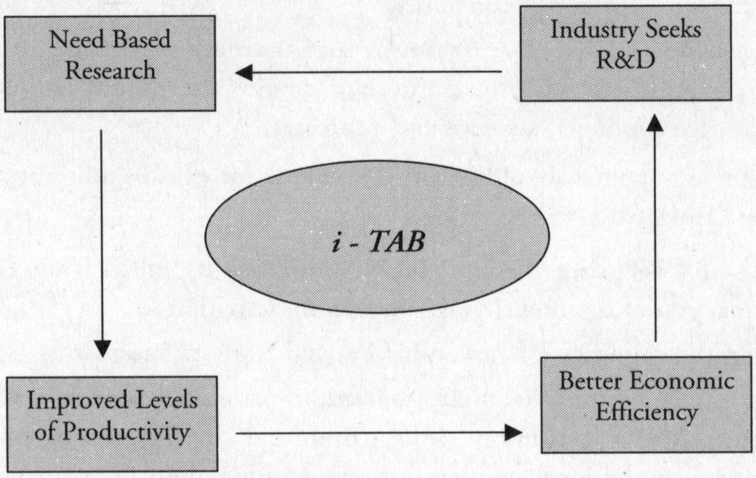

Functions of the Technology Assessment Board

- Liaison with the various academic and research institutions for providing a database of industry consultants and research scientists.
- Facilitate interaction with the experts / consultants and the industries by providing a forum for working out the contract details.
- Commercializing the transfer of technology and expertise while ensuring that the costs are kept to manageable levels.
- Maintain offices in at least 500 locations distributed uniformly in the country to network on a constant basis with SMEs.

10

CHAPTER TEN

Draft - Manufacturing Policy

THE purpose of this policy is to create a strong manufacturing sector with enhanced investment and technological capabilities. The manufacturing policy aims at reducing country costs and enhancing the transaction efficiency for business, including the necessary macro-economic, regulatory and business environment, the physical infrastructure, and the educated and skilled labour force under which manufacturing firms can invest and grow. The policy will focus on a free entry and exit policy from the manufacturing industry with supportive institutional and economic policies. The rationale behind the policy is to enable development of manufacturing industry in the context of competition enforced through liberalization, privatization and globalization. The policy will propose suitable amendments in regulations to accommodate a free market to replace the policies that were originally formulated with a socialistic perspective. The policy framework is given in terms of large, medium and small-scale industries and the reforms required in the basic factors for development such as infrastructure sector, education sector, manpower development and the policy initiatives in labour laws, direct and indirect taxes, EXIM policy and Institutional support to be created for supplementing the growth and development of manufacturing industry. The thrust of the policy will be in the areas of technology, export, incentives for investment in R&D, infrastructural development and establishment of clusters and specialized agencies for supporting industrial development.

10.1 Definition of Manufacturing

"Manufacturing comprises of mechanical, physical, or chemical transformation of materials, substances, or components into new products"

The following identified Industry Groups constitute manufacturing sector: -

- Food Products
- Cotton Textiles
- Wool, Silk, Textiles Products etc
- Jute Textiles
- Chemicals, etc
- Non-metallic products
- Non-Electrical Machinery, Tools & Parts
- Basic Metals

- Electrical Machinery
- Transport Equipments.

10.2 Developmental Policies for Large Scale Industries

Large scale enterprises will be encouraged to focus on industries where economies of scale is essential such as in paper, petrochemicals, steel, etc. Earlier policies did not have such a focused approach; for the purpose of facilitating SSIs they had often reserved industries where economies of scale simply do not exist. The new approach will reflect market sense rather than political sense, which is more essential in making a brand out of Indian manufacturing.

10.2.1 Sector Specific Policies

The following decelerating growth industries will be provided incentives to acquire new technology or to upgrade technology. Special attention is needed in these cases for retaining comparative advantage.

- Rubber, plastic, petroleum, and coal products
- Other manufacturing industries
- Leather and fur products
- Basic chemicals & chemical products (Except petro-products & coal)
- Wool, silk and man-made fibre textiles
- Textile products (including wearing apparel)
- Machinery and equipment other than transport equipment

The following negative growth industries will be provided facilities to increase the manufacturing capabilities in a non-discriminatory manner including the adoption of total quality and productivity programmes, technological upgrading, and investment in plant and industrial skills, using a network of private-sector led productivity centres, technical institutes and industrial extension services.

- Food products
- Cotton products
- Jute and other vegetable fibre textiles (except cotton)
- Metal products and parts (except machinery and equipment)
- Wood and wood products, furniture & fixtures

Food and Beverage Industries

- Village and farm owners to be educated on latest technologies used in food production.
- Incentives will be given for those operating in Special Economic Zones, by providing exemption from Central and State government tax on production.
- Special Economic Zones (Food) will provide facilities for large-scale storage and refrigeration. Such facilities can establish a strong channel for exports through intensive promotional schemes. They will also come up with training programmes to equip the manufacturers on frozen food technology and provide necessary support for development of frozen food industry.
- Multinational investment in food processing will be allowed with a maximum of 100% investment and are entitled to export incentives given by the government.
- Additional depreciation of 20 per cent will be provided for investments in new plant and machinery acquired after 1st April 2003 for new investments and expansion of capacity by additional 25 per cent.

Textile Industry

- Collaborative research will be undertaken in customization and development of IT tools like supply chain management for production of quality raw materials in textile sector like cotton, wool, etc.
- Creation of Special Textile Zones for assisting members to upgrade technology and automate processes and to support industries in exports through research and promotional activities in new potential markets. Market assistance schemes will be introduced.
- Export incentives will be extended for handloom and power loom products; financial assistance will be provided for upgrading technology.
- Additional depreciation of 20 per cent will be provided for investments in new plant and machinery acquired after 1st April 2003 for new investments and expansion of capacity by additional 25 per cent.

Automobile Sector

- Export of automotive components and automobiles cars, trucks, two wheelers and buses will be entitled to proportionate exemption in excise duty.
- Automobile Association of India should initiate the process of partnering with automotive component manufacturers for technological research initiatives and provide support on international developments in technology and changing needs of global markets to automotive manufacturers.

Electrical Equipments Industry

- Creation of Special Export Oriented Zones for electrical industry.
- Export oriented SMEs will be provided assistance to modernize technology.
- The policy will enable the accreditation of testing laboratories in India by overseas agencies.
- EOUs will be provided assistance for brand promotion, ware-housing facilities in overseas markets, etc.

Copper Industry

- Additional depreciation of 20 per cent will be provided for investments in new plant and machinery acquired after 1st April 2003 for new investments and expansion of capacity by additional 25 per cent.

Gems & Jewellery

- Additional depreciation of 20% will be provided for investments in new technology acquired after 1st April 2003.
- Create a cluster of players in the industry for promotion of export of jewellery.

Machine Tools

- Creation of Industry Clusters to promote export of machine tools.

Paper Industry

- Incentives for investment in backward integration by firms by additional depreciation of 20% on investments in plant and machinery acquired after 1st April 2003.

Petrochemicals industry

- Setting up of an institutional machinery to enable collaborative research and identify areas of focus for Indian firms for acquiring patents.

Petroleum & Energy sector

- A separate energy sector policy encompassing all segments is required for the furtherance of the sector to a sustainable level.

Rubber industry

- Creation of Industry Clusters to promote export of rubber products.

Steel Industry

- Investments in R&D will yield tax rebate of 20% on the investment.

10.2.2 State-Focused Policies

- Special incentives and concessions for manufacturing industries started in States other than Maharashtra, Gujarat, Karnataka, Andhra Pradesh and Tamil Nadu for the next five to ten years - Concessional rates for using power, flat income tax rate of 30%, exemptions from duty on imports for the first five years.
- State government approvals for starting up of industries should be completed within 90 days.

States with investments from manufacturing contributing more than 30 per cent will be provided additional 30% grant in the budget during that year or 20% of the tax collected from the manufacturing industries will be provided to the respective state governments.

10.2.3 Technology Policy

A 'National Institute of Technology' has to be created with focus on the development of technology initiatives. So far, technology progress was containerized into certain pockets of interest like selected academic institutions, areas such as space, agriculture etc. This often drives R&D away from the ground realities. The new technology policy will give a better perspective to the R&D efforts in the country. 'Partnering Innovation' will be the credo of this new approach. The chief facets of the new policy will be as follows:

- Partnering research
  - ⊙ Government - Private sector

    Government aid for projects on areas of concern for both parties.
  - ⊙ Government - Educational sector

    Government aid for projects on areas of concern for the government.
  - ⊙ Private sector - Educational sector

    Private sector sponsored research on areas of interest of the sponsor.
- Small Business Innovation research - sponsor projects on areas identified by the government.
- Advanced Technology programme - collaborative research on high risk enabling technology on cost sharing basis.
- It is proposed to put a ban on importing second hand technology like used machine tools.
- Incentives and concessions will be offered on technology upgradation by existing units.

- Government will focus more on basic research and will partner and collaborate for applied and development research initiatives.
- Venture capital funding will be provided for high technology research initiatives through the venture capital funding agencies.

Clustering for R&D

Firms in a particular industry will be given institutional support to identify and utilize possible synergies in taking a collaborative approach. Such an approach will enable them to reduce redundant costs of firm level R&D as well as firm-confined pollution control efforts. The policy suggests measures to successfully adopt clustering as a strategy for Indian firms.

10.2.4 Disinvestment Policy

Dis-invest loss making units first and then other units owned by the government with a disinvestment plan that is more transparent and comes with a package of measures to provide autonomy in operations, permit reduction of excessive workforce and take up measures to instill competition in the industry, as disinvestment without competitive environment will only result in furthering inefficiency.

- Privatize all companies owned by government sector- banking & insurance, manufacturing & mining, power, telecommunications, - begin with large companies - task should include the creation of pressures for enterprises to be competitive, relax rules and give more managerial autonomy, create conditions that enable PSUs to compete in equal terms with others.

10.3 Overarching Policy Initiatives

Direct Taxes

- Standard taxation rate (20%) in Special Zones for industries located in industry parks, free trade zones, industries located in identified backward areas.
- Income tax rebate for investment in R&D will be provided to enable better focus on research and development activities and to enable domestic players to benchmark their product quality to world standards. Tax rebate will be proportional to amount of investment made in R&D.

Indirect taxes

- Sales tax and excise duties will be the same for all categories of players in a sector and strict enforcement measures will be taken by the authorities.

Trade Policy

Import Duty

Reduce import duties on both primary goods and capital goods to the levels of South Asian countries (10%) over three to five years.

- High customs duty of the order of 250 per cent levied on import of low technology used as factor input such as import of second hand machine tools.
- Import duty on petrochemicals to be reduced gradually to bring in better competition in the industry.
- Import duty on imported cars reduced to 10%.

- Import tariff on automotive components reduced to 10%.

Simplification of Import Procedures

- Time lag, defined, as the period between request for investment license and effective starting date of operation (e.g. in the case of import of equipment) will be brought down by simplification of procedures and by fixing dead-lines for approvals.

Export Policy

- Automated Export Oriented Zones will be created for specific sectors.
- Foreign investment in export-oriented units is encouraged in all manufacturing sectors focusing on export of high technology products.
- Foreign Investment Parks will be created in specific sectors (as specified by the government) in Special export zones.
- The Government will provide the necessary infrastructure such as power, roads, etc. in such Parks to facilitate the smooth running of export-oriented units.
- Priority will be given to those investments bringing in high technology.

Export Incentives

Special Export incentives and concessions will be provided for the export of the priority items to specific countries as given in the Table.

Environmental Policy

- Strict enforcement of environmental policy in industries such as lubricants, paper etc
- Emphasis on strict enforcement of environment standards

Foreign Investment Promotion Policy

- Instead of the existing regulatory mechanism, it is proposed to have a FDI Promotion Authority.
- The policy will focus on target companies and their requirements instead of a scattered approach towards attracting FDI
- Individual States will be made direct stake-holders of this measure by integrating their infrastructure policies with the central FDI Promotion policy
- Exit barriers to FDI investors should be removed. Equity transfer restrictions will be categorically taken off.

TABLE 30

The Export List

Priority items	Targeted countries
Electrical equipments	China
Automobile components	Malaysia, Taiwan & other South Asian countries
Steel (Galvanized)	Iran, Iraq, West Asia, the Philippines and Bangladesh. Middle east and Rest of Asia (except China). Focus on Middle East countries
Two wheelers, cars, trucks	Bangladesh
Basic Metals	China
Copper	China
Petrochemicals	Asian and African markets

Other Policy Initiatives

- Remove all licensing and quasi-licensing restrictions that limit number of players in affected industries such as banking, dairy processing etc.
- Competition policy should maintain and encourage competition in order to foster greater efficiency in resource allocation and maximize consumer welfare.

10.4 Developmental Policies for Small-Scale Sector

Sectors in which SMEs have better chances will be those with lower start-up costs. Examples are paper accessories, fruits and vegetable processing. The crux of the policy will be such that technology should no longer be either an entry barrier or a limiting factor for SMEs. The policy measures will be as follows

- The policy of reservation of items for SSIs will no more be applicable. The investment limit for SSIs hereafter will be Rs. 10 Crore.
- Setting up of Township and Village Industries (TVEs) with investment limit of Rs. 5 Crore. The Panchayati Raj system already provides lots of governance power to the villages for easy development of local firms and enterprises.
- The tiny units will have investment limit of Rs. 25 lakhs to further self-employment.
- A new category SMEs - Small and Medium Enterprises with investment limit ranging from Rs.10-15 Crore will be permitted to enable foreign participation in such units. The government will encourage grafting Joint ventures with foreign companies in labour-intensive projects. The government will have to be pro-active and incubate the model of local firms with global tie-ups.
- Bank sponsored SSIs will be encouraged to enable adequate financial support and better growth of SSIs.
- SSI will be located in clusters– SSI Clusters to train and support the use of IT-based manufacturing processes. The cluster will provide infrastructure facilities, low-interest capital, trained labour, shared Research and Development, shared cost of pollution norms, holistic strategies for the industry branding of clusters, savings in logistics and identifying markets for business.
- National SSI Manpower Training Board will be set up as Public-Private Partnership to train SSIs in up-dated technologies wherein infrastructure will be provided by the government while the management is left to either an NGO or an industry association or private sector.
- Government would provide infra-structure support for such Large and SSI to set up marketing consortia (showrooms, godowns, trade fairs and convention centres) and promote trade delegations.
- National Testing and Certification Lab for SSIs would be set up as Public-Private Partnership to offer specialized facilities which often require expensive and sophisticated instruments. SSI associations can also set up and operate testing laboratories by availing the capital subsidy of 50 per cent being offered by the government.
- Industrial estates with specialized infrastructure and services of international standards will be set up by the government and handed over to associations of local industries for maintenance along with the tax revenues.
- Industrial estates and Special Economic Zones for SSI will set up small power plants to cater to the needs of the members and thereby enable SSIs to become self-reliant in terms of power.

- The Associations/ SSI Institutional support organizations will be streamlined with coordinated focus for each unit and will implement development programmes and enforce regulations as in the case of environment protection, safety of workers and so on.
- The Associations/ SSI Institutional support organizations will also undertake infrastructural development initiatives for the sustainability of SSIs.
- Collaborative relations between the large, medium and small units through subcontracting arrangements are to be encouraged by the SSI Board for furthering the development of SSIs.
- Centers for entrepreneurship development with specific agenda will be set up in every State to promote development of SSIs.
- National level and State level Institutes set up for development of SSIs will transform into promotional agencies for development of SSIs instead of focusing on regulating and controlling SSIs. Overlapping and duplicating organizations at the state and central level will be scrapped in one year.
- SSIs will be exempt from the Indian labour laws pertaining to retrenchment of permanent workers and will be permitted to use contract labour.

10.5 Institutional Support for Manufacturing Sector Development

Creation of Special Economic Zones

The SEZs, both old and new will have the following features:

- Permitting foreign investment up to 100%.
- Tax exemption on income -100% for first 10 years and thereafter a flat rate of 10% for 100% export-oriented units.
- Special income tax status - flat rate of 20% for industries supporting the units in NRI SEZs.
- Tax exemption on income -100% for first 10 years and thereafter a flat rate of 10%.
- 50% of the Units in each zone will be reserved for companies with FDI.
- Invest at least 10% of sales in research & development activities in India.

Creation of NRI SEZs

The SEZs, both old and new will have the following features

- Permitting NRIs upto 100%.
- Tax exemption on income-100% for first 10 years and thereafter a flat rate of 10% Special income tax status-flat rate of 20% for industries supporting the units in NRI SEZs.

Entrepreneurship

- Encourage entrepreneurship to convert the potential of skillful persons opting for VRS to productive use.

First generation entrepreneurs will get their project cleared by the respective state governments within a period of 90 days.

Create panel of experts in each sector to monitor and evaluate the progress of the sector continuously and suggest timely reforms required in each sector.

Public-private partnerships will be created to formulate and implement industrial strategy to provide the necessary framework for the co-ordinated action of various government agencies involved in industrial development.

Government- Industry Associations will collaborate

- To organize programmes to raise the awareness of manufacturing industry across society; improve the image of manufacturing industry.
- To provide all school children and teachers with direct experience of the industry.
- To work with educators to provide on-the-job training for skill and knowledge based courses offered by higher educational institutions

Technology Assessment Board

- i-TAB, as discussed earlier will provide the integrated approach towards channelising academic and institurional research expertise to manufacturing firms.

Development of Technology Parks

Technology parks with participation from private sector industries and higher educational institutes will be set up to encourage innovation and foster development of research on simple and high technology areas with partial assistance from government.

Development of Industry Parks

Industry parks on every sector will be created for development of industries to provide the benefits of common infrastructure, storage and distribution facilities, marketing assistance, technology advisory services, and other benefits to enable firms to raise manufacturing to world standards. There will be facilities that will enable Research and Development activities, training on know how, share Information technology based solutions for operations and share logistics.

Creation of Industry Strategy Groups

Large companies in each sector accounting for more than 10%-25% of the market share will form into a strategic group to work together on Research and Development activities, domestic and international market research, etc. with equal rights to every member irrespective of their market share. Even companies with market share of 25% or more will be permitted to form such groups for alliances on Research and Development.

10.6 Supporting Policies - Education, Infrastructure, Manpower

Infrastructure

An infrastructure vision 2020 will be conceptualized, which will serve to give direction to infrastructure development. The vision will encompass developmental strategies in all sectors while effectively addressing the issues in each of them, as discussed below.

Telecommunication Sector

- Specialized facilities can be jointly funded by government and the user-industries and thereafter, handed over to expert private agencies for maintenance.

- Infrastructure projects can be structured to make them attractive to private investors by offering them as 'packages' - combining commercially viable projects with the not-so viable ones or by offering subsidies and tax concessions for the unviable segments.
- Supportive and competitive telecom policy will be enforced through the Convergence Bill.

A National Power sector policy will be evolved encompassing power sector development with a three to five year time bound plan.

- Unbundle power generation and distribution activities and provide for private sector participation in distribution activities. Consortium ownership will be used to make it viable for a network of generators and distributors to participate in power sector activities. BOO, BOOT, BOLT schemes will be used for partnering with the private sector.
- Public-Private partnership in setting up small power plants will be encouraged initially by focusing on industries that are willing to set up their own power plants for internal consumption by providing financial incentives for such initiatives. A self-reliant power generating network for every cluster of 1000 industries will be aimed at over a period of five to ten years.
- 100% metering of connections is envisaged over a period of 5 years.
- Uniform power tariff is envisaged over a period of 5 years.
- Special incentives will be provided for investments in hydel power projects.
- Every state should become self-reliant in power over a period of five years.
- State electricity boards will be permitted to retrench excessive workforce.

Port Sector

A 'National Policy on Ports' will be formulated for port sector development, incorporating the following dimensions with a three to five year time bound plan.

- Indian Ports Act will be made applicable to major as well as minor ports. Outdated clauses in the Indian Ports Act and Major Ports Act will be amended.
- Throughput targets for ports will be set according to world benchmarks.
- Long Gestation periods of port infrastructure projects due to random checking and verifications imposed on companies will be put to end by obtaining clearances from the relevant authorities before a project is handed over for development. The concession period would start after obtaining statutory clearances.
- A strong and independent regulatory authority to oversee the regulatory functions of independent ports will be set up.
- Public-Private Partnerships through BOO, BOOT and BOLT will be used for different activities of port projects.
- The tariff will be determined by market dynamics rather than administered by the government.
- World standard Ports would be developed over a period of five years.
- Turnaround time (Unloading cargo) will be reduced to a maximum of 2 days over a period of three years.

Roads Sector Policy

- The National Highways will focus on converting to cement concrete roads over a period of five years.

- Private sector particpation in road maintenance will be encouraged through an intensive promotional strategy.

Educational Sector Reforms

- Innovation - to develop into a culture of thinking imparted from school itself.
- Raising skills & education level.

Labour Policy

- Reform labour laws by repealing section 5- B of the Industrial Disputes Act, by introducing standard retrenchment-compensation norms and by allowing full flexibility in the use of contract labour and retrenchment of permanent workers.
- Strict enforcement of labour laws protecting child labour and penalizing factories using child labour with heavy fines and stringent norms discouraging them to do so again.

10.7 Assumptions

- Funding options for development - efficient taxation, savings from investment in unprofitable government enterprises, FDI.
- The economy will emphasize on low technology and then graduate to high technology inventions to capitalize and protect the employment potential in the country.
- A successful manufacturing strategy cannot happen unless and until the distortions in the infrastructure sector is cleared. It is assumed that the reforms in infrastructure will be in tune with the requirements of the manufacturing sector.

11

CHAPTER ELEVEN

Conclusion

Manufacturing Vision - Framework

The model for manufacturing sector in India can be broadly divided into four sections

1. Economic and Social Foundation
2. Small-scale sector
3. Large or medium-scale sector
4. Institutions

Instead of having all the elements of the manufacturing sector as one jumbo group, this model shows a focused strategy for the future of the manufacturing sector in India.

11.1 Economic and Social Foundation

For manufacturing sector as a whole a sound economic and social foundation is vital. Our study reveals that the elements of this foundation should be

- Infrastructure policy
- Education policy
- Labour policy

Infrastructure Policy

This policy will identify the lacunas in the Indian infrastructure that are directly related to creating bottlenecks and higher cost of production for the manufacturing sector. It seeks to diagnose all sectors of Indian infrastructure including Roads, Ports, Power, and Telecommunication. The development of the infrastructure is sought to be time bound and of international standards. Development of each of the infrastructural element needs to be monitored in terms of goals achieved. Moreover, time and cost overrun will be strictly monitored. Private sector participation in these kinds of development has been poor due to the lack of government initiatives. The policy targets a private-public participation model with the private sector partner being assigned projects, which are critical and time-bound. The policy also envisions scope for involving Small and medium enterprises at various stages of a project. Infrastructure funds for specific sectors can be constituted to rope in funds from NRIs.

TOWARDS SUSTAINABLE MANUFACTURING SUPERIORITY

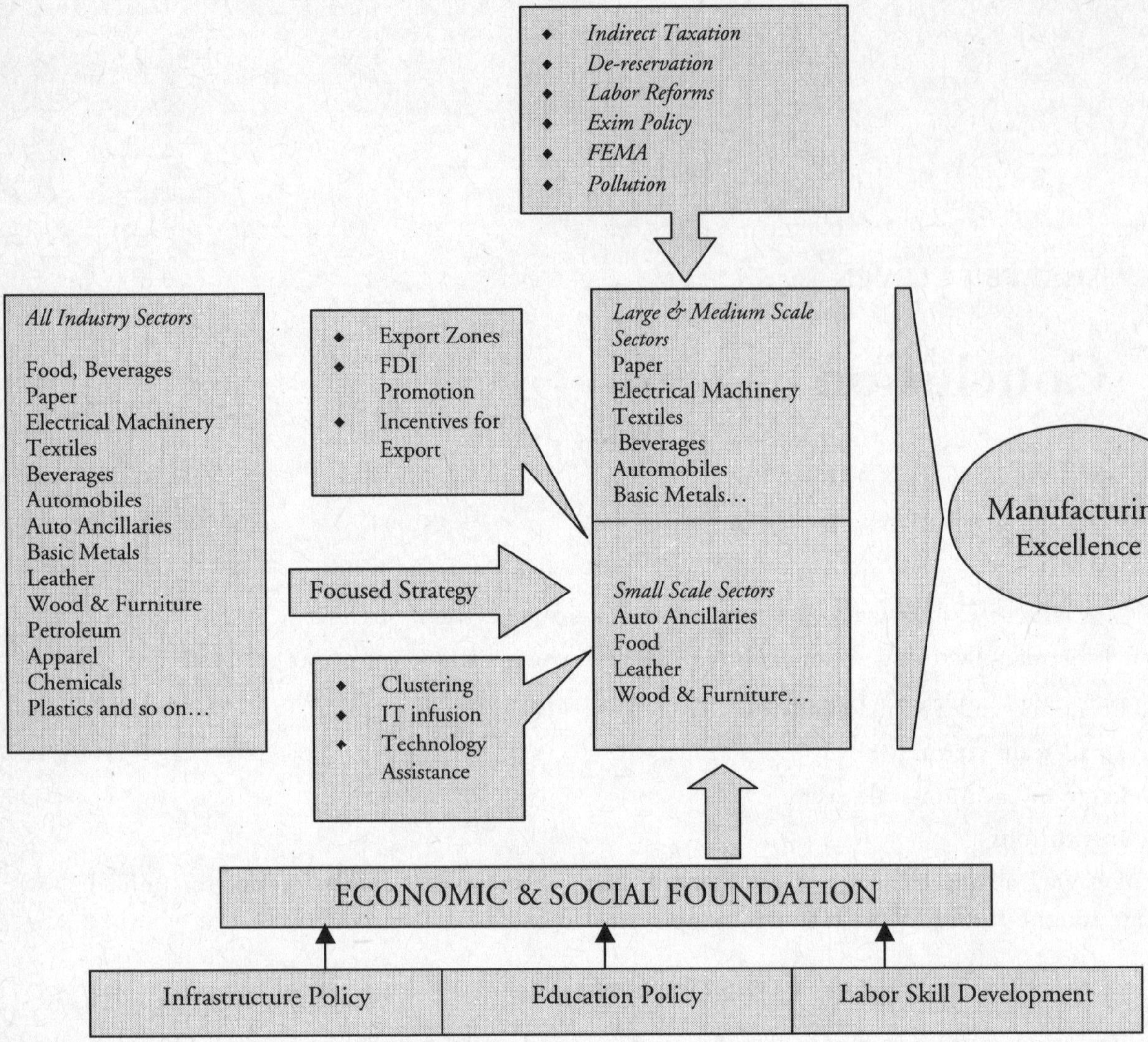

Education Policy

The education policy will address those issues that have been identified as limiting factors by the manufacturing sector. There has not been any visible encouragement from the manufacturing firms to attract young technologists to work on critical projects. Such initiatives will rarely come at firm level, due to huge establishment and sustenance charges. The education policy follows a proactive model towards attracting the youth towards a career in the sector. Manufacturing firms will pool resources for nurturing them to become solution providers to the industry. Firms in this sector will seek to establish "Manufacturing Technology Incubation Centers" (MTIC) to provide research support and IT solutions exclusively to manufacturing operations in collaboration with the top educational and software institutions. This tri-partite incubation centre will have students as the researcher, software firms as the technology provider and manufacturing firms as internal customers. Small manufacturing firms will also be able to use home-grown technology with lower customization costs.

Labour Skill Development

The proposed labour policy seeks to attract, train and retrain quality workers by providing standard working environment and identifiable value addition to them. The thrust of the policy is to help workers concentrate on self-development. Skills development will be by design and firms will provide time and facility for the workers to continuously upgrade their know-how. Cross-functional skill development among operational laborers will be promoted. This will prepare them to shift across departments and in the long run across industries. Firms need to create enthusiasm among the present lot of laborers for learning new technologies.

11.2 Small-scale sectors

◆ *Clustering*

A National Cluster Development Group will be formed, the functions of which will be to identify, analyze and nurture industry clusters in different locations in the country.

◆ *Technology Transfer*

Technology Assessment Board for eliminating import of second hand technology and the corresponding low levels of quality, productivity and sustainability of SMEs.

11.3 Large or medium-scale sector

◆ *FDI Promotion*

To promote foreign capital inflow, restrictions on FDI will be categorically taken off. This will ensure more competition in various sectors. As an externality, foreign capital will also bring in further infrastructural development of the country.

◆ *Trade Policy*

The trade policy will give stress to commodities with export potential. Incentives for export will be set up. This will include Industrial Parks, Export Zones, Clusters and so on.

◆ *Technology Policy*

11.4 Institutions

A new network of support institutions as discussed in various parts of the report will be formulated. Some of these will be:-

- Indian Infrastructure Development Mission, under which various sector regulatory authorities will function.
- National Cluster Development Group, which will take care of creating and marketing of successful industry clusters
- Indian Technology Assessment Board (i-TAB) to facilitate market driven research in the academic and research institutions while providing low-cost technology and expertise to the small scale sector

11.5 The Future

Future prospects of India as a manufacturing destination will have to be decided by the year 2010. Within that time frame we would have settled into any one of the following three paths.

Path 1 - With better, integrated and focused policies, India registers a cumulative annual GDP growth above 10 percent, poverty and unemployment becomes practically nil as the new surge in manufacturing matches the needs of a billion people.

Path 2 - In keeping with the tradition of chasing targets instead of actually achieving them, the current words of caution are ignored. Services also starts to feel the heat as we register a moderate growth rate in manufacturing. Results? Indian firms continue to import technology. Infrastructure still keeps the unholy reputation of adding to the cost of capital instead of helping to cut costs down. Ever chasing, ever missing, manufacturing firms keep on dragging their collective feet.

Path 3 - As part of a new development thinking, manufacturing is recognized as a negative sum game. Efforts to build up manufacturing prowess gives way to a complete service sector orientation. India strategically positions herself as the best among the cheapest locations for services outsourcing. The legacy of successful manufacturing is long forgotten.

Which way to go? Future is uncertain. But we can certainly say that Indian has immense potential in manufacturing. One important principle in Chess as well as Corporate Strategy is that the potential threat is more damaging than the threat in itself. Leveraging the potential to its full benefit is what the Government of India as well as Indian Manufacturing Firms should strive to achieve.

References

1. *Anil K Upadhyay (2000): 'Power Sector Reforms -Indian Experience and Global Trends', Economic and Political Weekly, pp 1023-28.*
2. *Sanjeev S Ahluwalia (2000): 'Power Tariff Reform in India', Economic and Political Weekly, pp 3407-3419.*
3. *A World Development Report , World Bank titled "Knowledge for Development"*
4. *CII, David Dollar, Taye Mengistae, Mary Hallward-Driemier, Giuseppe Iarossi*
5. *UNICTAD, World Investment Report 2001, The world Bank, World Development Report, 2000-01*
6. *A work of Kaufman, Kraay and Zoido-Lobaton (1999), combined indicators from 13 sources under five categories that measure broad dimensions of governance. World Bank, World Business Environment Survey 2000*
7. *FACS (Firm Analysis and competitiveness survey), 2002*
8. *David de Ferranti, Guillermo E. Perry, Daniel Lederman, William F. Maloney, "Knowledge and Natural Resources"*
9. *CSO (1999): Selected Socio-Economic Statistics, India - 1998, Central Statistical Organisation, Government of India, New Delhi, April.*
10. *EPW Research Foundation (1998): National Accounts Statistics of India - 1950-51 to 1996-97, Mumbai, October.*
11. *Greenaway, D (1997): 'Policy Forum: Competitiveness and Competitiveness Policy' in editorial note, The Economic Journal, 107, September 1484-85.*
12. *Kathuria, V (1999): 'Productivity Spillovers from Technology Transfer to Indian Manufacturing Firms', Journal of International Development, forthcoming.*
13. *Oughton, C (1997): 'Competitiveness Policy in the 1990s', The Economic Journal, 107, September 1486-1503.*
14. *Rao, S L (1998): 'Towards a National Competition Policy for India', Economic and Political Weekly, Vol 33 (9), M31-36.*
15. *Suresh Babu (30 March, 2000, EPW, "Competitiveness and competition across states")*
16. *Source: Adapted from momaya and Ajitabh (2000)*
17. *Ace global research on SEZs, 2002*
18. *Source: Arun Agrawal, Secretary, World Association for Small & Medium Enterprises (WASME), New Delhi, India Reference:*
19. *Porter, M. E. (1990): 'The Competitive Advantage of Nations.' (New York: The Free Press).*
20. *Cluster/Network Development Program, UNIDO*
21. *Per Lundequist, Dominic Power (2001): 'Putting Porter into Practice? Practices of Regional Cluster Building: Evidence from Sweden'*
22. *Accenture report "Making Indian Manufacturing Globally Competitive", 2001*
23. *World Bank Group, "India at a glance", September, 2002*
24. *ICRA Sector Research, "Exim Policy 2002: A guide to free trade", 2002*
25. *World Bank, "Improving the investment Climate in India, an investment climate assessment prepared by the World Bank Group in Collaboration with CII", February 2002*
26. *The World Peace Foundation and MIT, "the impact of ideas on trade policy: the origins of U.U agriculture and manufacturing policies", 1989 by Judith Goldstein*
27. *"Serving the world's poor", by C.K Prahald and Allen Hammond"*
28. *UK government manufacturing policy, "The government's Manufacturing Strategy", 2001*
29. *"UK Manufacturing: We can make it better",*
30. *"Implications of China's Entry into WTO: Some observations", T.K Bhaumik, Senior Adviser CII,*
31. *Economic Development Quarterly, Aug 92, Vol 6 Issue 3, p255*
32. *"Competition and Competitiveness among States", Suresh Babu, EPW, March 30, 2002*
33. *"Industrial Pollution control, Choosing the right option", Vinish Kathuria, G.S Haripriya, EPW, October 28, 2000*
34. *"Competition policy dilemmas", TT Ram Mohan, EPW commentary, July 15-21, 2000"*
35. *Case study of Apparel and Machine tools policies", Shuji Uchikawa, EPW commentary, November 27, 1999*
36. *"Institutional change and entrepreneurial development, SSI factor", Takshi Shinoda, EPW,Aug 26-Sept 2, 2000*

37. *"Enterprise Development for Rural Employment, Project Report", C.S Adhikari, EPW, March 27, 2000*

38. *Afsah, S, B Laplante and D Wheeler (1997): Regulation in the Information Age: Indonesia's Public Information Programme for Environmental Management, The World Bank, March.*

39. *Banerjee Bibek, G. Raghuram, and Narayan Rangaraj (2000), 'Konkan Railway Corporation Limited', Vikalpa-The Journal of Decision Makers, Jan.-Mar.*

40. *Button, Kenneth (1993), Transport Economics, Third Edition, Edward Elgar, London.*

41. *Central Institute of Road Transport (1986), Road Transport Taxation in India, CIRT Pune. (1995), Fare Policies: A Book of Readings, Central Institute of Road Transport, Pune. (1995),*

42. *HRTC: A Study of Its Financial and Operational Viability, Central Institute of Road Transport, Pune.*

43. *Gopalakrishnan, C. and A. Ramanathan (2000), 'A Model for Price-Induced Fare Revision in State Transport Undertakings in India', Journal of Transport Management, Vol. 19, No. 12, 699-704.*

44. *Gwilliam, Kenneth (1987), 'Market Failures, Subsidy and Welfare Maximisation', in Stephen Glaister (ed.), Transport Subsidy, Policy Journals, London.*

45. *Kessides, Ian and Robert Willig (1995), Restructuring Regulation of the Rail Industry for the Public Interest, Policy Research Working Paper, No.1506, The World Bank, Washington.*

46. *Mekoth, Nandakumar (1997), 'Passenger Road Transport in Goa: A Model of Unfair Competition', Indian Journal of Transport Management, Vol. 21, No. 1, 18-22.*

47. *Robert Goodland, Herman E. Daly, and Salah El Serafy, Population, Technology and Lifestyle: The Transition to Sustainability (Washington, D.C: Island Press, 1992), p5 Kaushik Deb - 'Private Investment and Policy - Developments in Transport Sector' Economic and Political Weekly 2000 pp 1277 - 1282 McKinsey Global Institute, (2001) 'India's Growth Potential'*

Websites Referred

1. www.foresight.gov.uk/manu2020
2. www.countrywatch.com
3. www.indiainfoline.com
4. www.epw.org.in
5. www.accenture.com
6. www.mckinsey.com
7. www.imd.ch
8. www.worldbank.org
9. www.mep.nist.gov
10. www.wef.org
11. www.3inetwork.org
12. www.unido.org
13. www.indianembassy.org

ANNEXURE-1

List of Jury Members

1. Shri Prabhat Kumar
   Former Cabinet Secretary, Government of India

2. Dr. A.S. Nigavekar
   Chairman, University Grants Commission,
   Government of India

3. Dr. R. Natarajan
   Chairman, All India Council for Technical Education (AICTE),
   Ministry of HRD, Government of India

4. Dr. P. Ghosh
   Addl. Secretary, Prime Minister's Office,
   Government of India

5. Padma Bhushan Prof. P.V. Indiresan
   Former Director, IIT, Madras

6. Shri Jagdish Khattar
   Managing Director, Maruti Udyog Ltd.

7. Dr. Arvind Virmani
   Director and Chief Executive, Indian Council for
   Research in International Economic Relations (ICRIER), New Delhi

8. Smt. K.J. Udeshi
   Executive Director, Reserve Bank of India

9. Shri R.K. Krishna Kumar
   Vice-Chairman, Tata Tea Ltd.;
   Managing Director, The Indian Hotels Company Ltd.
   (The Taj Group of Hotels)

10. Shri T. Joseph
    Consulting Editor, The Businessworld

11. Shri A.P. Gandhi
    Chairman, FAG Bearing (I) Ltd.
    and Former President, Hyundai India

ANNEXURE-2

List of Patrons / Members of WTI

Chairman:
Ranganath Misra
Former Chief Justice of India
MP (Rajya Sabha)

Vice Chairman:
S.S. Mahdi
Vice Chancellor, Jamia Millia Islamia

Abid Hussain
Former Ambassador of India to USA

N. Vittal
Former Central Vigilance Commissioner,
Government of India

T.S.R. Subramanian
Former Cabinet Secretary, Government of India

R.A. Mashelkar
Director General (CSIR), Government of India

Asis Dutta
Director (NCPGR),
Former Vice Chancellor, JNU

L.V. Saptharishi
Additional Secretary,
Ministry of Commerce, Government of India

S. Krishna Moorthy
Former Jt. Secretary, Ministry of
Civil Aviation & Tourism, Govt of India.

A.N. Saksena
Former Addl. Secretary, Ministry of Petroleum,
Government of India

P. Rajendran
Director & COO, NIIT

Gurnam Saran
President,
EMPI Group of Institutions

A.P. Gandhi
Chairman, FAG Bearings,
Former President, Hyundai India

S.P. Verma
Chief Editor, *IJPA,*
Indian Institute of Public Administration

Padma Seth
Consultant, Unicef

D.K. Batra
Head, Academic Affairs,
NIFT

Ashwin Kapur
Managing Director, Uni Products

C.P. Srivastava
Former Secretary, UGC

Pankaj Saran
Vice President,
EMPI Group of Institutions

V.A. Eshwar
Director, EMPI B-School

Ravi Kant Gupta
Chartered Accountant

Member Secretary:
Dhruba Jyoti Pati
Asst. Prof. (Media Studies)
EMPI B-School

ANNEXURE-3

List of Participants

(in order of registration)

★★ : equal among **top three** "policy drafts" and also included in this publication

★ : among the shortlisted nine "policy drafts"

	S. No.	Name of the Participants / Team	Organisation
	1.	Haridasan & Team	R.K. Consulting
	2.	M Victor Louis	St Joseph's College, Trichy
	3.	Sunil Dias & Team	IIM, Bangalore
	4.	R V Ramanamurthy & Team	Nalsar (University of Law), Hyderabad
	5.	Ravi Chadda & Team	Nalsar (University of Law), Hyderabad
	6.	Gautam Gupta	Arthur Andersen
	7.	Sudhakar Gadam & Team	Education & Development Research Centre
	8.	Amit Kumar & Team	Tata Infotech
	9.	Pradip Kumar Taluja	(individual)
	10.	Venkat Rangam & Team	IIM, Lucknow
★	11.	Gopal P Sinha & Team	CMERI, Durgapur
	12.	F C Mahajan & Team	Maharashtra University of Health Sciences
★	13.	Suniti Nagpurkar & Team	Vivekanand Education Society's College, Mumbai
	14.	Jyoti Lata Pandey & Team	NPL, New Delhi
	15.	C Satyadevi	St. Theresa College, Eluru
	16.	Varun Gupta & Team	IIM, Lucknow
	17.	N G Pendse & Team	R D University, Jabalpur
	18.	Koushik Panda & Team	National Institute of Science & Technology, Berhampur
	19.	P C Jain & Team	Sri Ram College of Commerce (SRCC), Delhi
	20.	Dilip Roy	Burdwan University, Burdwan
	21.	CTV Manickam & Team	Haldia Petrochemicals Ltd., Haldia
	22.	K Vanaja Reddy & Team	Sarojini Naidu Vanita Mahavidyalaya, Hyderabad
	23.	Prajapati Mahendra & Team	Diva Construction
	24.	V Ranganathan & Team	IIM, Bangalore
★	25.	Pratap Verma & Team	Federation of MP Chambers (MPCCI), Bhopal
	26.	C Ganesh & Team	Shanmugham Arts, Science, Technology & Research Academy
	27.	S Abdul Rasheed	(individual)
	28.	Noor Muhammad & Team	(individual)
	29.	Prakhar Mehra & Team	Dayalbagh Educational Institute, Dayalbagh, Agra
	30.	Vaishali Nandan	Society for Planning & Research in Sustainable Management
	31.	Milind Agarwal & Team	IIM, Lucknow
	32.	Sujeet Kumar & Team	Institute of Business Management & Research, Pune
	33.	Sandeep Chatterjee	IIM, Kozhikode
	34.	Sambit Kr Mishra & Team	(individual)
	35.	Shahid Ashraf & Team	Jamia Millia Islamia, New Delhi
	36.	Prashant Iyengar & Team	Nalsar (Univ of Law), Hyderabad
	37.	Sudhir Yadav & Team	D D Institute of Technology, Nadiad
	38.	Tarun Goyal	Geefcee Finance Ltd.
	39.	S Venkatachalapathy	Excel Textronics
	40.	Anup S & Team	Nalsar (Univ of Law), Hyderabad

41. N M Jadhav & Team .... Atul Hardware
42. Kingshuk Sen .... Distt. Nadia, West Bengal
43. D G Girdhari .... Marathwada University
44. Rajiv Kumar Govil & Team .... S M College, Chandausi (UP)
★★ 45. Atanu Chaudhuri & Team .... IIM, Lucknow
46. P G Deshmukh & Team .... IIM, Lucknow
47. R Rajesh & Team .... Shanmugha Arts, Science, Technology & Research Academy
48. V Mohanasundaram & Team .... PSG College, Coimbatore
49. K Abdul Ghani .... Technical Teachers Training Institute, Chennai
50. C Siva Rama Krishna Rao & Team .... Kakatiya University, Warangal
51. Hari Natarajan & Team .... IIM, Bangalore
52. R Shankar & Team .... IIT, Delhi
53. Aruna S & Team .... Avinashilingam Instt for Home Sc & Hr Education, Coimbatore
54. Gopal Iyengar & Team .... Kirloskar Instt. of Adv. Mgt. Studies, Mumbai
55. Kaustuva Barik & Team .... IGNOU, New Delhi
56. Krishan Khanna .... i-Watch, Mumbai
57. Anil Garg .... Vijay Nagar, Indore
58. Shantha Kurup & Team .... Avinashilingam Instt. for Home Sc. & Hr. Edn., Coimbatore
59. Srivatsan & Team .... Symbiosis Centre for Mgt. & HRD, Nashik
60. Shubhrangsu S Dutta .... RIT, Jamshedpur
61. Indranil Basu Ray .... (individual)
62. Prasanth V G & Team .... West Bengal National Univ. of Juridical Sc., Kolkata
63. Shahid Ahmad & Team .... Jamia Millia Islamia, New Delhi
64. J C Saboo & Team .... R J College, Mumbai
65. Swarup Ratan Sinha Ray .... (individual)
66. Yatendra Malhotra & Team .... FMS, University of Delhi, Delhi
67. S Saraswati & Team .... St Theresa's College, Eluru (AP)
68. Sudeepta Chaudhuri & Team .... Tata Services Limited
69. Laxmi Priya & Team .... Fatima College, Madurai
70. V Raj Kumar & Team .... Nalsar (University of Law), Hyderabad
71. Ravi Prakash Singh & Team .... Lucknow University
72. J Debroy & Team .... RCC Instt of Information Technology, Kolkata
73. Rudramoorthy & Team .... PSG College of Technology, Coimbatore
74. R C Tripathi & Team .... Indian Institute of Information Technology, Allahabad
75. Jayanshu B Gandhi .... (individual)
76. Promod S Dabrase .... Representing an NGO
77. Atanu Ghosh & Team .... SRCC, Delhi
78. P S Subbarao .... M R College, Vizianagaram
79. G Vijayan .... Hindusthan College of Engineering, Kanchipuram
80. Harsh Parekh & Team .... (individual)
81. Praveen Kumar Vecha & Kavitha Kumari .... (individual)
82. S N Kalani & Deepak Lele .... (individual)
83. S Narayanan & Team .... Vellore Institute of Technology, Vellore
84. Shilbhadra Dutt & Samrat Bose .... Jamia Millia Islamia, New Delhi
85. Hansmeet Chandok & Team .... Asian Traders
86. S Sreenivas Reddy & Team .... (individual)
87. Arun Kr. Misra .... Canara Bank Head Office, Bangalore
88. A C Srivastava .... Bharti Healthcare Ltd., New Delhi
89. P Jayaseela Pandian .... (individual)
90. Kiran Goel .... Panjab University, Chandigarh
★ 91. Abhijith Mukunda & Team .... BMS College, Bangalore
92. T Ramappa .... FKCCI, Bangalore
93. Sachin Nair & Team .... Gujarat Positra Port Infrastructure Ltd., Mumbai
94. Jagannath Mohanty .... NIST, Berhampur, Orissa,

95. N Thillainayagam .... Gandhigram Rural Institute, Gandhidham
96. S K Srivastava .... Computer Education Institute, Lucknow
97. Arun Kr Jagatramka .... Gujarat NRE Coke Ltd., Ahmedabad
98. Aditya K .... Saleem Nagar, Hyderabad
99. Narendra Kumar K & Pavan Kumar K N .... Srikrishna Nagar, Hyderabad
100. Swapna.S & Team .... Vikas Nagar, Hyderabad
101. Madhu M & Saritha K .... R B Nagar, Shamshabad
102. Rambabu Kodali & Team .... Birla Institute of Technology & Science, Pilani
103. K L Narayanan .... Vasant Vihar, Thane
104. Amol Kulkarni .... Pawan Nagar, Aurangabad
105. Subhajit Mitra Mazumder .... Doranda, Ranchi
106. Arvind Jajoo .... Tardeo Road, Mumbai
107. T J Joseph & Team .... Madras School of Economics, Chennai
108. Rajat Kathuria & Team .... International Management Institute, New Delhi
109. Abhay D Lidbe .... Tukaram Nagar, Dombirli
110. Padmabati Gahan .... Sambalapur University, Sambalapur
111. Rohit Pandey and Saurabh Agarwal .... Nanakpura, New Delhi
112. Gaurav Awasthi & Team .... IIM, Lucknow
113. V Venugopal Reddy & Team .... Annamacharya Institute of Technology & Science
114. Sanjit Dey .... Rishi Gases Pvt Ltd., Bilaspur
115. K S Chaube .... Panchkula, Haryana
116. Shobha Ahuja .... Yamuna Vihar, Delhi
117. Prof D Acharya & Team .... IIT, Kharagpur
★★ 118. Sanjay Upendram & Team .... Andersen Business Consulting (now KPMG consulting), Chennai
119. N V M Rao & Team .... BITS, Pilani
120. Ananthakrishnan Ganesan & Team .... Vasant Vihar, Thane
121. Saket Bihari .... MIT School of Management, Pune
122. Harnam Singh & Team .... OEF Group, Kanpur
123. Arun Kumar Singh & Team .... IIT, Delhi
124. Suresh Kamat G.B. & Team .... (individual)
125. Pankaj Rusia & Team .... (individual)
126. Vivek Patil .... (individual)
★ 127. B Ravi & Team .... IIT, Mumbai
★★ 128. M Thenmozhi & Team .... IIT-Madras, Chennai
129. Thamaraiselvan P .... (individual)
130. Soumendra N Bagchi & Team .... IIM, Ahmedabad
131. Basawaraj Kanta & Team .... Hyderabad
132. Sumit S Bhale & Team .... (individual)
133. Meghna Girotra .... (individual)
134. M S Shunmugam & Team .... IIT-Madras, Chennai
135. S S Jain & Team .... Shree Cement Ltd., Ajmer
136. H Ramakrishnan .... BHEL, Bangalore
137. Subhasish Sahoo & Team .... ICFAI Business School, Hyderabad
138. N Rajiv Menon & Team .... Guruvayurappan Institute of Management
139. Swaha Shome & Team .... ICFAI Business School, Gurgaon
140. Karthik & Team .... Dept of Management Studies, University of Madras
141. Rohan Kurian Varghese .... IIFT, New Delhi
142. Prof Chiplunkar & Team .... Chetana's R K Institute of Management & Research
143. Praveen Gokhale & Team .... Procter & Gamble India
144. Sumon Datta & Team .... IIT-Madras, Chennai
145. Deepak Goel & Team .... Ramjas College, Delhi
146. Prof Abhijit Nag & Team .... Symbiosis IBM, Pune
147. K Valavan .... (individual)
148. Gautam G & Team .... (individual)

149. Anurag Duggar & Team & Ms Bharti Sharma & Team .... Deepshikha College of Tech. Education
150. B M Ghodeshwar & Team .... NITIE, Mumbai
151. Pankaj Kumar M Joshi .... Narangpura, Ahmedabad
152. Sanjay Bhattacharya & Team .... IMT, Ghaziabad
153. M Guruprasad & Team .... AICAR Business School, Raigad
154. Rakesh Kupsad .... HBL NIFE Power Systems Ltd.
155. K Saravana Kumar & Team .... Bangalore
156. Sharad Chandra & Team .... Delhi
157. Tapobroto Chaterjee & Team .... (individual)
158. Laxmi Narashiman & Team .... ICFAIAN Business School, Bangalore
159. Elangcherian & Team .... Vellore Institute of Technology, Vellore
★ 160. Raj Kumar R & Team .... IIT, Delhi
161. Muthupandian K S .... Madurai
162. Praveen Kumar .... New Delhi
163. Khushdeep Dharni & Team .... Punjab College of Technical Education, Ludhiana
164. Haryana Electricity Regulatory Commission .... Panchkula
165. S.K. Srivastava .... PICUP, Lucknow
166. B. Ahmed .... BHEL, Hyderabad
167. S. Tiwari .... BHEL, Hardwar

Supporting Organisations

Punjab National Bank

HDFC Bank

MRF Limited

FICCI

Business World

Indian Airlines

Alkali Manufacturers Association of India

WTI logo credit : Ms. Lolita Dutta (Faculty, EMPI AdCom)